Advanced Photocatalytic Materials

Advanced Photocatalytic Materials

Editor

Vlassios Likodimos

MDPI • Basel • Beijing • Wuhan • Barcelona • Belgrade • Manchester • Tokyo • Cluj • Tianjin

Editor
Vlassios Likodimos
Section of Condensed Matter Physics, Department of Physics, National and Kapodistrian University of Athens, University Campus
Greece

Editorial Office
MDPI
St. Alban-Anlage 66
4052 Basel, Switzerland

This is a reprint of articles from the Special Issue published online in the open access journal *Materials* (ISSN 1996-1944) (available at: https://www.mdpi.com/journal/materials/special_issues/adv_Photocatal).

For citation purposes, cite each article independently as indicated on the article page online and as indicated below:

LastName, A.A.; LastName, B.B.; LastName, C.C. Article Title. *Journal Name* **Year**, *Volume Number*, Page Range.

ISBN 978-3-0365-1032-3 (Hbk)
ISBN 978-3-0365-1033-0 (PDF)

Contents

About the Editor

Vlassios Likodimos is Associate Professor at the Department of Physics of the National and Kapodistrian University of Athens (Greece), working on nanostructured metal oxides, carbon nanomaterials and composites for photoinduced- and solar-energy conversion applications. He has previously worked on scanning probe techniques and their application in ferroelectric and polymer materials, as well as on magnetic nanocomposites and strongly correlated systems, including manganites and high-Tc superconductors. He has participated in several EU and national research projects, with more than 160 papers in peer-reviewed journals (h index = 40 Scopus) and more than 130 communications in international conferences, and he serves as a reviewer in more than 60 peer reviewed journals, as well as being an Editorial Board member of *Materials* (MDPI).

Editorial

Advanced Photocatalytic Materials

Vlassis Likodimos

Section of Condensed Matter Physics, Department of Physics, National and Kapodistrian University of Athens, 15784 Panepistimiopolis, Greece; vlikodimos@phys.uoa.gr; Tel.: +30-2107-276-824

Received: 6 February 2020; Accepted: 9 February 2020; Published: 11 February 2020

Abstract: Semiconductor photocatalysts have attracted a great amount of multidiscipline research due to their distinctive potential for solar-to-chemical-energy conversion applications, ranging from water and air purification to hydrogen and chemical fuel production. This unique diversity of photoinduced applications has spurred major research efforts on the rational design and development of photocatalytic materials with tailored structural, morphological, and optoelectronic properties in order to promote solar light harvesting and alleviate photogenerated electron-hole recombination and the concomitant low quantum efficiency. This book presents a collection of original research articles on advanced photocatalytic materials synthesized by novel fabrication approaches and/or appropriate modifications that improve their performance for target photocatalytic applications such as water (cyanobacterial toxins, antibiotics, phenols, and dyes) and air (NO_x and volatile organic compounds) pollutant degradation, hydrogen evolution, and hydrogen peroxide production by photoelectrochemical cells.

Keywords: TiO_2 nanomaterials; visible light activated titania; heterojunction photocatalysts; photonic crystal catalysts; plasmonic photocatalysis; graphene-based photocatalysts; water and air purification; solar fuels

---

Semiconductor photocatalysis has been considered as a key technology to face the global concerns of environmental pollution and the ever increasing energy demands, through the utilization of environmentally benign earth-abundant materials and renewable energy sources, such as solar energy [1]. Photocatalytic materials have been attracting significant interest for diverse applications, ranging from sustainable water/air remediation as well as hydrogen and chemical fuel production by photocatalytic water splitting [2]. Their unique potential for solar powered technologies has been the stimulus for the development of nanostructured photocatalysts with improved structural, morphological, and electronic properties that could effectively evade the two main limitations of the process efficiency, i.e., the low quantum yield, stemming from the recombination of photogenerated charge carriers, and the poor visible light harvesting, pertinent mostly to wide band gap semiconductors such as the benchmark titanium dioxide (TiO_2) photocatalysts [3].

Research efforts have been accordingly focused on the design and fabrication of advanced photocatalytic materials relying on competent modification approaches such as coupling with plasmonic nanoparticles, surface engineering, and heterostructuring with other semiconducting and/or graphene-based nanomaterials, as well as tailoring the materials' structure and morphology (e.g., nanotubes, nanowires, and photonic crystals) in order to boost light harvesting and photon capture, charge separation, and mass transfer that play a pivotal role in photocatalytic environmental remediation and solar to chemical energy conversion applications [4,5].

This Special Issue consists of 10 original full-length articles on advanced photocatalytic materials fabricated by innovative synthetic routes and judicious compositional modifications, with diverse applications ranging from the degradation of hazardous water and air pollutants to hydrogen evolution and photoelectrocatalytic hydrogen peroxide production. T. M. Khedr at al. [6] reported on the degradation of microcystin-LR (MC-LR), a highly toxic and persistent hepatotoxin commonly detected in cyanobacterial algae blooms, by visible-light-activated C/N-co-modified mesoporous anatase/brookite TiO_2 photocatalysts, prepared by a one-pot hydrothermal method. The complete removal of MC-LR from aqueous solutions was achieved under visible light irradiation, related to the unique combination of visible light photogenerated electrons from anion-induced impurity states and interfacial charge transfer between the brookite and anatase phases.

N-modified TiO_2 was utilized by M. Janus et al. [7] as an additive for the technologically appealing application of photocatalytic active cement mortars that feature air purification by NO_x decomposition without compromising, and even improving, cement's mechanical properties. Laser pyrolysis was applied by K. Wang et al. [8] to synthesize novel C-modified titania/graphene nanocomposites with markedly high activity on different photocatalytic applications from acetic acid oxidative decomposition and methanol dehydrogenation (even without a Pt co-catalyst) to visible-light induced phenol degradation and Escherichia coli inactivation. Surface functionalization of TiO_2 photonic crystals by graphene oxide (GO) nanocolloids and subsequent thermal reduction was reported by Diamantopoulou et al. [9] as a promising approach for the development of efficient photocatalytic films that combine the unique slow photon-assisted light harvesting, surface area, and mass transport of macroporous photonic structures with the enhanced adsorption capability, surface reactivity, and charge separation of GO nanosheets.

Fluorine-doped tin oxide (FTO) inverse opals crystal were also exploited by X. Ke et al. [10] as macroporous photonic crystal substrates for the successive deposition of plate-like WO_3 and Ag_2S quantum dots with tunable photoelectrochemical response by the light incidence angle. Control over the shape and facet growth of TiO_2 nanocrystals was demonstrated by Y. Du et al. [11] by tuning the pH of the exfoliated metatitanic nanosheet solutions used as precursors in a simple fluorine-free microwave-assisted hydrothermal method, leading to enhanced photocatalytic and photovoltaic performance. The oriented morphology and enhanced surface area of TiO_2 nanowires, grown on TiO_2 nanotube arrays by electrochemical anodization, in combination with Au plasmonic nanoparticles were successfully applied by T.C.M.V. Do et al. [12] for the photocatalytic degradation of eight important antibiotics in model aquaculture wastewater from the Mekong Delta region.

Plasmonic Ag nanomaterials were also incorporated in silver-copper oxide heterostructures by H. Suarez et al. [13] to promote the full-spectrum photocatalytic-assisted volatile organic compound (VOC) oxidation in the gas phase, using n-hexane as a probe molecule, with LED illumination in the visible-NIR range. V. H. Nguyen et al. [14] reported on sulfate modification as an efficient means to improve the structural and photocatalytic properties of sol-gel synthesized $BiVO_4$ with annealing temperature, used as an alternative metal oxide photocatalyst beyond TiO_2, leading to enhanced decomposition of methylene blue as a model dye pollutant under LED visible light.

The sensitization of mesoporous Titania films by nanoparticulate CdS and CdSe, in combination with ZnS passivation was used by T. S. Andrade et al. [15] for the fabrication of ZnS/CdSe/CdS/TiO_2/FTO photoanodes, enabling broad visible light harvesting. These photoelectrodes combined with a Pt-free counter electrode, made of carbon cloth with deposited nanoparticulate carbon, were used for the assembly of a photoelectrochemical cell operating as a Photo Fuel Cell, i.e., without any external bias. These devices were demonstrated to photoelectrocatalytically produce substantial quantities of hydrogen peroxide with the Faradaic efficiency exceeding 100% in the presence of $NaHCO_3$ carbonate electrolyte.

All the published articles were thoroughly refereed through the standard single-blind peer-review process of Materials journal. As Guest Editor, I would like to acknowledge all of the authors for their excellent contributions and the reviewers for their valuable and prompt comments that greatly improved the quality of the papers. Finally, I would like to thank the staff members of Materials, in particular Ms. Floria Liu, Section Managing Editor, for her devoted efforts and kind assistance.

Conflicts of Interest: The author declares no conflict of interest.

References

1. *RSC Energy and Environment Series No. 14, Photocatalysis: Fundamentals and Perspectives*; Schneider, J.; Bahnemann, D.; Ye, J.; Li Puma, G.; Dionysiou, D.D. (Eds.) The Royal Society of Chemistry: Cambridge, UK, 2016.
2. Chen, X.; Selloni, A. Titanium Dioxide (TiO_2) Nanomaterials. *Chem. Rev.* **2014**, *114*, 9281–9282. [CrossRef] [PubMed]
3. Schneider, J.; Matsuoka, M.; Takeuchi, M.; Zhang, J.; Horiuchi, Y.; Anpo, M.; Bahnemann, D.W. Understanding TiO_2 Photocatalysis: Mechanisms and Materials. *Chem. Rev.* **2014**, *114*, 9919–9986. [CrossRef] [PubMed]
4. Kowalska, E.; Wei, Z.; Janczarek, M. Band-gap Engineering of Photocatalysts: Surface Modification versus Doping. In *Visible-Light-Active Photocatalysis: Nanostructured Catalyst Design, Mechanisms, and Applications*, 1st ed.; Ghosh, S., Ed.; Wiley-VCH Verlag GmbH & Co.: Weinheim, Germany, 2018; pp. 449–484.
5. Likodimos, V. Photonic Crystal-assisted Visible Light Activated TiO_2 Photocatalysis. *Appl. Catal. B Environ.* **2018**, *230*, 269–303. [CrossRef]
6. Khedr, T.M.; El-Sheikh, S.M.; Ismail, A.A.; Kowalska, E.; Bahnemann, D.W. Photodegradation of Microcystin-LR Using Visible Light-Activated C/N-co-Modified Mesoporous TiO_2 Photocatalyst. *Materials* **2019**, *12*, 1027. [CrossRef] [PubMed]
7. Janus, M.; Mądraszewski, S.; Zając, K.; Kusiak-Nejman, E.; Morawski, A.W.; Stephan, D. Photocatalytic Activity and Mechanical Properties of Cements Modified with TiO_2/N. *Materials* **2019**, *12*, 3756. [CrossRef] [PubMed]
8. Wang, K.; Endo-Kimura, M.; Belchi, R.; Zhang, D.; Habert, A.; Bouclé, J.; Ohtani, B.; Kowalska, E.; Herlin-Boime, N. Carbon/Graphene-Modified Titania with Enhanced Photocatalytic Activity under UV and Vis Irradiation. *Materials* **2019**, *12*, 4158. [CrossRef] [PubMed]
9. Diamantopoulou, A.; Sakellis, E.; Gardelis, S.; Tsoutsou, D.; Glenis, S.; Boukos, N.; Dimoulas, A.; Likodimos, V. Advanced Photocatalysts Based on Reduced Nanographene Oxide–TiO_2 Photonic Crystal Films. *Materials* **2019**, *12*, 2518. [CrossRef] [PubMed]
10. Ke, X.; Yang, M.; Wang, W.; Luo, D.; Zhang, M. Incidence Dependency of Photonic Crystal Substrate and Its Application on Solar Energy Conversion: Ag_2S Sensitized WO_3 in FTO Photonic Crystal Film. *Materials* **2019**, *12*, 2558. [CrossRef] [PubMed]
11. Du, Y.-E.; Niu, X.; Li, W.; An, J.; Liu, Y.; Chen, Y.; Wang, P.; Yang, X.; Feng, Q. Microwave-Assisted Synthesis of High-Energy Faceted TiO_2 Nanocrystals Derived from Exfoliated Porous Metatitanic Acid Nanosheets with Improved Photocatalytic and Photovoltaic Performance. *Materials* **2019**, *12*, 3614. [CrossRef] [PubMed]
12. Do, T.C.M.V.; Nguyen, D.Q.; Nguyen, K.T.; Le, P.H. TiO_2 and Au-TiO_2 Nanomaterials for Rapid Photocatalytic Degradation of Antibiotic Residues in Aquaculture Wastewater. *Materials* **2019**, *12*, 2434. [CrossRef] [PubMed]
13. Suarez, H.; Ramirez, A.; Bueno-Alejo, C.J.; Hueso, J.L. Silver-Copper Oxide Heteronanostructures for the Plasmonic-Enhanced Photocatalytic Oxidation of N-Hexane in the Visible-NIR Range. *Materials* **2019**, *12*, 3858. [CrossRef] [PubMed]
14. Nguyen, V.H.; Bui, Q.T.P.; Vo, D.-V.N.; Lim, K.T.; Bach, L.G.; Do, S.T.; Nguyen, T.V.; Doan, V.-T.; Nguyen, T.-D.; Nguyen, T.D. Effective Photocatalytic Activity of Sulfate-Modified $BiVO_4$ for the Decomposition of Methylene Blue Under LED Visible Light. *Materials* **2019**, *12*, 2681. [CrossRef] [PubMed]
15. Andrade, T.S.; Papagiannis, I.; Dracopoulos, V.; Pereira, M.C.; Lianos, P. Visible-Light Activated Titania and Its Application to Photoelectrocatalytic Hydrogen Peroxide Production. *Materials* **2019**, *12*, 4238. [CrossRef] [PubMed]

Article

Photodegradation of Microcystin-LR Using Visible Light-Activated C/N-co-Modified Mesoporous TiO_2 Photocatalyst

Tamer M. Khedr [1,2,3,*], Said M. El-Sheikh [1], Adel A. Ismail [1,4], Ewa Kowalska [3] and Detlef W. Bahnemann [2,5]

1 Nanomaterials and Nanotechnology Department, Central Metallurgical Research and Development Institute (CMRDI) P.O. Box: 87 Helwan, Cairo 11421, Egypt; selsheikh2001@gmail.com (S.M.E.-S.); aaismail@kisr.edu.kw (A.A.I.)
2 Institute of Technical Chemistry, Photocatalysis and Nanotechnology Research Unit, Leibniz Universität Hannover, Callinstr. 3, D-30167 Hannover, Germany; bahnemann@iftc.uni-hannover.de
3 Institute for Catalysis, Hokkaido University, N21, W10, Sapporo 001-0021, Japan; kowalska@cat.hokudai.ac.jp
4 Nanotechnology and Advanced Materials Program, Energy & Building Research Center (EBRC), Kuwait Institute for Scientific Research (KISR), P.O. Box 24885, Safat 13109, Kuwait
5 Laboratory "Photoactive Nanocomposite Materials" (Director), Saint-Petersburg State University, Ulyanovskaya str. 1, Peterhof, Saint-Petersburg 198504, Russia
* Correspondence: khedr.t@cat.hokudai.ac.jp or tamerkhedr56@gmail.com; Tel.: +81-11-706-9130

Received: 11 March 2019; Accepted: 22 March 2019; Published: 28 March 2019

Abstract: Microcystin-LR (MC-LR), a potent hepatotoxin produced by the cyanobacteria, is of increasing concern worldwide because of severe and persistent impacts on humans and animals by inhalation and consumption of contaminated waters and food. In this work, MC-LR was removed completely from aqueous solution using visible-light-active C/N-co-modified mesoporous anatase/brookite TiO_2 photocatalyst. The co-modified TiO_2 nanoparticles were synthesized by a one-pot hydrothermal process, and then calcined at different temperatures (300, 400, and 500 °C). All the obtained TiO_2 powders were analyzed by X-ray diffraction (XRD), Raman spectroscopy, transmission electron microscope (TEM), specific surface area (SSA) measurements, ultraviolet-visible diffuse reflectance spectra (UV-vis DRS), X-ray photoelectron spectroscopy (XPS), Fourier transform infrared (FTIR) spectroscopy, and photoluminescence (PL) analysis. It was found that all samples contained mixed-phase TiO_2 (anatase and brookite), and the content of brookite decreased with an increase in calcination temperature, as well as the specific surface area and the content of non-metal elements. The effects of initial pH value, the TiO_2 content, and MC-LR concentration on the photocatalytic activity were also studied. It was found that the photocatalytic activity of the obtained TiO_2 photocatalysts declined with increasing temperature. The complete degradation (100%) of MC-LR (10 mg L^{-1}) was observed within 3 h, using as-synthesized co-modified TiO_2 (0.4 g L^{-1}) at pH 4 under visible light. Based on the obtained results, the mechanism of MC-LR degradation has been proposed.

Keywords: anatase; brookite; C/N-TiO_2; microcystin-LR; photodegradation; visible light

1. Introduction

The desire to provide safe potable water has increased due to economic and population growth in recent years. Moreover, global freshwater resources, which might be used for the production of drinkable water, have been even diminished because of water pollution and climate change [1]. Both inorganic and organic compounds of natural, urban and industrial origins are considered as water pollutants. For example, cyanobacteria (commonly known as blue-green algae) are photosynthetic and microscopic organisms existing in aquatic systems, including fresh, brackish and marine water. It has been suggested that an increased content of cyanobacteria in water is caused by eutrophication, climatic changes, available nutrients of agricultural origin and waste disposal in water [1–4]. They can reproduce to compose harmful algal blooms (HABs) under convenient conditions (high temperatures and nutrient concentrations, and strong sunlight), and some cyanobacteria contain and release toxins as metabolic byproducts and during "cell death" (lysis). These toxic compounds, known as cyanotoxins, are considered a significant threat to human, animals, environment, and even ecosystems due to their high solubility in water, toxicity and chemical stability [1,2,5,6]. There are diverse species of cyanotoxins, detected in the water around the world, but the Environmental Protection Agency (EPA) clearly focused on three of them as being of the biggest threat to the environment: microcystin-LR (MC-LR), cylindrospermopsin (CYN), and anatoxin-a [1,7]. Out of all known cyanobacterial toxins, MC-LR (L for Leucine and R for Arginine) is considered as the most toxic of the microcystins, and the most abundant and harmful in aquatic systems [1,7,8]. MC-LR is a cyclic heptapeptide containing several moieties, including Adda, and other amino acids [7]. The toxicity of MC-LR is large due to the Adda chain, and thus the detoxification requires the removal of the Adda side chain or isomerization from trans to cis. However, the trans-to-cis photoisomerization demands UV irradiation, and thus is impossible under visible light (vis) [7]. Several technologies, including ozonation, chlorination, and phototransformation, have been tested to some extent for removing (and/or detoxifying) MC-LR from water [7,8]. Recently, great attention has been focused on promising green technologies, such as advanced oxidation technologies (AOTs) for removal of cyanotoxins, due to their ability to produce free radicals with strong oxidizing power (mainly hydroxyl radicals) resulting in possible complete degradation (mineralization) of cyanotoxins to CO_2 and H_2O. Among AOTs, TiO_2-based photocatalysis is probably the most widely used for the degradation of emerging pollutants because of promising characteristics, such as non-toxicity for both human and environment, low costs, strong oxidizing ability, photochemical stability, high efficiency, biocompatibility, and facile synthesis with various morphologies [9,10]. Indeed, Feitz et al. reported the decomposition of MC-LR (low concentration) in a natural organic matrix using conventional TiO_2/UV-based photocatalysis process [11]. Despite the considerable advancement in TiO_2-based photocatalysis that has been accomplished recently, there are some drawbacks that restrain the photocatalytic performance of TiO_2. The most serious drawback for TiO_2 is its large band gap (despite excellent redox ability), which requires UV irradiation (expensive, toxic, and with low content (ca. 4–5%) in the solar spectrum) for photo-excitation, thereby restricting its photocatalytic activity under vis (safe and cheap since natural solar radiation might be used (46% of the sun's electromagnetic radiation)). In addition, rapid electron–hole recombination (typical for all semiconductors) results in low quantum yields of photocatalytic reactions [7,9,10,12]. Therefore, various strategies, including chemical and structural modifications, have been applied to improve the photocatalytic performance of TiO_2 [12–16]. It should be noticed that the formation of mixed-phase TiO_2 nanostructure [9,10], and heterojunctions with other semiconductors [17] facilitate the charge facilitates the charge separation to overcome the electron–hole recombination, and thus enhancing the activity of TiO_2. The TiO_2 band gap can be modified by the aggregation state and by differences in particle size, as Reinosa et al. previously demonstrated [18]. Moreover, surface modification and doping (interstitial and substitutional) with metal and non-metal (ions and compounds) have demonstrated to be effective methods for increasing the photocatalytic efficiency of TiO_2 resulting from enhanced electron–hole separation and appearance of vis absorption [7–9,12,13,16]. Recently, research has focused on synthesis of co-modified TiO_2 because of their higher photocatalytic activity than that of single-modified TiO_2 [7–9,12,13,16]. However, there are only limited studies focusing on the synthesis of non-metal co-modified TiO_2 for degradation of MC-LR [7]. As far as we know, there is only one report by Liu et al. on the activity of C-N co-modified anatase-TiO_2 for degradation of MC-LR under visible light irradiation [19]. They prepared C-N co-modified anatase films through a sol-gel method from titanium (IV) isopropoxide (TIP), polyoxyethylene sorbitan monooleate (Tween 80, nonionic surfactant) and anhydrous ethylenediamine (EDA) as Ti, C, and N source, respectively. Unfortunately, titanium alkoxides (e.g., TIP) are very sensitive to moisture, thereby synthesis under inert gas and/or the multi-steps processes (resulting in low yields) must be applied [9,10]. Moreover, incomplete removal (65%) was only achieved for 0.5 mg L^{-1} MC-LR during 5-h vis irradiation (fluorescent lamps with a UV block filter, λ > 420 nm, pH 3). Therefore, in this study, commercially available $Ti_2(SO_4)_3$ and glycine were used as the TiO_2-precursor and non-metal ions source, respectively, to prepare C/N co-modified mesoporous A/B TiO_2 through one-pot surfactant-free hydrothermal approach. Additionally, the as-prepared TiO_2 powder was further calcined at different temperatures (300, 400, and 500 °C) to investigate the effect of thermal treatment on the properties, and thus photocatalytic activities of TiO_2 photocatalysts. The as-synthesized and calcined TiO_2 powders were characterized by advanced techniques, and used for MC-LR degradation using low-cost irradiation source (visible-LED lamp, λ = 420 nm). After 3 h irradiation, the complete removal of MC-LR (C_0 = 10 mg L^{-1}) was achieved using 0.4 g L^{-1} as-prepared C-N co-modified TiO_2 at pH 4.

2. Materials and Methods

2.1. Materials

Titanium(III) sulfate (TS, $Ti_2(SO_4)_3$, Fisher, Loughborough, Leicestershire, UK, 15%), sodium nitrate ($NaNO_3$, Koch-light laboratories Ltd., Haverhill, Suffolk, UK, 98%), glycine (GLY, H_2N-CH_2-COOH, Sigma-Aldrich, St. Louis, MO, USA, 99%), sodium hydroxide (NaOH, LobaChemie, Pellets, Mumbai, India, 98%), MC-LR (Cal-Biochem, Nottingham, UK, 99%), and absolute ethanol (CH_3CH_2OH, Sigma-Aldrich, Darmstadt, Germany, 99.8%) were used without any further purification. The water used for the experiment was deionized water (DI, 8.2 μ Ω).

2.2. Preparation of Photocatalyst

In a typical procedure, two different mixtures were prepared. For the first mixture (A), $NaNO_3$ (0.1 μmol L^{-1}) was added to TS solution (0.5 μmol L^{-1}) and then subjected to gentle stirring for 25 min till the formation of a transparent solution. The second mixture (B) consisting of an aqueous solution of GLY (1 mol L^{-1}) and NaOH (2 mol L^{-1}) was mixed with a solution (A) dropwise under continuous stirring, and then stirred for 30 min to complete the reaction, resulting in obtaining a milk suspension. Then, the obtained suspension (maintained at 40 mL) was poured into the 100-mL Teflon-lined tube. The sample was treated at 200 °C for 20 h, and then naturally cooled to room temperature, collected, washed several times with ethanol and water, and dried at 60 °C for 12 h. Finally, the as-prepared sample was calcined in a furnace open to air at three temperatures (300, 400, and 500 °C) for 1 h. The as-synthesized and calcined samples were named as CDT-0.00, CDT-300, CDT-400, and CDT-500, respectively.

2.3. Characterization of Photocatalyst

The crystal structure of TiO_2 was analyzed by the X-ray diffraction (XRD, D8, Bruker AXS X-ray diffractometer, Karlsruhe, Germany) in a 2θ range from 15 to 70°. Raman spectra were obtained on a Senterra Dispersive Micro-Raman (Bruker, Munich, Germany) under excitation with the 532-nm line of a doubled Nd:YAG laser at an incident power of 10 mW. Transmission electron microscope (TEM, JEOL-JEM-1230, Tokyo, Japan) was used to investigate the particle sizes and morphologies. Brunauer–Emmett–Teller (BET) specific surface areas were estimated based on nitrogen adsorption isotherms at 350 °C using Quanta Chrome Instruments (). All samples were degassed at 180 °C for 12 h before the N_2 physisorption measurements. The BET specific surface area was estimated using the adsorption data in the relative pressure (P/P_o). The Barrett–Joyner–Halenda (BJH) pore size distribution was estimated from adsorption data. UV-vis absorption spectra were measured using UV-vis spectrophotometer (UV-2501 PC, Shimadzu, Tokyo, Japan) with an integrating sphere and $BaSO_4$ as a reflectance standard. The reflectance data were converted to F(R) values according to the

Kubelka–Munk theory. The (F(R) hν)$^{1/2}$ versus energy of the exciting light was plotted to obtain the band-gap energy [10]. The X-ray photoelectron spectroscopy (XPS, Thermo Fisher Scientific, Waltham, MA, USA) measurements were performed to investigate the surface chemical composition. The room temperature FT-IR absorption spectra were recorded using JASCO 3600 spectrometer (Tokyo, Japan), in the spectral range from 400 to 4000 cm^{-1}. The PL spectrometer (Shimadzu RF-5301PC, Tokyo, Japan) was used to determine the photoluminescence properties.

2.4. Photocatalytic Experiments

A stock aqueous solution of MC-LR (30 mg L^{-1}) was prepared (pH 6.3). Prior the photocatalytic experiments, the standard calibration curve was made for MC-LR concentrations in the range of 5–30 mg L^{-1}. The photocatalytic degradation experiments were carried out in a double jacket round quartz reactor with a 50 mL volume. The temperature was maintained at 25 °C by circulation thermostated water around the reactor. Firstly, an aqueous solution of MC-LR (40 mL) was added to a round quartz reactor containing TiO_2 powder, and then sonicated to obtain a uniform suspension, which was then stirred for 180 min in the dark to achieve the adsorption equilibrium for MC-LR on the catalyst surface before turning the experiments [20]. Thereafter, the suspension was irradiated from the top by a visible-LED lamp (λ_{max} = 420 nm, intensity = 1 mW cm^{-2}, height 20 cm). The samples were dragged at different times, and then filtered using a 0.22-μm filter membrane. The residual MC-LR concentration at different durations was adopted using a High-Performance Liquid Chromatography (HPLC, 1260, Agilent, Hamburg, Germany) with a G1311C-1260 Quat pump and a G1365D-1260 MWD UV detector (Hamburg, Germany), set at 238 nm with a C18 column (100 mm Long × 4.6 mm i.d., 3.5 μm particles) and using the method reported before [21]. The reaction rates were estimated and fitted with the Langmuir–Hinshelwood first-order kinetic model. The degradation rate (r) was calculated using Equation (1) [3,9,10]:

$$r = \mathrm{K} \times \mathrm{C_0}^{\mathrm{n}}, \quad (1)$$

where K is the rate constant, C_0 is the initial MC-LR concentration, and n is the order of the reaction. The MC-LR photodegradation efficiency (E %) using the as-prepared and calcined TiO_2 samples under visible light was determined using Equation (2) [9,10]:

$$E\% = (1 - (\mathrm{C}/\mathrm{C_0})) \times 100, \quad (2)$$

where C_0 and C are the MC-LR concentrations before and after irradiation, respectively.

3. Results and Discussion

3.1. Characteristics of Photocatalyst

The crystallographic structure of TiO_2 powders, obtained by thermal hydrolysis of an aqueous solution of $Ti_2(SO_4)_3$, which was pre-oxidized by sodium nitrate in the presence of 1 mol L^{-1} glycine at pH 10 and then calcined at different temperatures (300, 400, and 500 °C), was investigated by XRD. The XRD patterns of CDT-0.00, CDT-300, CDT-400, and CDT-500 samples are displayed in Figure 1. The diffraction peaks of all samples showed mixed-phase titania of anatase and brookite. The peaks at 25.27° (101), 36.96° (103), 37.76° (004), 38.56° (112), 47.84° (200), 53.76° (105), 55.04° (121), 62.56° (204), and 68.64° (116) were well attributed to anatase-TiO_2 (JCPDS No. 84-1286) [9,10]. Additionally, the formation of brookite TiO_2 phase (JCPDS no.15-0875) could be confirmed by the peaks at 25.27° (120), 30.72° (121), 37.76° (201), 47.84° (231), 53.76° (320), 55.04° (241), and 68.64° (400) [9,10]. Although almost all brookite peaks overlapped with those of anatase, the most intensive brookite peak at 30.72° clearly indicated its presence. It should be pointed that brookite (metastable) is not a common titania polymorph, whereas anatase (metastable) and rutile (stable) have been frequently reported. Therefore, these results indicate the role of glycine, Na^+, and pH value in the formation of anatase-brookite TiO_2. It has been proposed that insoluble titanium hydroxide ($Ti(OH)_4$) is firstly obtained by the direct hydrolysis of a mixture of aqueous $Ti_2(SO_4)_3$ and $NaNO_3$ [9,10,22]. Then, by adding sodium hydroxide (to adjust pH to be 10) and glycine, the insoluble sodium titanate ($Na_{2-x}H_2Ti_2O_5.H_2O$) and soluble five-membered ring complex $[Ti(OH)_x(C_2H_5NO_2)_y]^{z-}$ are formed during the hydrothermal treatment [10,22–25]. The sodium titanate is transformed into brookite at a high temperature in highly alkaline medium, whereas $[Ti(OH)_x(C_2H_5NO_2)_y]^{z-}$ is transferred into anatase to obtain anatase-brookite heterojunction titania [10,22]. The content of the anatase and brookite phases were calculated using Zhang formula [26]. It was observed that an increase in calcination temperature resulted in a decrease in brookite content. Therefore, it was found that brookite was gradually converted to anatase over this temperature range (see Figure 1), and/or amorphous titania was transformed to anatase (as suggested by more intensive anatase peaks). Although some reports suggest the direct transition of brookite into rutile (not via anatase) [27] and an accelerated anatase to rutile transition (starting at 500 °C) in the presence of brookite [28], this study showed that the brookite presence does not necessarily cause fast and direct anatase-to-rutile transition (even at 500 °C). The Scherrer's equation was applied to estimate the average crystallite size for the anatase and brookite phases from the broadening of peaks 25.27° (120) and 30.72° (121), respectively (see Table 1) [10]. Clearly, both anatase and brookite crystallite sizes showed a typical upward trend with increasing calcination temperature. The change in phase structure and particle size of TiO_2 with increasing calcination temperature is corroborated by some previous reports [29–31]. For example, Allen et al. studied the influence of calcination temperature on the morphology (crystalline phase and size) and photocatalytic performance of TiO_2. They noticed a similar trend for the calcination of anatase-brookite TiO_2 samples, where anatase content increased gradually with increasing temperature, in contrast to the decrease in brookite content. Moreover, the crystal size of anatase and brookite increased with increasing calcination temperature [31].

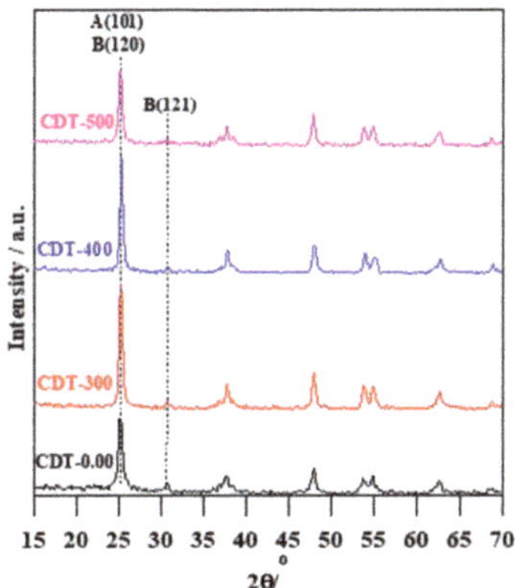

Figure 1. XRD patterns for CDT-0.00, CDT-300, CDT-400, and CDT-500 samples.

Table 1. Crystal size, BET surface area, pore volume, pore size and band gap of CDT-0.00, CDT-300, CDT-400, and CDT-500 samples.

Samples	Crystal Size (nm)		BET Surface Area ($m^2\ g^{-1}$)	Pore Volume ($cm^3\ g^{-1}$)	Pore Size (nm)	Band Gap (eV)
	A	B				
CDT-0.00	16	32	72.41	0.29	8.6	2.79
CDT-300	17	35	66.00	0.27	9.3	2.82
CDT-400	20	42	55.22	0.26	9.8	2.86
CDT-500	25	50	20.51	0.23	10.7	3.00

To confirm the phase structure of the obtained TiO_2, Raman spectroscopy was used. In Figure 2a and Table 2, it is explicit that all the TiO_2 samples exhibited certain peaks matching to the vibration modes of anatase and brookite [9,32]. Figure 2b shows the main Raman peak of anatase (Eg mode, peak position at around 145 cm^{-1}). The peak was red-shifted and its width increased with increasing temperature, and these observations are consistent with the literature data [33]. The redshift and broadening of Raman peaks of Eg mode can be assigned to thermal expansion, intrinsic anharmonicity and phonon confinement effects [33]. Heating results in expanding of material leading to redshift, and the intrinsic anharmonicity is stronger with temperature [33,34]. The phonon–phonon interactions, which may be due to the O-Ti-O bond vibrational type, were very obvious in the Eg mode, hence the phonon confinement effects should cause an increase in the shift and asymmetric broadening of Raman peaks [33,34].

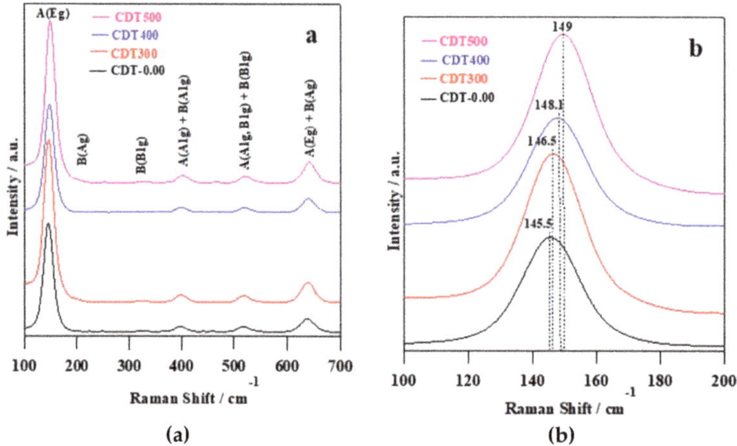

Figure 2. (**a**) Raman spectra of CDT-0.00, CDT-300, CDT-400, and CDT-500 samples; and (**b**) enlargement of the main peak around 145 cm^{-1}.

Table 2. Raman vibration bands and their assignments in CDT-0.00, CDT-300, CDT-400, and CDT-500 samples at 532 nm excitation wavelength.

Samples	Assignment [Peak Position (cm^{-1})]	
	Anatase	**Brookite**
CDT-0.00	Eg [145.5, 636], B1g [517.5], A1g [396, 517.5]	Ag [199, 636], A1g [396], B1g [517.5]
CDT-300	Eg [146.5, 639], B1g [518], A1g [397, 518]	Ag [199, 639], A1g [397], B1g [323, 518]
CDT-400	Eg [148.1, 641.5], B1g [518.1], A1g [399, 518.1]	Ag [199, 641.5], A1g [399], B1g [518.1]
CDT-500	Eg [149, 644], B1g [521], A1g [400, 521]	Ag [199, 644], A1g [400], B1g [323, 521]

Figure 3 shows exemplary TEM images and the corresponding electron diffraction (ED) patterns of all samples obtained at different calcination temperatures. All samples contained nano-rod-like particles and small nano-quasi-spherical-like particles. Since brookite has orthorhombic structure, in which TiO_6 octahedron shares three edges and corners, the rod-like particles of brookite have frequently been reported [35–39]. Therefore, it was assumed that the small nanoparticles Were related to the anatase phase, whereas brookite formed the rod-like particles. Similar to crystallite size, the particle sizes increased with an increase in the calcination temperature. According to the insets in Figure 3a–d, the interface, as revealed by the electron diffraction (ED) patterns, consisted of anatase/brookite, being consistent with XRD data.

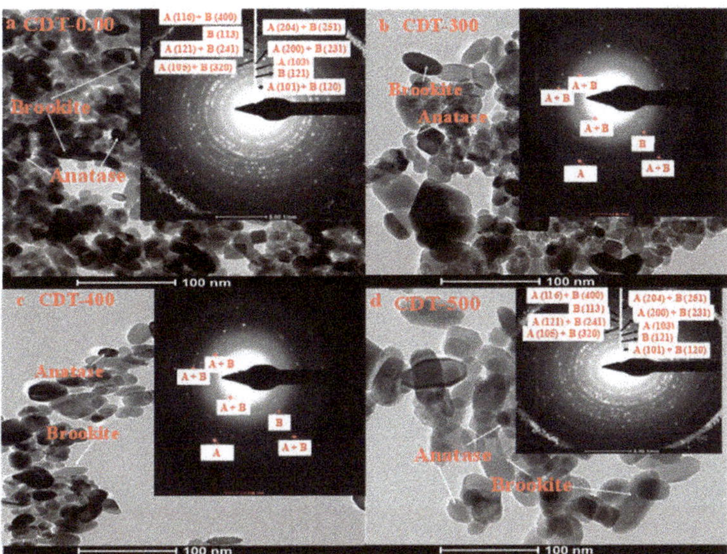

Figure 3. (**a–d**) TEM images of CDT-0.00, CDT-300, CDT-400, and CDT-500 samples, respectively.

Figure 4a shows the nitrogen adsorption–desorption isotherms and pore-size distribution (inset) of CDT-0.00, CDT-300, CDT-400, and CDT-500 samples. It was found that all TiO_2 samples exhibited IV-type isotherm according to IUPAC classification, proving the formation of a mesoporous structure [32,40]. The broad hysteresis loops (0.35–0.96 P/P_o) for CDT-0.00 and CDT-300 samples and narrow ones (0.53–0.96 P/P_o) for CDT-400 and CDT-500 samples indicate that thermal treatment caused a shift for hysteresis loops to higher relative pressure, and this might correlate with the gradual loss of mesoporous structure with increasing temperature [41]. Moreover, it should be mentioned that, with increasing temperature, the particle and pore sizes of the TiO_2 photocatalyst increased, resulting in narrowing the hysteresis loops because the large particles could be considered as irregular voids [42]. Additionally, it must be pointed that thermal treatment influenced all surface properties, e.g., pore sizes increased from 8.6 to 10.7 nm, whereas pore volume and specific surface area decreased from 0.29 to 0.23 $cm^3\ g^{-1}$ and 72.4 to 20.5 $m^2\ g^{-1}$, respectively, by thermal treatment at 500 °C (as summarized in Table 1). Figure 4b presents the diffuse reflectance spectra and band gap estimations (inset) of as-prepared and calcined TiO_2 photocatalysts. The increase in calcination temperature resulted in the hypochromic effect, i.e., shift of the absorption edge towards shorter wavelengths, demonstrating the broadening of the band gap (from 2.79 to 3.00 eV), as shown in Figure 4b and Table 1. Generally, brookite shows broader bandgap than anatase, ca. 3.3 eV, but the modification resulted in bandgap narrowing; however, temperature again caused an increase in the band gap. Accordingly, it is proposed that, by applying the thermal treatment, the non-metal elements released gradually, hence the visible light response of the samples was diminished, resulting in redshift of absorption edge and broadening of band gap, similarly to previous reports [43,44].

To determine the nature of incorporated elements, the modified TiO_2 photocatalysts were also characterized by XPS and FT-IR (see Figures 5–7). Figure 5a shows XPS spectra of the samples for titanium (Ti 2p). All samples exhibited two Ti components with the core level binding energies at 458.5, and 464.2 eV for Ti $2p_{3/2}$ and Ti $2p_{1/2}$, respectively. These peaks are attributed to TiO_2 lattice (Ti-O-Ti) [7,8,10,19,32]. Figure 5b gives the XPS of O 1s spectra of the TiO_2 samples. The samples displayed two O components with the O 1s binding energies at 529.7 and 531.6 eV. The low binding energy is ascribed to oxygen in TiO_2 lattice (Ti−O−Ti), and the higher one to the surface hydroxyl groups, resulting from chemisorbed water [7,8,10,19,32]. It was observed that the content of surface hydroxyl groups decreased with increasing the calcination temperature, and this is consistent with the FTIR results. The decrease in surface hydroxyl groups might lead to reducing the possibility for trapping photogenerated holes, and thus enhancing the possibility of the electron–hole recombination, which agrees with the PL results [7,8,10,19,32,45]. The XPS spectra for C 1s are presented in Figure 6a. Two peaks were observed with the binding energies of 284.4 and 288.7 eV, which were assigned to C-C, and C-H and C-O, C=O, O=C-O, Ti-O-C, and C-N bonds, respectively [10,19,32]. For nitrogen (N 1s), one peak with binding energy of 401.7 eV might correspond to the chemisorbed N species (NO, N_2O, NO^{2-}, and NO^{3-}), hyponitrite species, interstitial N-doping (Ti-O-N and Ti-N-O linkage) and substitutional N-doping (O-Ti-N linkage) (see Figure 6b) [10,19,32]. The surface compositions of C/N co-modified TiO_2 samples are summarized in Table 3. It indicates that the atomic concentration of Ti and O increased with increasing temperature, while O/Ti ratio remained almost unchanged. On the other hand, the content of N and C decreased with increasing calcination temperature, which might be explained by the thermal release of surface modifiers (non-metal elements), as already reported [43,44].

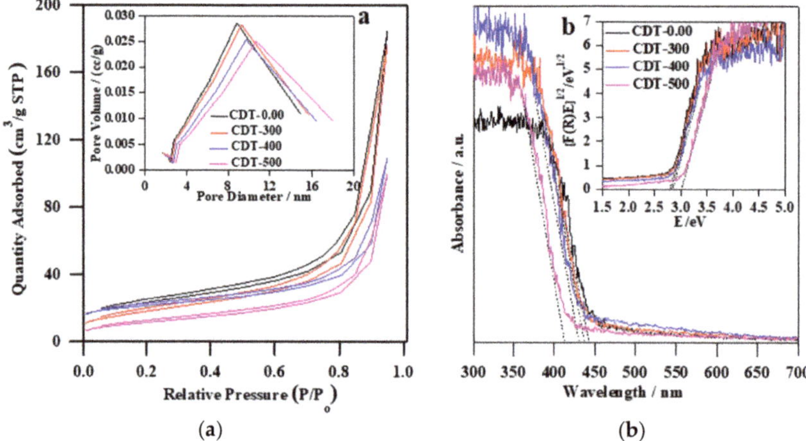

Figure 4. (**a**) N_2 sorption isotherm and pore size distribution (inset); and (**b**) diffuse reflectance spectra and Tauc plots (inset) for all obtained samples.

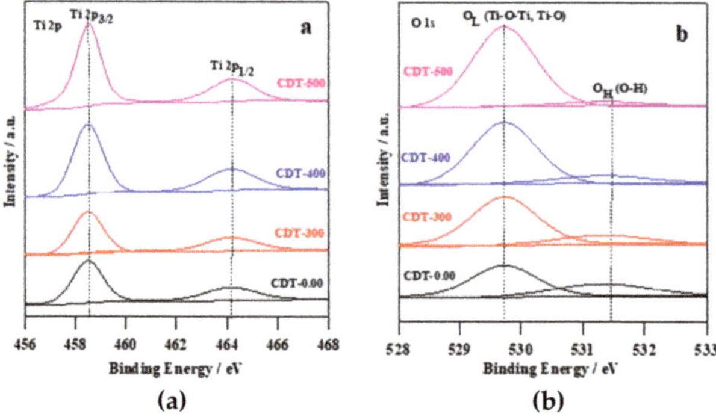

Figure 5. XPS detailed scans in the energy regions of: (**a**) Ti 2p; and (**b**) O 1s of all samples.

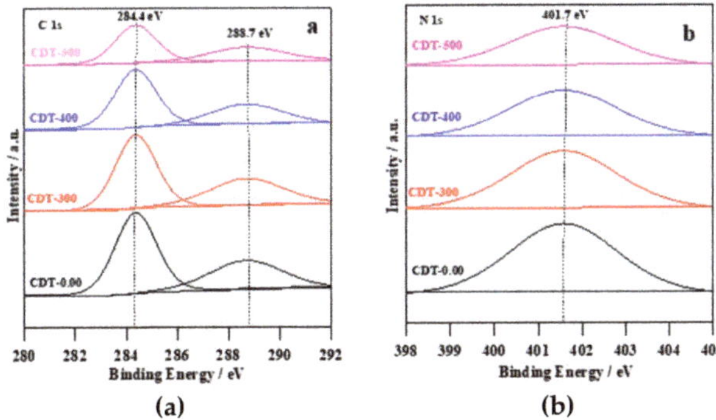

Figure 6. XPS detailed scans in the energy regions of: (**a**) C 1s; and (**b**) N 1s of the obtained photocatalysts.

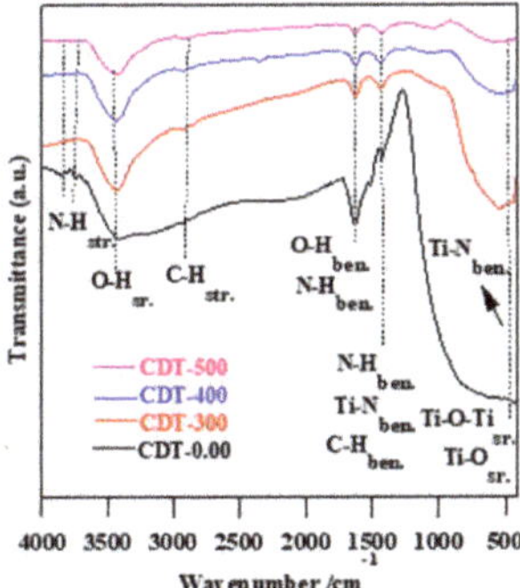

Figure 7. FT-IR of CDT-0.00, CDT-300, CDT-400, and CDT-500 samples.

Table 3. The surface composition of Ti, O, C, and N obtained from XPS.

Samples	Content (at. %)				Ratio
	Ti 2p	O 1s	C 1s	N 1s	O/Ti
CDT-0.00	20.20	56.20	21.70	1.90	2.8
CDT-300	20.64	57.60	20.10	1.66	2.8
CDT-400	21.05	58.10	19.40	1.45	2.8
CDT-500	21.98	58.60	18.20	1.22	2.7

The existence of carbon and nitrogen for the obtained TiO_2 powders was further confirmed by FTIR spectroscopy, as shown in Figure 7. It was found that prepared TiO_2 samples displayed vibration modes at the 4000–400 cm^{-1} range, indicating the possible incorporation of C and N in the TiO_2 lattice (see Figure 7 and Table 4) [9,10,32,40]. The Ti–O bond and Ti–O–Ti bridge stretching are located at 900–400 cm^{-1} [9,10]. The low FTIR bands located at 480 cm^{-1} are attributed to Ti-$N_{str.}$ bonds [10,32]. The bands located at 1416 cm^{-1} are related to Ti-$N_{ben.}$, C-$H_{ben.}$, and N-$H_{ben.}$ [10,32]. N-$H_{ben.}$ and O-$H_{str.}$ bonds are located at 1630 cm^{-1} [9,10,32,40]. The FTIR bands located at 2922 cm^{-1} are assigned to C-$H_{str.}$ [10]. The broad bands located at 3418 cm^{-1} are characterized to O-$H_{str.}$ [9,10,32,40]. The weak FTIR bands located at 3700–3800 cm^{-1} are due to N-H_{str}. [10,32]. It is worth noting that the intensities of O-H, N-H, C-H and Ti-N vibration bands decreased with increasing calcination temperature, confirming XPS data, i.e., the thermal treatment caused a release of non-metal elements and/or chemosorbed H_2O [46–48].

Table 4. FTIR vibration bands and their locations for CDT-0.00, CDT-300, CDT-400, and CDT-500 samples.

Vibration Mode	Bands Location (cm^{-1})
Ti-O; Ti-O-Ti (stretching)	400–900
O-H (stretching)	3418
O-H (bending)	1630
N-H (stretching)	3700–3800
N-H (bending)	1630; 1416
C-H (stretching)	2922
C-H (bending)	1416
Ti-N (bending)	1416, 480

3.2. *Photocatalytic Activity of Photocatalyst*

3.2.1. Degradation of MC-LR over TiO_2 Photocatalyst

To compare the catalytic activity of as-synthesized (CDT-0.00) and calcined (CDT-300, CDT-400, and CDT-500) C-N co-modified TiO_2 for MC-LR removal, three kinds of experiments were performed: MC-LR adsorption in the dark (blank test), MC-LR photolysis (without photocatalyst), and MC-LR photodegradation (visible light + photocatalyst). In the initial experiments, the pH value and the content of TiO_2 were 6.3 and 0.3 g L^{-1}, respectively. Before the photocatalytic reaction, the adsorption of MC-LR on the surface of co-modified TiO_2 samples in the dark was performed (see Figure 8a). It was found that 3-h stirring resulted in 40%, 39%, 33% and 20% adsorption of MC-LR on the surface of CDT-0.00, CDT-300, CDT-400 and CDT-500, respectively. More efficient adsorption of MC-LR on the surface of as-synthesized co-modified TiO_2 than that on the calcined samples could result from larger specific surface area, as shown in inset of Figure 8a. Blank tests (in the absence of photocatalyst and irradiation) indicated that MC-LR is stable and irradiation was necessary for its efficient decomposition (Figure 8a,b). It was found that 3-h vis irradiation resulted in almost complete degradation of MC-LR, reaching 95% (0.0172 mni^{-1}, r = 1.04 × 10^{-4} mmol L^{-1}), 90% (0.0131 mni^{-1}, r = 0.81 × 10^{-4} mmol L^{-1}), 86% (0.0113 min^{-1}, r = 0.76 × 10^{-4} mmol L^{-1}), and 79% (0.0084 mn^{-1}, r = 0.68 × 10^{-4} mmol L^{-1}) removal efficiencies for CDT-0.00, CDT-300, CDT-400, and CDT-500 samples, respectively, as shown in Figure 8b. Therefore, it was confirmed that all co-modified titania samples displayed a high photocatalytic performance. Several reasons could be responsible for this high photocatalytic activity, such as formation of mixed-phase structure (anatase/brookite (A/B)), mesoporous morphology, and non-metal co-modification [9–19,22,25,32,48]. The formation of A/B mixed-phase TiO_2 might facilitate the migration of electrons from brookite to anatase, and holes from anatase to brookite, and hence reducing the possibility of electron–hole recombination [9,10,22,25]. The high photocatalytic activity of the mesoporous TiO_2 can be explicated by: (i) more active sites for photocatalytic reaction; (ii) the cumulation of hydroxyl radicals inside the pores; (iii) the high dispersion of mesoporous TiO_2 in the aqueous solution; and (iv) the rapid diffusion of MC-LR to the reactive sites on the surface of the mesoporous TiO_2 photocatalyst [49,50]. The high activity of the photocatalysts under visible light can be attributed to narrowing of band gap resulting from non-metal modification, as shown in our previous report on non-modified titania [32]. Moreover, it was observed that the degradation rate was decreased with an increase in calcination temperature, which might be explained by: (i) the increases in crystal and pore sizes; and (ii) the decreases in specific surface area and the content of non-metal elements. It is worth noting that the charge-carriers recombination was gradually increasing with increasing the calcination temperature, probably due to a decrease in the amount of adsorbed water molecules, which play an essential role in the photocatalytic activity since they can react with photogenerated holes forming reactive oxygen species (ROS), thereby inhibiting charge recombination [19]. These hypotheses were confirmed by changes in the XPS spectra (a decrease in hydroxyl groups), FTIR bands (O-H vibration) and PL intensities (Figures 5, 7 and 9, respectively) with an increase in calcination temperature. For example, the intensity of the bands located at 3418 and 1630 cm^{-1}, attributed to the O-H vibration modes of adsorbed water molecules, decreased with increasing the calcination temperature (Figure 7).

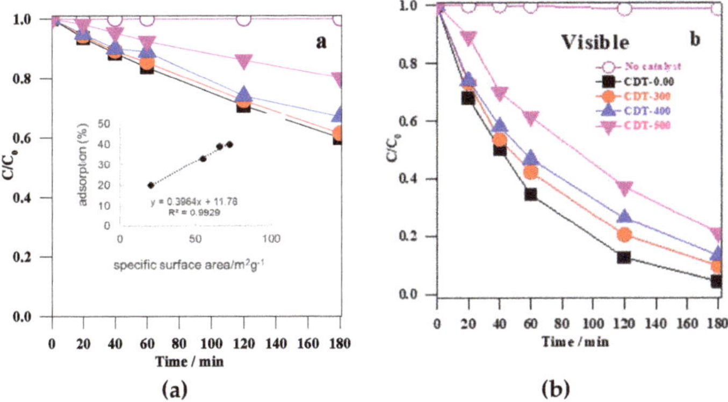

Figure 8. Change of the MC-LR concentration with initial concentration 10 mg L^{-1} versus reaction time: (**a**) under dark condition; and (**b**) under visible light illumination for 3 h using CDT-0.00, CDT-300, CDT-400, CDT-500 catalysts (0.3 g L^{-1}) at pH 6.3.

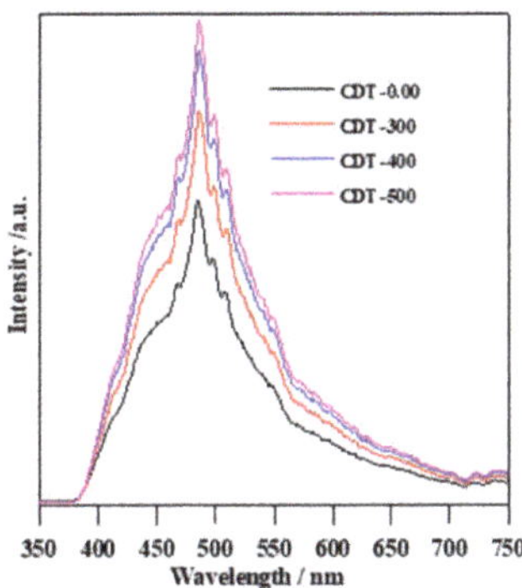

Figure 9. PL spectra of CDT-0.00, CDT-300, CDT-400, and CDT-500 catalysts.

The photoluminescence (PL) spectroscopy has been used widely to investigate the charge separation/recombination of photogenerated charges. Additionally, the PL emission intensity is related directly to the recombination rate of the e^-/h^+ pairs, i.e., the lower is the PL emission intensity, the lower is the recombination rate, hence an increase in the photocatalytic activity of the materials [10,33]. Indeed, higher PL intensity of the sample CDT-0.00 than those of calcined samples (Figure 9) correlates well with photocatalytic activity (Figure 8), confirming the lower charge carriers' recombination. The results indicate that PL peaks of all samples are located at the same position around 485 nm, which can be attributed to the free electron recombination process from the conduction band (CB) to the ground state recombination center [32]. The PL intensities increased with increasing the calcination temperature, i.e., the as-synthesized TiO_2 catalyst (CDT-0.00) exhibited the lowest PL intensity indicating the lowest charge carriers' recombination, and thus the highest photocatalytic activity for MC-LR degradation.

3.2.2. Improvement of Photocatalytic Degradation

To attain the highest efficiency of MC-LR degradation on co-modified TiO_2 under vis irradiation, the condition-dependent activity was investigated, i.e., the influence of the initial pH value, the initial content of TiO_2, and the initial concentration of MC-LR. Firstly, the pH effect on the degradation of MC-LR (10 mg L^{-1}) on CDT-0.00 photocatalyst (0.4 g L^{-1}) was investigated for three different pH values (4, 6.3, and 8), and the data are shown in Figure 10a. The complete MC-LR degradation (100%, K = 0.034 min^{-1}, and r = 2.01 × 10^{-4} mmol L^{-1}) was achieved at the acidic pH 4, whereas at more neutral (pH 6.3) and basic (pH 8) conditions, the total degradation efficiencies were lower reaching 97% (K = 0.019 min^{-1}, and r = 1.16 × 10^{-4} mmol L^{-1}) and 81% (K = 0.009 min^{-1}, and r = 0.59 × 10^{-4} mmol L^{-1}), respectively. The enhanced photocatalytic degradation of MC-LR in acidic conditions has already been reported [3,4]. It is well known that the TiO_2 surface is positively/negatively charged below/above the point of zero charge (p_{zc} = 6–8 depending on titania sample) [51–55]. Hence, the TiO_2 surface is positively charged (Ti-OH_2^+) at pH 4. It is important to note that MC-LR is negatively charged above pH 2.10, and positively charged below this value [56]. Therefore, obtained results, i.e., best activity at pH 4, suggest that attractive forces between the positively charged TiO_2 ($TiOH_2^+$) and the negatively charged MC-LR (MC-LRH^-) enhance the adsorption of MC-LR on the surface of C/N co-modified TiO_2 photocatalyst, thus improving its photocatalytic degradation.

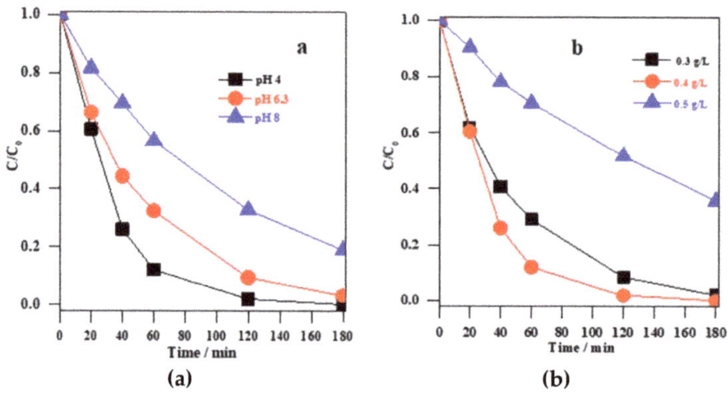

Figure 10. (**a**) Photodegradation of 10 mg L^{-1} MC-LR in the presence of CDT-0.00 (0.4 g L^{-1}) under visible irradiation at different pH values (4, 6.3, and 8); and (**b**) photodegradation of 10 mg L^{-1} MC-LR using different dosages of CDT-0.00 (0.3, 0.4, and 0.5 g L^{-1}) under visible irradiation at pH 4.

Next, the influence of photocatalyst dose was investigated, and obtained data are shown in Figure 10b for three different values (0.3, 0.4, and 0.5 g L^{-1}) at acidic conditions (pH 4) and for initial concentration of MC-LR of 10 mg L^{-1}. It was found that the photocatalytic efficiency increased with an increase in the photocatalyst content from 0.3 to 0.4 g L^{-1} (from 98% (K = 0.021 min^{-1}, and r = 1.27 × 10^{-4} mmol L^{-1}) to 100% (K = 0.034 min^{-1}, and r = 2.01 × 10^{-4} mmol L^{-1})), due to more efficient photon absorption, similar to reported data [21,39,57,58]. However, further increase in the photocatalyst dose (0.5 g L^{-1}) resulted in a decrease in the photocatalytic efficiency to 64% (K = 0.006 min^{-1}, and r = 0.37 × 10^{-4} mmol L^{-1}), due to the "inner filter" effect (reduced penetration depth and increased scattering of the incident light beam) [20,56]. Therefore, for reliable comparison between different photocatalysts the photoirradiation, experiments should be performed at maximum photon absorption, i.e., after reaching plateau in photocatalyst dose-activity dependence, as clearly discussed by Kisch [58]. Although, usually for titania powders of specific surface areas of about 50–200 m^2g^{-1}, the optimum dose in the range of 0.5–3.0 g L^{-1} [58] has been proposed, our data indicate that even 0.5 g L^{-1} could be too large (probably due to intensive coloration by C/N-modification).

The influence of initial MC-LR concentration on the photocatalytic performance of as-synthesized C/N co-modified TiO_2 with an initial TiO_2 dose of 0.4 g L^{-1} at pH 4 is shown in Figure 11. It was found that an increase in MC-LR concentration resulted in a decrease in degradation efficiency. This obvious behavior is caused by the fact that higher concentration means higher competition (more molecules) for same content of reactive oxygen species, as well as the competition between oxidation by-products and original compound (MC-LR) for both adsorption (on the photocatalyst surface) and degradation. The MC-LR degradation presented a first-order behavior (as confirmed by logarithmic plot (Figure 11b)) with the reaction rates of 2.01×10^{-4} mmol L^{-1} (K = 0.034 min^{-1}), 1.03×10^{-4} mmol L^{-1} (K = 0.017 min^{-1}), and 0.73×10^{-4} mmol L^{-1} (K = 0.011 min^{-1}) and degradation efficiency after 180-min vis irradiation of 100%, 95% and 83% at the initial MC-LR concentration of 10, 20, and 30 mg L^{-1}, respectively.

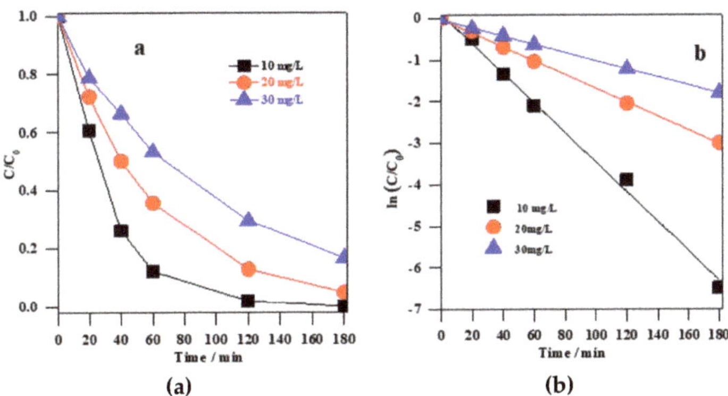

Figure 11. (**a**) Photodegradation of MC-LR with different initial concentration (10, 20, and 30 mg L^{-1}), using 0.4 g L^{-1} CDT-0.00 under visible irradiation at pH 4; and (**b**) plot of C/C_0 vs irradiation time.

3.2.3. Proposed Mechanism for MC-LR Removal

(i) Adsorption:

Before irradiation, an aqueous solution of MC-LR was stirred with C/N co-modified TiO_2 powder in the dark for 180 min to attain the adsorption equilibrium for MC-LR on the photocatalyst surface. Scheme 1 shows the proposed mechanism of MC-LR adsorption on the TiO_2 surface in the dark at acidic conditions, i.e., fast adsorption due to the attractive forces between the positively charged TiO_2 ($TiOH_2^+$) and the negatively charged MC-LR (MC-LRH$^-$).

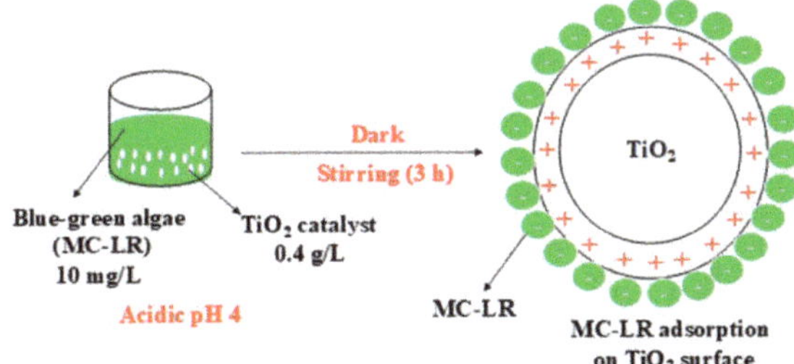

Scheme 1. Proposed mechanism of MC-LR adsorption on co-modified TiO_2 at acidic pH 4.

(ii) Photodegradation:

The proposed mechanism of the MC-LR photodegradation on C/N co-modified TiO_2 photocatalyst is shown in Scheme 2. Firstly, vis irradiation could excite semiconductor only from the impurity energy levels (non-metal levels, considering surface modification with doping-like nature, e.g., surface oxygen replaced by nitrogen and carbon) to the conduction band, and leaving photo-generated holes at the impurity levels in both anatase and brookite phase, as presented in Equation (3).

$$TiO_2 + h\nu \xrightarrow{Photo-excitation} h^+ + e^-, \quad (3)$$

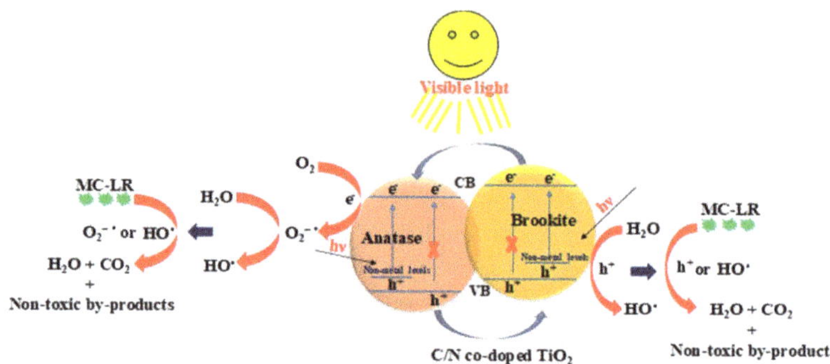

Scheme 2. Possible mechanism of MC-LR degradation over C/N co-modified A/B TiO_2 photocatalyst.

Moreover, it is possible (and has already been proposed in our previous and other reports [9,10,22,25,32,40,47,48]) that photogenerated electrons could transfer from brookite to anatase, whereas the holes could migrate inversely (from anatase to brookite), thereby enhancing the electron–hole separation. Following that, these separated charges can oxidize and reduce the adsorbed H_2O and O_2, respectively, to form high reactive oxygen species (ROS) by the following reactions:

$$H_2O + h^+ \xrightarrow{Oxidation} {}^\bullet OH, \quad (4)$$

$$O_2 + e^- \xrightarrow{Reduction} {}^{\bullet -}O_2, \quad (5)$$

$${}^{\bullet -}O_2 + H_2O \longrightarrow {}^\bullet OH \quad (6)$$

Finally, MC-LR adsorbed on the TiO_2 surface can be degraded either directly by h^+ or indirectly by formed ROS (Equations (4)–(6)) to be converted finally into non-toxic products, H_2O and CO_2, as shown in Equation (7).

$$\text{MC-LR} + h^+,\ {}^\bullet OH,\ (\text{and/or})\ {}^{\bullet -}O_2 \xrightarrow{Degradation} H_2O + CO_2 + \text{non-toxic by-products} \quad (7)$$

4. Summary and Conclusions

MC-LR toxic pollutant was removed from aqueous solution over visible light responsive C/N co-modified, mesoporous, mixed-phase (A/B) TiO_2 photocatalyst, synthesized through one-pot hydrothermal approach using glycine as C/N source. The as-synthesized and calcined TiO_2 at different temperatures (300, 400, and 500 °C) were prepared, and characterized by XRD, Raman, TEM, BET-surface area, UV-vis-DRS, XPS, FTIR, and PL. The important parameters, including the initial pH value, the photocatalyst content, and MC-LR concentration, were studied to reach the best conditions for photocatalytic degradation of MC-LR using the co-modified TiO_2 under visible irradiation. The results reveal that all mesoporous co-modified A/B TiO_2 photocatalysts exhibited high photocatalytic activity toward MC-LR degradation, probably because of their mesoporous structure and non-metal modification. The photocatalytic efficiency decreased with increasing calcination temperature, as a result of decreased specific surface area and a release of non-metal modifiers. The complete degradation of 10 mg L^{-1} MC-LR over 0.4 g L^{-1} as-synthesized C/N co-modified TiO_2 at pH 4 was achieved under vis irradiation for 180 min. It is proposed that mixed-phase formation and non-metal co-modification cause the high photocatalytic activity under visible light irradiation, as shown in Scheme 2. Therefore, it is concluded that this photocatalyst is suitable for efficient water/wastewater treatment under natural solar radiation.

Author Contributions: Conceptualization, T.M.K. and S.M.E.-S.; methodology, T.M.K.; experimental research, T.M.K.; data analysis, T.M.K. and E.K.; writing—original draft preparation, T.M.K.; writing—review and editing, T.M.K., S.M.E.-S., A.A.I. and E.K.; resources, D.W.B., S.M.E.-S., A.A.I. and E.K.; publication fees, E.K; and supervision, S.M.E.-S., A.A.I. and D.W.B.

Funding: This work was supported by the Egyptian Ministry of Higher Education (Cultural Affairs and Missions Sector). A part of this work was also supported by US-Egypt Joint project Cycle 17 no. 229 and short-term fellowship Cycle 6, Science & Technological Development Fund in Egypt (STDF-STF) under Grant No. 25503.

Acknowledgments: T.M. Khedr acknowledges Institute of Technical Chemistry, Photocatalysis and Nanotechnology Research Unit, Leibniz Universität Hannover, Germany for hosting him during this research work.

Conflicts of Interest: The author declares no conflict of interest.

References

1. Chae, S.; Noeiaghaei, T.; Oh, Y.; Kim, I.S.; Park, J.-S. Effective Removal of Emerging Dissolved Cyanotoxins from Water using Hybrid Photocatalytic Composites. *Water Res.* **2019**, *149*, 421–431. [CrossRef]
2. Carmichael, W.W. The cyanotoxins. *Adv. Bot. Res.* **1997**, *27*, 211–256.
3. Nawaz, M.; Moztahida, M.; Kim, J.; Shahzad, A.; Jang, J.; Miran, W.; Lee, D.S. Photodegradation of microcystin-LR using graphene-TiO_2/sodium alginate aerogels. *Carbohydr. Polym.* **2018**, *199*, 109–118. [CrossRef] [PubMed]
4. Khadgi, N.; Upreti, A.R. Photocatalytic degradation of Microcystin-LR by visible light active and magnetic, $ZnFe_2O_4$-Ag/rGO nanocomposite and toxicity assessment of the intermediates. *Chemosphere* **2019**, *221*, 441–451. [CrossRef]
5. Antoniou, M.G.; de la Cruz, A.A.; Dionysiou, D.D. Cyanotoxins: New Generation of Water Contaminants. *J. Environ. Eng.* **2005**, *131*, 1239–1243. [CrossRef]
6. Guo, Q.; Li, H.; Zhang, Q.; Zhang, Y. Fabrication, characterization and mechanism of a novel Z-scheme Ag_3PO_4/NG/Polyimide composite photocatalyst for microcystin-LR degradation. *Appl. Catal. B Environ.* **2018**, *229*, 192–203. [CrossRef]
7. Andersen, J.; Han, C.; O'Shea, K.; Dionysiou, D.D. Revealing the degradation intermediates and pathways of visible light-induced NF-TiO_2 photocatalysis of microcystin-LR. *Appl. Catal. B Environ.* **2014**, *154–155*, 259–266. [CrossRef]
8. Wang, X.; Wang, X.; Zhao, J.; Song, J.; Zhou, L.; Wang, J.; Tong, X.; Chen, Y. An alternative to in situ photocatalytic degradation of microcystin-LR by worm-like N, P co-doped TiO_2/expanded graphite by carbon layer (NPT-EGC) floating composites. *Appl. Catal. B Environ.* **2017**, *206*, 479–489. [CrossRef]
9. El-Sheikh, S.M.; Khedr, T.M.; Zhang, G.; Vogiazi, V.; Ismail, A.A.; O'Shea, K.; Dionysiou, D.D. Tailored synthesis of anatase-brookite heterojunction photocatalysts for degradation of cylindrospermopsin under UV-Vis light. *Chem. Eng. J.* **2017**, *310*, 428–436. [CrossRef]
10. Khedr, T.M.; El-Sheikh, S.M.; Ismail, A.A.; Bahnemann, D.W. Photodegradation of 4-aminoantipyrine over Nano-Titania Heterojunctions Using Solar and LED Irradiation Sources. *J. Environ. Chem. Eng.* **2019**, *17*, 102797. [CrossRef]
11. Feitz, A.J.; Waite, D.T.; Jones, G.J.; Boyden, B.H.; Orr, P.T. Photocatalytic degradation of the blue green algal toxin microcystin-LR in a natural organic aqueous matrix. *Environ. Sci. Technol.* **1999**, *33*, 243–249. [CrossRef]
12. Devi, L.G.; Kavitha, R. A review on non-metal ion doped titania for the photocatalytic degradation of organic pollutants under UV/solar light: Role of photogenerated charge carrier dynamics in enhancing the activity. *Appl. Catal. B Environ.* **2013**, *140–141*, 559–587. [CrossRef]
13. Hu, X.; Hu, X.; Tang, C.; Wen, S.; Wu, X.; Long, J.; Yang, X.; Wang, H.; Zhou, L. Mechanisms underlying degradation pathways of microcystin-LR with doped TiO_2 photocatalysis. *Chem. Eng. J.* **2017**, *330*, 355–371. [CrossRef]
14. Fiorenza, R.; Bellardita, M.; Scirè, S.; Palmisano, L. Effect of the addition of different doping agents on visible light activity of porous TiO_2 photocatalysts. *Mol. Catal.* **2018**, *455*, 108–120. [CrossRef]

15. Pedrosa, M.; Pastrana-Martínez, L.M.; Pereira, M.F.R.; Faria, J.L.; Figueiredo, J.L.; Silva, A.M.T. N/S-doped graphene derivatives and TiO_2 for catalytic ozonation andphotocatalysis of water pollutants. *Chem. Eng. J.* **2018**, *348*, 888–897. [CrossRef]
16. Abdelraheem, W.H.M.; Patil, M.K.; Nadagouda, M.N.; Dionysiou, D.D. Hydrothermal synthesis of photoactive nitrogen- and boron- codoped TiO_2 nanoparticles for the treatment of bisphenol A in wastewater: Synthesis, photocatalytic activity, degradation byproducts and reaction pathways. *Appl. Catal. B* **2019**, *241*, 598–611. [CrossRef]
17. Reinosa, J.J.; Álvarez Docio, C.M.; Zapata-Ramírez, V.; Fernández, J.F. Hierarchical nano ZnO-micro TiO_2 625 composites: High UV protection yield lowering photodegradation in sunscreens. *Ceram. Int.* **2018**, *44*, 2827–2834. [CrossRef]
18. Reinosa, J.J.; Leret, P.; Álvarez Docio, C.M.; del Campo, A.; Fernández, J.F. Enhancement of UV absorption behavior in ZnO-TiO_2 composites. *Bol. Soc. Esp. Ceram. Vid.* **2016**, *55*, 55–62. [CrossRef]
19. Liu, G.; Han, C.; Pelaez, M.; Zhu, D.; Liao, S.; Likodimos, V.; Kontos, A.G.; Falaras, P.; Dionysiou, D.D. Enhanced visible light photocatalytic activity of C N-codoped TiO_2 films for the degradation of microcystin-LR. *J. Mol. Catal. A Chem.* **2013**, *372*, 58–65. [CrossRef]
20. Peng, G.; Fan, Z.; Wang, X.; Sui, X.; Chen, C. Photodegradation of microcystin-LR catalyzed by metal phthalocyanines immobilized on TiO_2-SiO_2 under visible-light irradiation. *Water Sci. Technol.* **2015**, *72*, 1824–1831. [CrossRef]
21. He, X.; Pelaez, M.; Westrick, J.A.; O'Shea, K.E.; Hiskia, A.; Triantis, T.; Kaloudis, T.; Stefan, M.I.; de la Cruz, A.A.; Dionysiou, D.D. Efficient removal of microcystin-LR by UV-C/H_2O_2 in synthetic and natural water samples. *Water Res.* **2012**, *46*, 1501–1510. [CrossRef] [PubMed]
22. Shen, X.; Zhang, J.; Tian, B.; Anpo, M. Tartaric acid-assisted preparation and photocatalytic performance of titania nanoparticles with controllable phases of anatase and brookite. *J. Mater. Sci.* **2012**, *47*, 5743–5751. [CrossRef]
23. Nagase, T.; Ebina, T.; Iwasaki, T.; Hayashi, H.; Onodera, Y.; Chatterjee, M. Hydrothermal synthesis of brookite. *Chem. Lett.* **1999**, *9*, 911–912. [CrossRef]
24. Yamashita, H.; Harada, M.; Misaka, J.; Takeuchi, M.; Neppolian, B.; Anpo, M. Photocatalytic degradation of organic compounds diluted in water using visible light-responsive metal ion-implanted TiO_2 catalysts: Fe ion-implanted TiO_2. *Catal. Today* **2003**, *84*, 91–196. [CrossRef]
25. Shen, X.; Tian, B.; Zhang, J. Tailored preparation of titania with controllable phases of anatase and brookite by an alkalescent hydrothermal route. *Catal. Today* **2013**, *201*, 151–158. [CrossRef]
26. Zhang, H.; Banfield, J.F. Understanding polymorphic phase transformation behavior during growth of nanocrystalline aggregates: Insights from TiO_2. *J. Phys. Chem. B* **2000**, *104*, 3481–3487. [CrossRef]
27. Perego, C.; Clemençon, I.; Rebours, B.; Revel, R.; Durupthy, O.; Cassaignon, S.; Jolivet, J.-P. Thermal Stability of Brookite-TiO_2 Nanoparticles with Controlled Size and Shape: In-situ studies by XRD. *Mater. Res. Soc. Symp. Proc.* **2009**, *1146*, NN04.
28. Hu, Y.; Tsai, H.-L.; Huang, C.-L. Effect of brookite phase on the anatase–rutile transition in titania nanoparticles. *J. Eur. Ceram. Soc.* **2003**, *23*, 691–696. [CrossRef]
29. Ye, X.S.; Sha, J.; Jiao, Z.K.; Zhang, L.D. Thermoanalytical characteristic of nanocrystalline brookite-based titanium dioxide. *Nanostruct. Mater.* **1997**, *8*, 919–927. [CrossRef]
30. Kandiel, T.A.; Robben, L.; Alkaim, A.; Bahnemann, D. Brookite versus anatase TiO_2 photocatalysts: Phase transformations and photocatalytic activities. *Photochem. Photobiol. Sci.* **2013**, *12*, 602–609. [CrossRef]
31. Allen, N.S.; Mahdjoub, N.; Vishnyakov, V.; Kelly, P.J.; Kriek, R.J. The effect of crystalline phase (anatase, brookite and rutile) and size on the photocatalytic activity of calcined polymorphic titanium dioxide (TiO_2). *Polym. Degrad. Stab.* **2018**, *150*, 31–36. [CrossRef]
32. El-Sheikh, S.M.; Khedr, T.M.; Hakki, A.; Ismail, A.A.; Badawy, W.A.; Bahnemann, D.W. Visible Light Activated Carbon and Nitrogen Co-doped Mesoporous TiO_2 as Efficient Photocatalyst for Degradation of Ibuprofen. *Sep. Purif. Technol.* **2017**, *173*, 258–268. [CrossRef]
33. Gao, K. Strongly intrinsic anharmonicity in the low-frequency Raman mode in nanocrystalline anatase TiO_2. *Physica B* **2007**, *398*, 33–37. [CrossRef]
34. Ohsaka, T.; Izumi, F.; Fujiki, Y. Raman spectrum of anatase, TiO_2. *J. Raman Spectrosc.* **1978**, *7*, 321. [CrossRef]

35. Lin, H.; Li, L.; Zhao, M.; Huang, X.; Chen, X.; Li, G.; Yu, R. Synthesis of High-Quality Brookite TiO_2 Single-Crystalline Nanosheets with Specific Facets Exposed: Tuning Catalysts from Inert to Highly Reactive. *J. Am. Chem. Soc.* **2012**, *134*, 8328–8331. [CrossRef] [PubMed]
36. Di Paola, A.; Bellardita, M.; Palmisano, L. Brookite, the Least Known TiO_2 Photocatalyst. *Catalysts* **2013**, *3*, 36–73. [CrossRef]
37. Zhao, H.; Liu, L.; Andino, J.M.; Li, Y. Bicrystalline TiO_2 with controllable anatase–brookite phase content for enhanced CO_2 photoreduction to fuels. *J. Mater. Chem. A* **2013**, *1*, 8209–8216. [CrossRef]
38. Zhu, G.; Lin, T.; Lü, X.; Zhao, W.; Yang, C.; Wang, Z.; Yin, H.; Liu, Z.; Huang, F.; Lin, J. Black brookite titania with high solar absorption and excellent photocatalytic performance. *J. Mater. Chem. A* **2013**, *1*, 9650–9653. [CrossRef]
39. Xu, J.; Wu, S.; Ri, J.H.; Jin, J.; Peng, T. Bilayer film electrode of brookite TiO_2 particles with different morphology to improve the performance of pure brookite-based dye sensitized solar cells. *J. Power Sources* **2016**, *327*, 77–85. [CrossRef]
40. Khedr, T.M.; El-Sheikh, S.M.; Ismail, A.A.; Bahnemann, D.W. Highly Efficient Solar Light-Assisted TiO_2 Nanocrystalline for Photodegradation of Ibuprofen Drug. *Opt. Mater.* **2019**, *88*, 117–127. [CrossRef]
41. Jiang, R.; Zhu, H.-Y.; Chen, H.-H.; Yao, J.; Fu, Y.-Q.; Zhang, Z.-Y.; Xu, Y.-M. Effect of calcination temperature on physical parameters and photocatalytic activity of mesoporous titania spheres using chitosan/poly (vinyl alcohol) hydrogel beads as a template. *Appl. Surf. Sci.* **2014**, *319*, 189–196. [CrossRef]
42. Robben, L.; Ismail, A.A.; Lohmeier, S.J.; Feldhoff, A.; Bahnemann, D.W.; Buhl, J.-C. Facile synthesis of highly ordered mesoporous and well crystalline TiO_2: Impact of different gas atmosphere and calcinations temperature on structural properties. *Chem. Mater.* **2012**, *24*, 1268–1275. [CrossRef]
43. Matsumoto, T.; Iyi, N.; Kaneko, Y.; Kitamura, K.; Ishihara, S.; Takasu, Y.; Murakami, Y. High visible-light photocatalytic activity of nitrogen-doped titania prepared from layered titania/isostearate nanocomposite. *Catal. Today* **2007**, *120*, 226–232. [CrossRef]
44. Zhang, G.; Zhang, Y.C.; Nadagouda, M.; Han, C.; O'Shea, K.; El-Sheikh, S.M.; Ismail, A.A.; Dionysiou, D.D. Visible light-sensitized S, N and C co-doped polymorphic TiO_2 for photocatalytic destruction of microcystin-LR. *Appl. Catal. B Environ.* **2014**, *144*, 614–621. [CrossRef]
45. Dong, F.; Zhao, W.; Wu, Z. Characterization and photocatalytic activities of C, N and S co-doped TiO_2 with 1D nanostructure prepared by the nano-confinement effect. *Nanotechnology* **2008**, *19*, 365–607. [CrossRef]
46. Ma, D.; Xin, Y.; Gao, M.; Wu, J. Fabrication and photocatalytic properties of cationic and anionic S-doped TiO_2 nanofibers by electrospinning. *Appl. Catal. B Environ.* **2014**, *147*, 49–57. [CrossRef]
47. El-Sheikh, S.M.; Zhang, G.; El-Hosainy, H.M.; Ismail, A.A.; O'Shea, K.E.; Falaras, P.; Kontos, A.G.; Dionysiou, D.D. High performance sulfur, nitrogen and carbon doped mesoporous anatase–brookite TiO_2 photocatalyst for the removal of microcystin-LR under visible light irradiation. *J. Hazard. Mater.* **2014**, *280*, 723–733. [CrossRef]
48. Khedr, T.M.; El-Sheikh, S.M.; Hakki, A.; Ismail, A.A.; Badawy, W.A.; Bahnemann, D.W. highly active non-metals doped mixed-phase TiO_2 for photocatalytic oxidation of ibuprofen under visible light. *Photochem. Photobiol. A Chem.* **2017**, *346*, 530–540. [CrossRef]
49. Ismail, A.A.; Bahnemann, D.W. Mesoporous titania photocatalysts: Preparation, characterization and reaction mechanisms. *J. Mater. Chem.* **2011**, *21*, 11686–11707. [CrossRef]
50. Atitar, M.F.; Ismail, A.A.; Al-Sayari, S.A.; Bahnemann, D.; Afanasev, D.; Emeline, A.V. Mesoporous TiO_2 nanocrystals as efficient photocatalysts: Impact of calcination temperature and phase transformation on photocatalytic Performance. *Chem. Eng. J.* **2015**, *264*, 417–424. [CrossRef]
51. Leroy, P.; Tournassat, C.; Bizi, M. Influence of surface conductivity on the apparent zeta potential of TiO_2 nanoparticles. *J. Colloid Interface Sci.* **2011**, *356*, 442–453. [CrossRef] [PubMed]
52. Suttiponparnit, K.; Jiang, J.; Sahu, M.; Suvachittanont, S.; Charinpanitkul, T.; Biswas, P. Role of Surface Area, Primary Particle Size, and Crystal Phase on Titanium Dioxide Nanoparticle Dispersion Properties. *Nanoscale Res. Lett.* **2011**, *6*, 27–34. [CrossRef] [PubMed]
53. Daou, I.; Chfaira, R.; Zegaoui, O.; Aouni, Z.; Ahlafi, H. Physico-Chemical Characterization and Interfacial Electrochemical Properties of Nanoparticles of Anatase-TiO_2 Prepared by the Sol-Gel Method. *Med. J. Chem.* **2013**, *2*, 569–582.
54. Zeng, M. Influence of TiO_2 Surface Properties on Water Pollution Treatment and Photocatalytic Activity. *Bull. Korean Chem. Soc.* **2013**, *34*, 953–956. [CrossRef]

55. Wang, P. Aggregation of TiO_2 Nanoparticles in Aqueous Media: Effects of pH, Ferric Ion and Humic Acid. *Int. J. Environ. Sci. Nat. Res.* **2017**, *1*, 555575. [CrossRef]
56. Wu, S.; Lv, J.; Wang, F.; Duan, N.; Li, Q.; Wang, Z. Photocatalytic degradation of microcystin-LR with a nanostructured photocatalyst based on upconversion nanoparticles@TiO_2 composite under simulated solar lights. *Sci. Rep.* **2017**, *7*, 14435. [CrossRef] [PubMed]
57. Zanjanchi, M.A.; Ebrahimian, A.; Arvand, M. Sulphonated cobalt phthalocyanine-MCM-41: An active photocatalyst for degradation of 2,4-dichlorophenol. *J. Hazard. Mater.* **2010**, *175*, 992–1000. [CrossRef]
58. Kisch, H. On the Problem of Comparing Rates or Apparent Quantum Yields in Heterogeneous Photocatalysis. *Angew. Chem. Int. Ed.* **2010**, *49*, 9588–9589. [CrossRef]

Article

TiO_2 and Au-TiO_2 Nanomaterials for Rapid Photocatalytic Degradation of Antibiotic Residues in Aquaculture Wastewater

Tho Chau Minh Vinh Do [1], Duy Quoc Nguyen [1], Kien Trung Nguyen [2] and Phuoc Huu Le [3,*]

1 Department of Drug Quality Control – Analytical Chemistry – Toxicology, Faculty of Pharmacy, Can Tho University of Medicine and Pharmacy, 179 Nguyen Van Cu Street, Can Tho City 94000, Vietnam
2 Department of Physiology, Faculty of Medicine, Can Tho University of Medicine and Pharmacy, 179 Nguyen Van Cu Street, Can Tho City 94000, Vietnam
3 Department of Physics and Biophysics, Faculty of Basic Sciences, Can Tho University of Medicine and Pharmacy, 179 Nguyen Van Cu Street, Can Tho City 94000, Vietnam
* Correspondence: lhuuphuoc@ctump.edu.vn; Tel.: +84-292-3-739730

Received: 24 June 2019; Accepted: 27 July 2019; Published: 31 July 2019

Abstract: Antibiotic residues in aquaculture wastewater are considered as an emerging environmental problem, as they are not efficiently removed in wastewater treatment plants. To address this issue, we fabricated TiO_2 nanotube arrays (TNAs), TiO_2 nanowires on nanotube arrays (TNWs/TNAs), Au nanoparticle (NP)-decorated-TNAs, and TNWs/TNAs, which were applied for assessing the photocatalytic degradation of eight antibiotics, simultaneously. The TNAs and TNWs/TNAs were synthesized by anodization using an aqueous NH_4F/ethylene glycol solution. Au NPs were synthesized by chemical reduction method, and used to decorate on TNAs and TNWs/TNAs. All the TiO_2 nanostructures exhibited anatase phase and well-defined morphology. The photocatalytic performance of TNAs, TNWs/TNAs, Au-TNAs and Au-TNWs/TNAs was studied by monitoring the degradation of amoxicillin, ampicillin, doxycycline, oxytetracycline, lincomycin, vancomycin, sulfamethazine, and sulfamethoxazole under ultraviolet (UV)-visible (VIS), or VIS illumination by LC-MS/MS method. All the four kinds of nanomaterials degraded the antibiotics effectively and rapidly, in which most antibiotics were removed completely after 20 min treatment. The Au-TNWs/TNAs exhibited the highest photocatalytic activity in degradation of the eight antibiotics. For example, reaction rate constants of Au-TNWs/TNAs for degradation of lincomycin reached 0.26 min^{-1} and 0.096 min^{-1} under UV-VIS and VIS irradiation, respectively; and they were even higher for the other antibiotics. The excellent photocatalytic activity of Au-TNWs/TNAs was attributed to the synergistic effects of: (1) The larger surface area of TNWs/TNAs as compared to TNAs, and (2) surface plasmonic effect in Au NPs to enhance the visible light harvesting.

Keywords: TiO_2 nanomaterials; Au nanoparticles; anodization; photocatalytic degradation of antibiotics; LC-MS/MS

---

1. Introduction

Titanium dioxide (TiO_2) is one of the most widely studied materials for applications in solar cells [1–3], pollutant degradation [4–6], photolysis of water [7], gas sensing [8], and bio-applications [9,10], due to its excellent photocatalytic reactivity, high chemical stability, non-toxicity, biocompatibility, and low cost [11–13]. However, the large band gap of TiO_2 (3.2 eV) limits it light absorption to only 5% of the solar spectrum [14–16]. Considerable effort has been made to improve the light absorption of TiO_2 by doping with non-metals (N, F, S) [17–19] or chemical modification to narrow the band gap [20]. In addition, visible-light absorption can also be achieved by coupling TiO_2 to small-band-gap quantum dots [21]. Recently, a new approach involving metal nanostructures in enhancing the visible-light photoactivity of TiO_2 via plasmonic effect has received much attention [22–26]. Moreover, metal nanoparticles (NPs) have demonstrated good photo-stability [14].

TiO_2 nanomaterials are of great interest because of their large surface area and high light absorption capability [27–31]. In this study, TiO_2 nanotube arrays (TNAs) and TiO_2 nanowires on nanotube arrays (TNWs/TNAs) are of interest, because they can provide a large surface-to-volume ratio and unidirectional electrical channel [32,33]. TNWs/TNAs presented a better photocatalytic degradation of methylene blue than that of TNAs, which was attributed to the presence of partial coverage of TNWs on the surface of TNAs for the enhanced surface area [6]. By using the anodic oxidation, the nanostructures of TNAs and TNWs/TNAs can be fabricated on immobilized Titanium folds that allows retrieval of the photocatalysts from the reaction solution after treatment, so they can be reused for many times.

The aquaculture production sector of the Mekong Delta (Vietnam) has reached an annual production of 1.14 million tons in 2012 [34,35]. Antibiotics are commonly used in aquaculture for the prevention and treatment of diseases. However, Vietnam has very little enforced regulation pertaining to antibiotic usage in domestic aquaculture. Consequently, antibiotic residues in aquaculture wastewater of the Mekong Delta region are considered as an emerging environmental problem, due to their adverse effects on ecosystems, the aquaculture production and its economy [36,37], and human health [38–42]. Indeed, antibiotic residues in the environment have been found at low levels, usually in the $ng \cdot L^{-1}$–$\mu g \cdot L^{-1}$ range [38–40]. The antibiotic residues can result in bacterial antibiotic resistance [41,42], which in turn can be a serious risk to humans and other animals [37]. To address this environmental issue, photocatalysis has received tremendous attention, owing to its great potential in removing antibiotics from aqueous solutions via a green, economic, and effective process [43,44]. Indeed, photocatalytic degradation of tetracycline using nanosized titanium dioxide in an aqueous solution has been studied. Also, the degradation of paracetamol in aqueous solutions by TiO_2 photocatalysis in powder and immobilized forms have been studied [45]. Y. He et al. studied the degradation of pharmaceuticals (i.e., Propranolol, Diclofenac, Carbamazepine, and Ibuprofen) in wastewater using immobilized TiO_2 photocatalysis under simulated solar irradiation [46]. Therefore, the hypothesis of this study is that the antibiotic residues in aquaculture wastewater can be degraded effectively and rapidly by using nanostructured TiO_2 and Au-TiO_2 photocatalysts.

In this study, we fabricated TNAs, TNWs/TNAs, Au NP-decorated TNAs, and Au NP-decorated TNWs/TNAs and utilized them to degrade antibiotic residues in aquaculture wastewater of the Mekong Delta (Vietnam) via the photocatalysis process. Indeed, for the first time, we successfully developed a sensitive, specified and repeatable analytic procedure to assess the photocatalytic removal efficiency of important classes of antibiotics, including amoxicillin (AMOX), ampicillin (AMPI), doxycycline (DXC), oxytetracycline (OTC), lincomycin (LCM), vancomycin (VCM), sulfamethazine (SMT), and sulfamethoxazole (SMZ) simultaneously in aquaculture wastewater, using liquid chromatography/tandem mass spectrometry (LC-MS/MS) analysis. LC-MS/MS is the combination of liquid chromatography (LC) with mass spectrometry (MS). Structural-morphological, and photocatalytic degradation kinetics of the eight antibiotics under UV and/or VIS irradiation are discussed in detail.

2. Experimental Details

TNAs and TNWs/TNAs were grown on Titanium (Ti) foil substrates (99.9% purity, 1 cm × 2.5 cm size, 0.4 mm thickness) by anodic oxidation. Prior to anodization, the substrate was first ultrasonically cleaned using acetone, methanol, and deionized water, followed by drying in a N_2 gas flow. The anodization was performed using a two-electrode system with the Ti foil as an anode and a stainless steel foil (SS304) as a cathode. The electrolyte contained 0.3 wt % NH_4F (SHOWA, Tokyo, Japan) in ethylene glycol (EG) solution with 2 vol % water. The Ti foil was anodized at 30 V for 1 h and 5 h to grow TNAs and TNWs/TNAs, respectively. The samples were then annealed at 400 °C for 1 h to induce sample crystallization. Au nanoparticles were prepared by chemical reduction method in which water (100 mL) containing $HAuCl_4$. $4H_2O$ (0.2 mM) and citric acid (0.5 mM) was stirred at 120 °C. The Au-TNAs and Au-TNWs/TNAs were prepared by immersing the samples in the Au solution for 6 h at room temperature. The samples were then annealed at 400 °C for 1h to improve the crystallinity and Au-TiO_2 interfaces.

The crystal structures of the nanomaterials were characterized by X-ray diffraction (XRD, Bruker D2, Bruker, Billerica, MA, USA) using Cu Kα radiation (λ = 1.5406 Å). Morphologies of the samples were characterized by scanning electron microscopy (SEM, JEOL JSM-6500, Pleasanton, CA, USA). An antibiotic solution was designed and prepared to reflect the practical aquaculture wastewater samples, collected at Dam Doi district of Ca Mau province, which is one of the large aquaculture areas of Mekong Delta, Vietnam. The aquaculture wastewater had a biochemical oxygen demand (BOD) of 10.7 mg/L, chemical oxygen demand (COD) of 19.6 mg/L, and low concentration of organic matter. The spiked mixture solution of standard eight antibiotics with an initial concentration of 500 ng/mL was dissolved in blank wastewater samples containing 0.1% (v/v) formic acid. Photocatalytic reactions were carried out by immersing a sample into a 30 ml antibiotic solution under UV-VIS at approximately 120 mW·cm^{-2} or VIS illumination at approximately 95mW·cm^{-2} using a 100 W Xenon lamp. Prior to illumination, the catalyst was immersed into the solution and magnetic stirring followed for 20 min in the dark, to ensure absorption-desorption equilibrium between the photocatalyst (sample) and antibiotic solution. A band-pass filter for $\lambda \geq 400$ nm was used to select the VIS spectrum region from the Xenon lamp. The reaction temperature was kept at 32–33 °C for all photocatalytic reactions. After a certain photocatalytic reaction time, qualitative and quantitative analysis of antibiotics was determined by LC-MS/MS technique. We used ultra performance liquid chromatography (Acquity H-Class, Waters, Milford, MA, USA) coupled with a triple quadrupole mass detector (Xevo-TQD, Waters, Milford, MA, USA), and equipped with an electrospray ionization (ESI) interface. Mass analysis was in positive and multiple-reaction monitoring (MRM) and daughter ion mode. The Agilent Poroshell 120 Phenyl-hexyl (4.6 × 150 mm; 2.7 μm) column was used, and the mobile phase included acetonitrile-methanol-aqueous formic acid 0.1% in gradient program [47]. The results were evaluated using the degradation percentage of each antibiotic at various reaction times, starting at 0 and followed by 2, 5, 9, 14, and 20 min, as the ratio between the initial peak area of antibiotic solution (without photocatalytic treatment) and peak area of treated antibiotic solution. It was possible to follow the degradation progress of every antibiotic by calculating these areas with Masslynx Software 4.1.

3. Results and Discussion

Figure 1 shows the XRD patterns of TNAs, TNWs/TNAs, Au-TNAs, and Au-TNWs/TNAs. All the samples exhibited the anatase phase of TiO_2 with preferred orientations of (004), (101) and (105) lattice planes at 37.8°, 25.1°, and 53.8°, respectively (JCPDS No. 21–1272). Also, there were no rutile peaks, indicating that the TiO_2 nanomaterials in this study possessed a pure anatase phase. This result agreed with those reported in [4,5,13,19,48,49]. A closer inspection of the (004) peaks revealed that Au (111) component was found in the (004) peaks of Au-TNAs and Au-TNWs/TNAs, as demonstrated in Figure 1c, confirming the presence of crystalline Au NPs in these samples.

The grain sizes (D) of the samples were estimated by using the Scherrer equation: $D = 0.9\lambda/\beta\cos\theta$, where λ, β, and θ are the X-ray wavelength, full width at half maximum of the anatase phase TiO_2 (004)-oriented peak, and Bragg diffraction angle, respectively [50]. Clearly, the estimated grain size varied in a narrow range between 21.3 nm and 24.7 nm, and the full width at half maximum (FWHM) of the (004) peak remained almost constant (Figure 1b). Those results confirmed that the grain size and the crystallinity of four nanomaterials were almost the same.

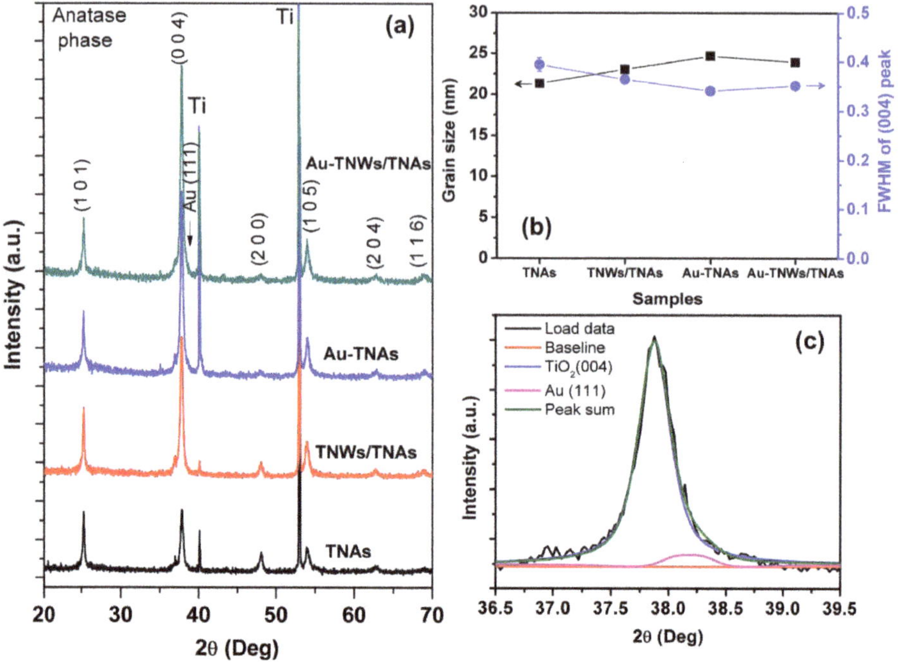

Figure 1. (**a**) The XRD patterns of TiO_2 nanotube arrays (TNAs), TiO_2 nanowires on nanotube arrays (TNWs/TNAs), Au-TNAs, and Au-TNWs/TNAs. (**b**) Grain size and the full width at half maximum (FWHM) of (004) peaks of the four nanomaterials. (**c**) The (004) peak of Au-TNAs shows two components of TiO_2 (004) and Au (111).

Figure 2 shows the morphology of TNAs, TNWs/TNAs, Au-TNAs, and Au-TNWs/TNAs. Clearly, the TNAs exhibited a highly ordered, uniformed, and clean surface. The TNAs had tube diameter of 75 nm and thickness of 5.4 μm (Figure 2a inset). In Figure 2b, TNWs/TNAs exhibited a TNWs (length of 6 μm) covering on the TNAs. The thickness of TNWs/TNAs film was 8.6 μm, as shown in the inset of Figure 2b. The inset in Figure 2c shows the morphology of as-synthesized Au nanoparticles with size of 20 ± 10 nm. For Au-TNAs samples, Au nanoparticles distributed relatively uniformly on the surface of TNAs (Figure 2c). In addition, a typical energy-dispersive X-ray spectroscopy (EDS) spectrum of Au-decorated TiO_2 samples in this study is shown in the inset of Figure 2c. Obviously, Ti, O, Au peaks were observed, confirming the successful fabrications for Au-TNAs and Au-TNWs/TNAs samples. Finally, the morphology of Au-TNWs/TNAs can be observed in Figure 2d.

During the anodization process, TNA growth is driven by the anodic-oxidation reaction (to form TiO_2 from Ti) and the chemical dissolution of the TiO_2 layer under the presence of electric field [19,51–53]. The reactions are given below:

Anodic reaction: $Ti + 2H_2O - 4e \rightarrow TiO_2 + 4H^+$

Cathodic reaction: $4H^+ + 4e \rightarrow 2H_2$

Chemical etching (dissolution) reaction: $TiO_2 + 6F^- + 4H_4^+ \rightarrow TiF_6^{2-} + 2H_2O$

The current density (j) changes with anodizing time (t) in an anodic oxidation process [53,54]. Initially, the j rapidly decreases, then slightly increases, and finally remains a constant [54]. According to the j–t characteristics, the TNAs growth process can be divided into three stages. In the early stage, the formation of a non-conductive thin oxide layer, associated with the decrease of j (Figure 3a). Next, there is the local growth of pits as evidenced by the slight increase of j (Figure 3b). Finally, the nanotube arrays are grown from the initial pits when j remains a constant (Figure 3c). When the dissolution rate of the wall of the nanopores is slower than that of the growth rate of nanopores, the diameter and length of the nanotubes will gradually increase. And, these sizes will remain unchanged when the growth rate is equal to the dissolution rate [53,55].

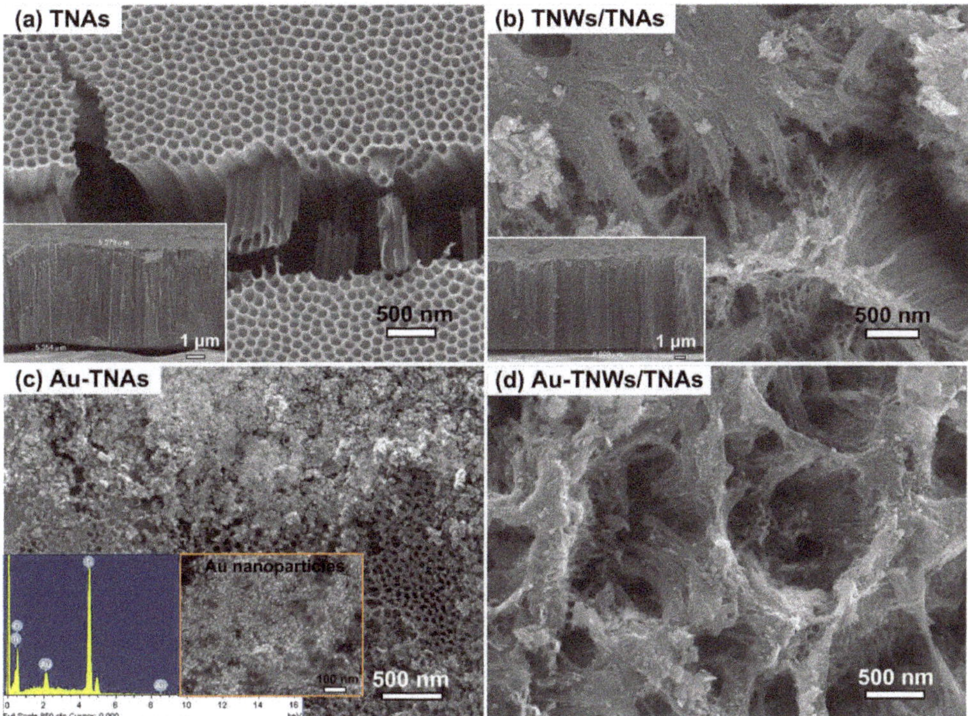

Figure 2. SEM images of (**a**) TNAs, (**b**) TNWs/TNAs, (**c**) Au-TNAs, and (**d**) Au-TNWs/TNAs. The insets in (**c**) show a typical EDS spectrum for Au-TNAs and Au-TNWs/TNAs, and the morphology of as-synthesized Au nanoparticles.

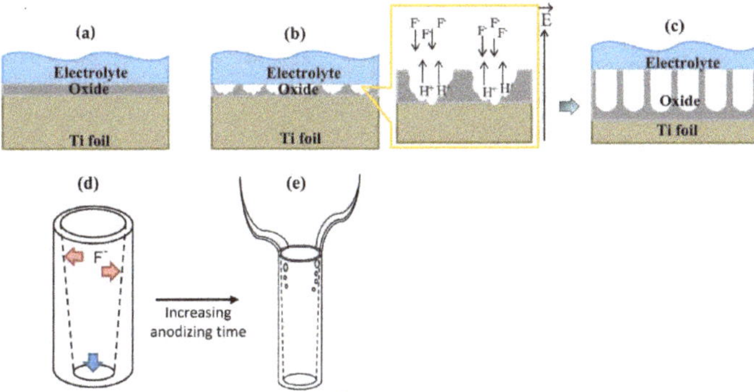

Figure 3. The growth process of TiO_2 nanotube arrays (TNAs): (**a**) non-conductive thin oxide layer forming, (**b**) local growth of the pits, (**c**) growth of the semicircle pores and developed nanotube arrays, (**d**) The shape and wall thickness profile of TNAs prior to the emergence of nanowires (TNWs), (**e**) Schematic of the TNWs/TNAs structure.

In the EG/H_2O solution containing NH_4F electrolyte, the migration of F^- toward the electric field at the bottom electrode is inhibited by the highly viscous solution. Thus, the F^- concentration at the tube mouth is much higher than it is at the tube bottom [6], while the chemical dissolution reaction is enhanced under the presence of H^+ ions from water. Consequently, the tube wall thickness near the tube mouth was thinner than the lower sections, as shown in Figure 3d. By increasing anodizing time, strings of through holes are formed on the tube wall and they would initiate and propagate downward from the top to the bottom of TNAs (or along the F^- migration direction). Meanwhile, the holes near the top expand and connect to each other, and finally split into nanowires (Figure 3e) [6].

The photocatalytic degradation kinetic of LCM is used to evaluate the photocatalytic performance of the four nanomaterials. The pseudo-first-order rate constants were determined by fitting the data with the Langmuir–Hinshelwood kinetics rate model [56,57]. Figure 4a,b shows photocatalytic degradation of LCM using five reaction conditions, namely photolysis (UV-VIS or VIS), and photocatalysis with TNAs, TNWs/TNAs, Au-TNAs, and Au-TNWs/TNAs nanomaterials. Both photolysis and photocatalysis reactions generally follow the exponential decay, $C_t = C_0 \times e^{-kt}$, where C_t is the concentration of antibiotic at time t (ng/mL), C_0 is the initial concentration (ng/mL), and *k* is the reaction rate constant (min^{-1}). By performing the linear fitting on the plot of $-\ln(C_t/C_0)$ versus reaction time t, the *k* is yielded, and the fittings are shown in Figure 4c,d. Specifically, the *k* values of LCM were 4.8×10^{-2} min^{-1} and 0.93×10^{-2} min^{-1} under UV-VIS and VIS irradiation, respectively. This indicates that UV irradiation degrades the antibiotics better than VIS, due to the higher photon energy via the photolysis effect [46,58,59]. As shown in Figure 4a,b, the photocatalysis shows significantly better performance in eliminating LCM than photolysis. The *k* values for LCM were in ranges of 14.8×10^{-2}–26×10^{-2} min^{-1} under UV-VIS illumination and 7.2×10^{-2}–9.5×10^{-2} min^{-1} under VIS illumination (Figure 5a). That means that the reaction rates of photocatalysis were 3.1–5.5 times and 7.6–10.3 times higher than those of UV-VIS photolysis and VIS photolysis, respectively.

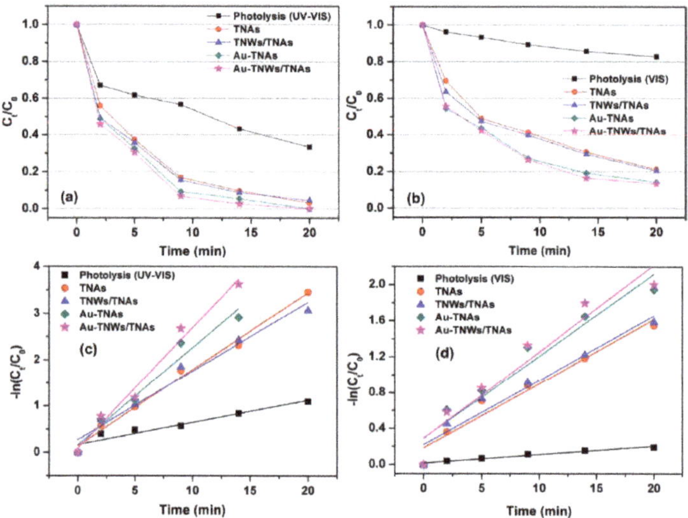

Figure 4. (**a**) Photocatalytic degradation of lincomycin (LCM, 500 ng/mL) using five reaction conditions of photolysis (UV-VIS), and photocatalysis (with TNAs, TNWs/TNAs, Au-TNAs, and Au-TNWs/TNAs). (**b**) Photocatalytic degradation of LCM under photolysis of the visible light ($\lambda \geq 400$ nm of Xenon lamp) and the photocatalysis conditions. (**c**,**d**) LCM degradation kinetic curves of the five reaction conditions under UV-VIS illumination (**c**) and VIS illumination (**d**).

Figure 5a shows the *k* values of the four kinds of nanomaterials under UV-VIS and VIS irradiation. Generally, the *k* of TNWs/TNAs is higher than that of TNAs, which is primarily attributed to the presence of partial coverage of TNWs on the surface of TNAs for the enhanced surface area [6,53]. There was a significant enhancement in the k values by decorating TNAs and TNWs/TNAs with Au NPs, because of the enhancement of the visible-light photoactivity of TiO_2 via the localized surface plasmon resonance (LSPR) effect [14,22,60] (Figure 5a). The LSPR of spherical Au NPs (20 ± 10 nm diameter) in this study was suggested by the absorption peak at 529 nm (Figure 5b), which was well consistent with the LSPR-peaks of Au nanoparticles in [61,62]. In addition, the absorption enhancement in VIS region for Au-TiO_2 was confirmed by the UV-VIS absorption spectra in [61,63]. LSPR can be described as the local electromagnetic fields near the surface of Au NPs being strongly enhanced when the electromagnetic field of the incident light becomes associated with the oscillations of the conduction electrons of Au NPs. Indeed, optical simulations clearly presented LSPR-enhanced electric fields at the interface of Au-TiO_2, owing to photo-excited Au nanoparticles [64]. Herein, a proposed mechanism for enhanced photocatalytic activity of Au-TiO_2 is that the LSPR-absorption of Au NPs generate photoexcited electrons and holes under VIS irradiation, and then the energetic electrons can inject into the conduction band of TiO_2 and trigger photocatalytic reactions (Figure 5c) [61,62,65,66]. Therefore, Au-TNWs/TNAs possessed the highest photocatalytic performance amongst the four kinds of nanomaterials, due to the synergistic effects of large surface area and the LSPR effect.

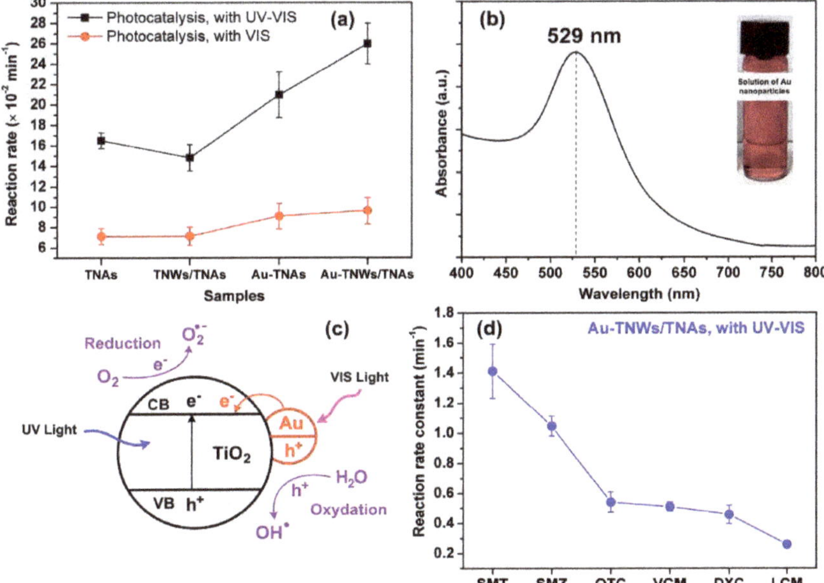

Figure 5. (**a**) Reaction rate constant (k) of various nanomaterials in photocatalytic degradation of Lincomycin (500 ng/mL) under UV-VIS and VIS irradiation. (**b**) Absorption spectrum of the solution of Au nanoparticles, showing the LSPR peak at 529 nm; and the inset image is a photograph of the Au nanoparticle solution. (**c**) A proposed mechanism for the photocatalytic activity of Au-TiO_2 upon the excitation of the Au surface plasmon band. (**d**) Reaction rate constant of various antibiotics under photocatalysis using Au-TNWs/TNAs under UV-VIS irradiation. The antibiotic abbreviations: SMT, sulfamethazine; VCM, vancomycin; OTC, oxytetracycline; SMZ, sulfamethoxazole; DXC, doxycycline; LCM, lincomycin.

Figure 5d summarizes the k values of various antibiotics treated using photocatalytic reaction of the Au-TNWs/TNAs (the best nanomaterial in this study) under UV-VIS irradiation. Here, the k is determined by the intrinsic photocatalytic property of the nanomaterial and the photolysis of antibiotics. AMOX and AMPI with β lactam ring structures decomposed rapidly by photolysis reaction with UV-VIS illumination [67]. In addition, the reaction rate of SMT and SMZ reached high values of 1.41 min^{-1} and 1.05 min^{-1}, respectively; meanwhile, it was only 0.26 min^{-1} for LCM. That is because the former has amine bond structure [68], while LCM has amide bond structure [68]. Similarly, all the molecule structures with amide bonds of VCM, DXC, and OTC are more resistant to photolysis. Consequently, VCM, DXC, and OTC exhibited lower k values (1.05, 0.46, and 0.54 min^{-1}) and needed a reaction time above 20 min to completely degrade. For comparison, the photocatalytic degradation rate of OTC using the Au-TNWs/TNAs (i.e., 0.54 min^{-1}) was far higher the k of 0.032 min^{-1} using TiO_2 nanobelts loading Au NPs [63].

For the typical LC-MS/MS analysis in more detail, Figure 6a illustrated photocatalytic kinetic analysis of OTC at various reaction times of 0, 2, 5, 9, 14, and 20 min using Au-TNWs/TNAs and UV-VIS irradiation. As a result, removal percentage of OTC increased dramatically as a function of reaction time, and obtained 100% at 20 min. This indicates that antibiotics can completely degrade using the photocatalytic reaction with TiO_2-based nanomaterials. Additionally, the UV-VIS photolysis or photocatalysis of antibiotics can produce potentially harmful substances [47,68]. Figure 6b shows the mass spectra of intermediates of OTC after 9 and 14 min of photocatalytic reaction. It is observed that intermediates separate at retention times of 4.58, 5.65, 10.97 min, respectively. At first, the OTC derived molecule 460.01 m/z is observed with a precursor ion $[M\text{-}H]^+$ 461.01 in positive mode for the pristine blank sample. In monitoring reaction mode, there are only three product ions with the transition of m/z 461 → 426, 443 and 201 m/z. Meanwhile, after exposure to UV-VIS and Au-TNWs/TNAs, new product impurity ions with 126, 114, 126 m/z appeared at retention times of 4.58 min; ions 230, 106, 92 m/z at a retention time of 5.65 min, and 123.98, 92 m/z at 10.07 min also appeared. These results suggested the presence of decomposed products of the investigated antibiotics.

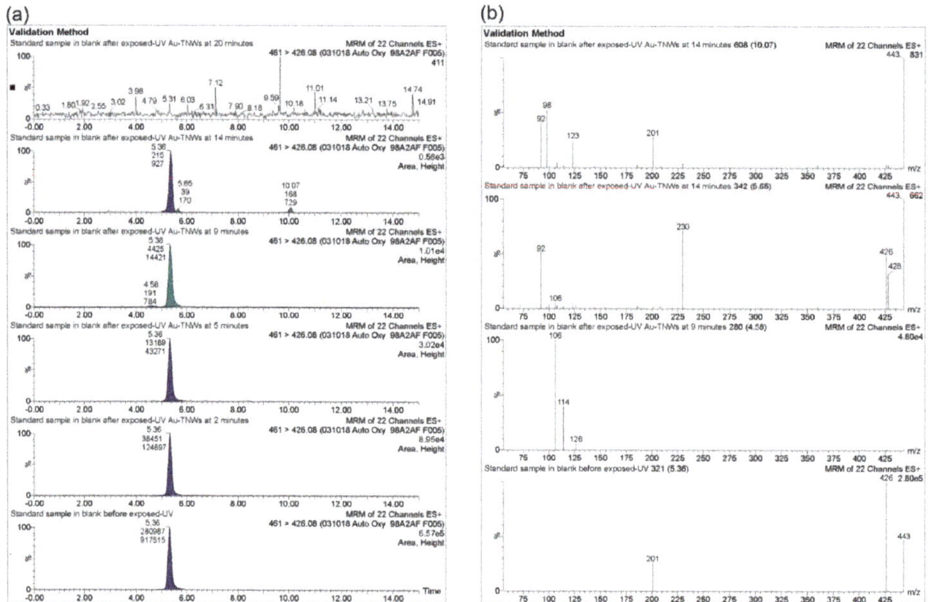

Figure 6. (**a**) Chromatogram ultra performance liquid chromatography (UPLC)-MS/MS photocatalytic degradation kinetic model of oxytetracycline (OTC, 500 ng/mL) with exposure to UV-VIS and Au-TNWs/TNAs. (**b**) Chromatogram UPLC-MS/MS impurities of OTC decomposition with exposed-UV-VIS and Au-TNWs/TNWs at reaction times of 9 and 14 min.

4. Conclusions

In this study, TiO_2-based nanomaterials (i.e., TNAs, TNWs/TNAs, Au-TNAs, and Au-TNWs/TNAs) were developed toward the end of enhanced photocatalytic degradation of popular antibiotics. All the four kinds of nanomaterials exhibited the anatase phase with (004) and (101)-preferred orientation, grain size of 21.3–24.7 nm, and a similar crystallinity. The morphology of the samples was highly uniform and well-defined, which is promising for enhanced photocatalytic activity. In addition, we proposed and shed light on the formation mechanisms of TNAs and TNWs/TNAs. The nanomaterials were utilized for evaluating the photocatalytic degradation of antibiotics in model aquaculture wastewater by an LC-MS/MS method. The photocatalytic activity of TNWs/TNAs was higher than that of TNAs, primarily owing to the larger surface area of the former than the latter. By decorating Au NPs onto TNAs or TNWs/TNAs, the photocatalytic activity of Au-TNAs and Au-TNWs/TNAs was enhanced significantly compared to that of TNAs and TNWs/TNAs, because of the local surface plasmon resonance effect. Consequently, the Au-TNWs/TNAs achieved the highest activity for decomposition of antibiotics under UV-VIS or VIS irradiation. Based on the photocatalysis's kinetic results, the photolysis of the eight antibiotics is of great concern. It was found that the photolysis of antibiotics depends on the stability of their structures. Indeed, the beta-lactam group (AMOX, AMPI) is more sensitive to photolysis than the sulfonamides group (SMT, SMZ) under UV-VIS irradiation. The photo-degradation pattern of more stable antibiotics (i.e., LCM, DXC, OTC, and VCM) followed pseudo-first order kinetics well, and their reaction rate constants were 0.26, 0.46, 0.54, and 0.51 min^{-1}, respectively. Furthermore, the appearance of transformation products of the investigated antibiotics was evident after the chromatographic analyses, whose identification is of interest for future studies.

Author Contributions: T.C.M.V.D., D.Q.N., and P.H.L. performed the experiments and analyzed the data; T.C.M.V.D. and P.H.L. wrote and edited the paper; K.T.N. supervised the project.

Funding: This research is funded by Vietnam National Foundation for Science and Technology Development (NAFOSTED) under grant number 103.99-2016.75.

Conflicts of Interest: The authors declare no conflict of interest.

References

1. Roy, P.; Kim, D.; Lee, K.; Spiecker, E.; Schmuki, P. TiO_2 nanotubes and their application in dye-sensitized solar cells. *Nanoscale* **2010**, *2*, 45–59. [CrossRef]
2. Gunawan, B.; Musyaro'ah; Huda, I.; Indayani, W.; Endarko, S.R. The influence of various concentrations of N-doped TiO_2 as photoanode to increase the efficiency of dye-sensitized solar cell. *AIP Conf. Proc.* **2017**, *1788*, 030128.
3. Mor, G.; Shankar, K.; Paulose, M.; Varghese, O.K.; Grimes, C.A. Use of Highly-Ordered TiO_2 Nanotube Arrays in Dye-Sensitized Solar Cells. *Nano Lett.* **2006**, *6*, 215–218. [CrossRef]
4. Lai, Y.K.; Huang, J.Y.; Zhang, H.F.; Subramaniam, V.P.; Tang, Y.X.; Gong, D.G.; Sundar, L.; Sun, L.; Chen, Z.; Lin, C.J. Nitrogen-doped TiO_2 nanotube array films with enhanced photocatalytic activity under various light sources. *J. Hazard Mater.* **2010**, *184*, 855–863. [CrossRef]
5. Yang, G.; Jiang, Z.; Shi, H.; Xiao, T.; Yan, Z. Preparation of highly visible-light active N-doped TiO_2 photocatalyst. *J. Mater. Chem.* **2010**, *20*, 5301–5309. [CrossRef]
6. Hsu, M.-Y.; Hsu, H.-L.; Leu, J. TiO_2 Nanowires on Anodic TiO_2 Nanotube Arrays (TNWs/TNAs): Formation Mechanism and Photocatalytic Performance. *J. Electrochem. Soc.* **2012**, *159*, H722–H727. [CrossRef]
7. Mor, G.K.; Shankar, K.; Paulose, M.; Varghese, O.K.; Grimes, C.A. Enhanced photocleavage of water using titania nanotube arrays. *Nano Lett.* **2005**, *5*, 191–195. [CrossRef]
8. Nisar, J.; Topalian, Z.; De Sarkar, A.; Osterlund, L.; Ahuja, R. TiO_2-based gas sensor: A possible application to SO_2. *ACS Appl. Mater. Interfaces* **2013**, *5*, 8516–8522. [CrossRef]
9. Abbasi, A.; Sardroodi, J.J. A theoretical investigation of the interaction of Immucillin-A with N-doped TiO_2 anatase nanoparticles: Applications to nanobiosensors and nanocarriers. *Nanomed. Res. J.* **2017**, *2*, 7–17.
10. Roy, S.C.; Paulose, M.; Grimes, C.A. The effect of TiO_2 nanotubes in the enhancement of blood clotting for the control of hemorrhage. *Biomaterials* **2007**, *28*, 4667–4672. [CrossRef]

11. Ansari, S.A.; Khan, M.M.; Ansari, M.O.; Cho, M.H. Nitrogen-doped titanium dioxide (N-doped TiO_2) for visible light photocatalysis. *New J. Chem.* **2016**, *40*, 3000–3009. [CrossRef]
12. Zhang, P.; Fujitsuka, M.; Majima, T. TiO_2 mesocrystal with nitrogen and fluorine codoping during topochemical transformation: Efficient visible light induced photocatalyst with the codopants. *Appl. Catal. B Environ.* **2016**, *185*, 181–188. [CrossRef]
13. Preethi, L.K.; Antony, R.P.; Mathews, T.; Loo, S.C.J.; Wong, L.H.; Dash, S.; Tyagi, A.K. Nitrogen doped anatase-rutile heterostructured nanotubes for enhanced photocatalytic hydrogen production: Promising structure for sustainable fuel production. *Int. J. Hydrogen Energy* **2016**, *41*, 5865–5877. [CrossRef]
14. Pu, Y.C.; Wang, G.; Chang, K.D.; Ling, Y.; Lin, Y.K.; Fitzmorris, B.C.; Liu, C.M.; Lu, X.; Tong, Y.; Zhang, J.Z. Au nanostructure-decorated TiO_2 nanowires exhibiting photoactivity across entire UV-visible region for photoelectrochemical water splitting. *Nano Lett.* **2013**, *13*, 3817–3823. [CrossRef]
15. Asahi, R.; Morikawa, T.; Irie, H.; Ohwaki, T. Nitrogen-doped titanium dioxide as visible-light-sensitive photocatalyst: Designs, developments, and prospects. *Chem. Rev.* **2014**, *114*, 9824–9852. [CrossRef]
16. Mazierski, P.; Nischk, M.; Golkowska, M.; Lisowski, W.; Gazda, M.; Winiarski, M.J.; Klimczuk, T.; Zaleska-Medynska, A. Photocatalytic activity of nitrogen doped TiO_2 nanotubes prepared by anodic oxidation: The effect of applied voltage, anodization time and amount of nitrogen dopant. *Appl. Catal. B Environ.* **2016**, *196*, 77–88. [CrossRef]
17. Devi, L.G.; Kavitha, R. Review on modified N–TiO_2 for green energy applications under UV/visible light: selected results and reaction mechanisms. *RSC Adv.* **2014**, *4*, 28265–28299. [CrossRef]
18. Asahi, R.; Morikawa, T.; Ohwaki, T.; Taga, Y. Visible-Light Photocatalysis in Nitrogen-Doped Titanium Oxides. *Science* **2001**, *293*, 269–271. [CrossRef]
19. Le, P.H.; Hieu, L.T.; Lam, T.-N.; Hang, N.T.N.; Truong, N.V.; Tuyen, L.T.C.; Phong, P.T.; Leu, J. Enhanced Photocatalytic Performance of Nitrogen-Doped TiO_2 Nanotube Arrays Using a Simple Annealing Process. *Micromachines* **2018**, *9*, 618. [CrossRef]
20. Chen, X.; Liu, L.; Yu, P.Y.; Mao, S.S. Increasing Solar Absorption for Photocatalysis with Black Hydrogenated Titanium Dioxide Nonocrystals. *Science* **2011**, *331*, 746–751. [CrossRef]
21. Selinsky, R.S.; Ding, Q.; Faber, M.S.; Wright, J.C.; Jin, S. Quantum dot nanoscale heterostructures for solar energy conversion. *Chem. Soc. Rev.* **2013**, *42*, 2963–2985. [CrossRef]
22. Zhang, Z.; Zhang, L.; Hedhili, M.N.; Zhang, H.; Wang, P. Plasmonic gold nanocrystals coupled with photonic crystal seamlessly on TiO_2 nanotube photoelectrodes for efficient visible light photoelectrochemical water splitting. *Nano Lett.* **2013**, *13*, 14–20. [CrossRef]
23. Hou, W.; Cronin, S.B. A review of surface plasmon resonance-enhanced photocatalysis. *Adv. Funct. Mater.* **2013**, *23*, 1612–1619. [CrossRef]
24. Yen, Y.C.; Chen, J.A.; Ou, S.; Chen, Y.S.; Lin, K.-J. Plasmon-Enhanced Photocurrent using Gold Nanoparticles on a Three-Dimensional TiO_2 Nanowire-Web Electrode. *Sci. Rep.* **2017**, *7*, 1–8. [CrossRef]
25. Tahir, M.; Tahir, B.; Amin, N.A.S. Gold-nanoparticle-modified TiO_2 nanowires for plasmon-enhanced photocatalytic CO_2 reduction with H_2 under visible light irradiation. *Appl. Surf. Sci.* **2015**, *356*, 1289–1299. [CrossRef]
26. Duan, Y.; Zhou, S.; Chen, Z.; Luo, J.; Zhang, M.; Wang, F.; Xu, T.; Wang, C. Hierarchical TiO_2 nanowire/microflower photoanode modified with Au nanoparticles for efficient photoelectrochemical water splitting. *Catal. Sci. Technol.* **2018**, *8*, 1395–1403. [CrossRef]
27. Yu, Y.; Zhang, P.; Guo, L.; Chen, Z.; Wu, Q.; Ding, Y.; Zheng, W.; Cao, Y. The design of TiO_2 nanostructures (nanoparticle, nanotube, and nanosheet) and their photocatalytic activity. *J. Phys. Chem. C* **2014**, *118*, 12727–12733. [CrossRef]
28. Verma, R.; Gangwar, J.; Srivastava, A.K. Multiphase TiO_2 nanostructures: A review of efficient synthesis, growth mechanism, probing capabilities, and applications in bio-safety and health. *RSC Adv.* **2017**, *7*, 44199–44224. [CrossRef]
29. Tian, J.; Zhao, Z.; Kumar, A.; Boughton, R.I.; Liu, H. Recent progress in design, synthesis, and applications of one-dimensional TiO_2 nanostructured surface heterostructures: A review. *Chem. Soc. Rev.* **2014**, *43*, 6920–6937. [CrossRef]
30. Ahn, C.; Park, J.; Kim, D.; Jeon, S. Monolithic 3D titania with ultrathin nanoshell structures for enhanced photocatalytic activity and recyclability. *Nanoscale* **2013**, *5*, 10384–10389. [CrossRef]

31. Cho, S.; Ahn, C.; Park, J.; Jeon, S. 3D nanostructured N-doped TiO_2 photocatalysts with enhanced visible absorption. *Nanoscale* **2018**, *10*, 9747–9751. [CrossRef]
32. Huang, J.; Zhang, K.; Lai, Y. Fabrication, Modification, and Emerging Applications of TiO_2 Nanotube Arrays by Electrochemical Synthesis: A Review. *Int. J. Photoenergy* **2013**, *2013*, 761971. [CrossRef]
33. Li, S.; Zhang, G.; Guo, D.; Yu, L.; Zhang, W. Anodization Fabrication of Highly Ordered TiO_2 Nanotubes. *J. Phys. Chem. C* **2009**, *113*, 12759–12765. [CrossRef]
34. Andrieu, M.; Rico, A.; Phu, T.M.; Huong, D.T.T.; Phuong, N.T.; Van den Brink, P.J. Ecological risk assessment of the antibiotic enrofloxacin applied to Pangasius catfish farms in the Mekong Delta, Vietnam. *Chemosphere* **2015**, *119*, 407–414. [CrossRef]
35. Uddin, G.M.N.; Larsen, M.H.; Christensen, H.; Aarestrup, F.M.; Phu, T.M.; Dalsgaard, A. Identification and Antimicrobial Resistance of Bacteria Isolated from Probiotic Products Used in Shrimp Culture. *PLoS ONE* **2015**, *10*, e0132338. [CrossRef]
36. Thuy, H.T.T.; Nguyen, T.D. The potential environmental risks of pharmaceuticals in Vietnamese aquatic systems: Case study of antibiotics and synthetic hormones. *Environ. Sci. Pollut. Res.* **2013**, *20*, 8132–8140. [CrossRef]
37. Cabello, F.C. Heavy use of prophylactic antibiotics in aquaculture: a growing problem for human and animal health and for the environment. *Environ. Microbiol.* **2006**, *8*, 1137–1144. [CrossRef]
38. Giang, C.N.D.; Sebesvari, Z.; Renaud, F.; Rosendahl, I.; Minh, Q.H.; Amelung, W. Occurrence and dissipation of the antibiotics sulfamethoxazole, sulfadiazine, trimethoprim, and enrofloxacin in the Mekong Delta, Vietnam. *PLoS ONE* **2015**, *10*, 1–24.
39. Managaki, S.; Murata, A.; Takada, H.; Bui, C.T.; Chiem, N.H. Distribution of macrolides, sulfonamides, and trimethoprim in tropical waters: Ubiquitous occurrence of veterinary antibiotics in the Mekong Delta. *Environ. Sci. Technol.* **2007**, *41*, 8004–8010. [CrossRef]
40. Hoa, P.T.P.; Managaki, S.; Nakada, N.; Takada, H.; Shimizu, A.; Anh, D.H.; Viet, P.H.; Suzuki, S. Antibiotic contamination and occurrence of antibiotic-resistant bacteria in aquatic environments of northern Vietnam. *Sci. Total Environ.* **2011**, *409*, 2894–2901. [CrossRef]
41. Sy, N.V.; Harada, K.; Asayama, M.; Warisaya, M.; Dung, L.H.; Sumimura, Y.; Diep, K.T.; Ha, L.V.; Thang, N.N.; Hoa, T.T.T.; et al. Chemosphere Residues of 2-hydroxy-3-phenylpyrazine, a degradation product of some β-lactam antibiotics, in environmental water in Vietnam. *Chemosphere* **2017**, *172*, 355–362. [CrossRef]
42. Brunton, L.A.; Desbois, A.P.; Garza, M.; Wieland, B.; Mohan, C.V.; Häsler, B.; Tam, C.C.; Le, P.N.T.; Nguyen, T.P.; Van, P.T.; et al. Identifying hotspots for antibiotic resistance emergence and selection, and elucidating pathways to human exposure: Application of a systems-thinking approach to aquaculture systems. *Sci. Total Environ.* **2019**, *687*, 1344–1356. [CrossRef]
43. Li, D.; Shi, W. Recent developments in visible - light photocatalytic degradation of antibiotics. *Chin. J. Catal.* **2016**, *37*, 792–798. [CrossRef]
44. Klavarioti, M.; Mantzavinos, D.; Kassinos, D. Removal of residual pharmaceuticals from aqueous systems by advanced oxidation processes. *Environ. Int.* **2009**, *35*, 402–417. [CrossRef]
45. Teixeira, S.; Gurke, R.; Eckert, H.; Kuhn, K.; Fauler, J.; Cuniberti, G. Photocatalytic degradation of pharmaceuticals present in conventional treated wastewater by nanoparticle suspensions. *J. Environ. Chem. Eng.* **2016**, *4*, 287–292. [CrossRef]
46. He, Y.; Sutton, N.B.; Rijnaarts, H.H.H.; Langenhoff, A.A.M. Degradation of pharmaceuticals in wastewater using immobilized TiO_2 photocatalysis under simulated solar irradiation. *Appl. Catal. B Environ.* **2016**, *182*, 132–141. [CrossRef]
47. Ambrosetti, B.; Campanella, L.; Palmisano, R. Degradation of Antibiotics in Aqueous Solution by Photocatalytic Process: Comparing the Efficiency in the Use of ZnO or TiO_2. *J. Environ. Sci. Eng. A* **2015**, *4*, 273–281.
48. Sun, L.; Cai, J.; Wu, Q.; Huang, P.; Su, Y.; Lin, C. N-doped TiO_2 nanotube array photoelectrode for visible-light-induced photoelectrochemical and photoelectrocatalytic activities. *Electrochim. Acta* **2013**, *108*, 525–531. [CrossRef]
49. Preethi, L.K.; Antony, R.P.; Mathews, T.; Walczak, L.; Gopinath, C.S. A Study on Doped Heterojunctions in TiO_2 Nanotubes: An Efficient Photocatalyst for Solar Water Splitting. *Sci. Rep.* **2017**, *7*, 1–15. [CrossRef]

50. Tuyen, L.T.C.; Jian, S.-R.; Tien, N.T.; Le, P.H. Nanomechanical and Material Properties of Fluorine-Doped Tin Oxide Thin Films Prepared by Ultrasonic Spray Pyrolysis: Effects of F-Doping. *Materials* **2019**, *12*, 1665. [CrossRef]
51. Wang, D.; Yu, B.; Wang, C.; Zhou, F.; Liu, W. A novel protocol toward perfect alignment of anodized TiO_2 nanotubes. *Adv. Mater.* **2009**, *21*, 1964–1967. [CrossRef]
52. Yan, J.; Zhou, F. TiO_2 nanotubes: Structure optimization for solar cells. *J. Mater. Chem.* **2011**, *21*, 9406. [CrossRef]
53. Le, P.H.; Leu, J. Recent Advances in TiO_2 Nanotube-Based Materials for Photocatalytic Applications Designed by Anodic Oxidation. In *Titanium Dioxide-Material for Sustainable Environmen*; Yang, D., Ed.; Intechopen: London, UK, 2018.
54. Roy, P.; Berger, S.; Schmuki, P. TiO_2 nanotubes: Synthesis and applications. *Angew. Chemie Int. Ed.* **2011**, *50*, 2904–2939. [CrossRef]
55. Nie, X.; Chen, J.; Li, G.; Shi, H.; Zhao, H.; Wong, P.K.; An, T. Synthesis and characterization of TiO_2 nanotube photoanode and its application in photoelectrocatalytic degradation of model environmental pharmaceuticals. *J. Chem. Technol. Biotechnol.* **2013**, *88*, 1488–1497. [CrossRef]
56. Hoffmann, M.R.; Martin, S.T.; Choi, W.Y.; Bahnemann, D.W. Environmental Applications of Semiconductor Photocatalysis. *Chem. Rev.* **1995**, *95*, 69–96. [CrossRef]
57. Sharma, S.D.; Saini, K.K.; Kant, C.; Sharma, C.P.; Jain, S.C. Photodegradation of dye pollutant under UV light by nano-catalyst doped titania thin films. *Appl. Catal. B Environ.* **2008**, *84*, 233–240. [CrossRef]
58. Topkaya, E.; Konyar, M.; Yatmaz, H.C.; Öztürk, K. Pure ZnO and composite ZnO/TiO_2 catalyst plates: A comparative study for the degradation of azo dye, pesticide and antibiotic in aqueous solutions. *J. Colloid Interface Sci.* **2014**, *430*, 6–11. [CrossRef]
59. Yang, L.; Yu, L.E.; Ray, M.B. Degradation of paracetamol in aqueous solutions by TiO_2 photocatalysis. *Water Res.* **2008**, *42*, 3480–3488. [CrossRef]
60. Chen, Y.; Tian, G.; Pan, K.; Tian, C.; Zhou, J.; Zhou, W.; Ren, Z.; Fu, H. In situ controlled growth of well-dispersed gold nanoparticles in TiO_2 nanotube arrays as recyclable substrates for surface-enhanced Raman scattering. *Dalt. Trans.* **2012**, *41*, 1020–1026. [CrossRef]
61. Chen, Y.; Bian, J.; Qi, L.; Liu, E.; Fan, J. Efficient Degradation of Methylene Blue over Two-Dimensional Au/TiO_2 Nanosheet Films with Overlapped Light Harvesting Nanostructures. *J. Nanomater.* **2015**, *16*, 1–10. [CrossRef]
62. Linic, S.; Christopher, P.; Ingram, D.B. Plasmonic-metal nanostructures for efficient conversion of solar to chemical energy. *Nat. Mater.* **2011**, *10*, 911–921. [CrossRef]
63. Chen, Q.; Wu, S.; Xin, Y. Synthesis of Au-CuS-TiO_2 nanobelts photocatalyst for efficient photocatalytic degradation of antibiotic oxytetracycline. *Chem. Eng. J.* **2016**, *302*, 377–387. [CrossRef]
64. Liu, Z.; Hou, W.; Pavaskar, P.; Aykol, M.; Cronin, S.B. Plasmon Resonant Enhancement of Photocatalytic Water Splitting Under Visible Illumination. *Nano Lett.* **2011**, *11*, 1111–1116. [CrossRef]
65. Kowalska, E.; Mahaney, O.O.P.; Abe, R.; Ohtani, B. Visible-light-induced photocatalysis through surface plasmon excitation of gold on titania surfaces. *Phys. Chem. Chem. Phys.* **2010**, *12*, 2344–2355. [CrossRef]
66. Wang, P.; Huang, B.; Dai, Y.; Whangbo, M.-H. Plasmonic Photocatalysts: Harvesting Visible Light with Noble Metal Nanoparticles. *Phys. Chem. Chem. Phys.* **2012**, *14*, 9813–9825. [CrossRef]
67. Timm, A.; Borowska, E.; Majewsky, M.; Merel, S.; Zwiener, C.Z.; Bräse, S.; Horn, H. Photolysis of four β lactam antibiotics under simulated environmental conditions: Degradation, transformation products and antibacterial activity. *Sci. Total Environ.* **2019**, *651*, 1605–1612. [CrossRef]
68. Hu, A.; Zhang, X.; Luong, D.; Oakes, K.D.; Servos, M.R.; Liang, R.; Kurdi, S.; Peng, P.; Zhou, Y. Adsorption and photocatalytic degradation kinetics of pharmaceuticals by TiO_2 nanowires during water treatment. *Waste Biomass Valorization* **2012**, *3*, 443–449. [CrossRef]

Article

Advanced Photocatalysts Based on Reduced Nanographene Oxide–TiO_2 Photonic Crystal Films

Angeliki Diamantopoulou [1], Elias Sakellis [2], Spiros Gardelis [1], Dimitra Tsoutsou [2], Spyridon Glenis [1], Nikolaos Boukos [2], Athanasios Dimoulas [2] and Vlassis Likodimos [1,*]

1 Section of Solid State Physics, Department of Physics, National and Kapodistrian University of Athens, Panepistimiopolis, 15784 Ilissia, Greece

2 Institute of Nanoscience and Nanotechnology, National Center for Scientific Research "Demokritos", 15310 Agia Paraskevi, Greece

* Correspondence: vlikodimos@phys.uoa.gr; Tel.: +30-2107276824

Received: 30 June 2019; Accepted: 5 August 2019; Published: 7 August 2019

Abstract: Surface functionalization of TiO_2 inverse opals by graphene oxide nanocolloids (nanoGO) presents a promising modification for the development of advanced photocatalysts that combine slow photon-assisted light harvesting, surface area, and mass transport of macroporous photonic structures with the enhanced adsorption capability, surface reactivity, and charge separation of GO nanosheets. In this work, post-thermal reduction of nanoGO–TiO_2 inverse opals was investigated in order to explore the role of interfacial electron transfer vs. pollutant adsorption and improve their photocatalytic activity. Photonic band gap-engineered TiO_2 inverse opals were fabricated by the coassembly technique and were functionalized by GO nanosheets and reduced under He at 200 and 500 °C. Comparative performance evaluation of the nanoGO–TiO_2 films on methylene blue photodegradation under UV-VIS and visible light showed that thermal reduction at 200 °C, in synergy with slow photon effects, improved the photocatalytic reaction rate despite the loss of nanoGO and oxygen functional groups, pointing to enhanced charge separation. This was further supported by photoluminescence spectroscopy and salicylic acid UV-VIS photodegradation, where, in the absence of photonic effects, the photocatalytic activity increased, confirming that fine-tuning of interfacial coupling between TiO_2 and reduced nanoGO is a key factor for the development of highly efficient photocatalytic films.

Keywords: TiO_2; photonic crystals; graphene oxide nanocolloids; reduced graphene oxide; photocatalysis

1. Introduction

Over the last decades, TiO_2 has been widely studied as a low cost, nontoxic, and highly stable photocatalyst, which can degrade a great number of gaseous and aqueous pollutants for environmental remediation [1]. Despite its unique advantages, the photocatalytic efficiency of TiO_2 is, however, limited by the recombination of photogenerated charge carriers [2] and its wide energy band gap (3.0–3.2 eV), which requires ultraviolet light for electron excitation [3]. Extensive research has accordingly been focused on tailoring titania's structural and morphological characteristics as well as its compositional and electronic properties by doping, defect engineering, and heterostructuring in order to enhance charge separation and visible-light harvesting [1]. Particular interest has been recently placed on TiO_2 photonic crystals, a promising titania modification featuring a periodic and mesoporous structure that can manipulate light propagation at specific wavelengths [4]. Because of their periodicity, photonic crystals—the most common being the inverse opal structure fabricated by the self-assembly of colloidal opal templates [5,6]—exhibit photonic band gaps (PBGs), which constitute frequency regions within the crystal where electromagnetic irradiation cannot propagate. At wavelengths near the PBG edges, slow photon effects appear, i.e., light propagation at reduced group velocity, leading to an increase in the optical path length and an enhancement of light–matter interactions [7–10].

Among the diverse heterostructured TiO_2 photocatalysts, nanocomposites comprising titania with graphene or its derivatives have emerged as a distinct material class of topical interest because of graphene materials' exceptional electron acceptor action and charge transport, together with their unique two-dimensional morphology and broadband light absorption [11–13]. In particular, graphene oxide (GO) has attracted much attention since 2004, when graphene research first emerged [14], as a solution-processable graphene precursor. It consists of graphene sheets with randomly distributed oxygenated groups on each plane and thus contains both sp^2 and sp^3 carbon atoms [15]. Owing to its abundant oxygen groups, GO demonstrates rich surface chemistry and reactivity, which can be favorably exploited for the assembly of highly efficient GO–TiO_2 composite photocatalysts with strong interfacial coupling and improved charge separation under optimal conditions [16,17] as well as exceptional pollutant adsorption ability [18]. Significant efforts have been recently devoted to determine and control the size and thickness variation of micrometer-sized GO sheets [19], which are crucial for their surfactant properties and emulsion ability [20,21] as well as for efficient GO incorporation and electrical performance as filler in polymer composites [22]. In the case of composite photocatalysts, GO nanosheets may act as scavenger and shuttle of TiO_2-photogenerated electrons, inhibiting electron-hole recombination depending on the oxidation state of GO and the nature of the pollutant [23–26], although GO may also operate as a sensitizer of TiO_2 via the formation of p–n junctions that enable electron transfer from GO to TiO_2 under visible light [27], depending on the variation of the π–π^* localized states of the inhomogenously distributed aromatic sp^2 domains, especially in reduced GO (rGO) [23,28]. Moreover, electrodeposition of GO coatings on anodized TiO_2 nanotube arrays and post-thermal reduction was recently shown to enhance the photocatalytic performance of pristine GO–TiO_2 films due to the enhanced charge separation and adsorption ability of the GO sheets [29]. Controlled synthesis of stable aqueous GO nanocolloids consisting of nanometer-sized GO nanosheets (nanoGO) was recently demonstrated by the chemical exfoliation of graphite nanofibers [30]. Compared to regular GO sheets, which extend over several microns, the smaller size of nanoGO sheets increases the oxygen-containing functional groups in the basal planes and the sheet edges, leading to higher colloidal stability and hydrophilicity as well as more available surface reaction and anchoring sites and enhanced interfacial coupling with TiO_2 nanomaterials [31]. Very recently, surface functionalization of PBG-engineered TiO_2 photonic films by GO nanocolloids was demonstrated as an effective approach to enhance the photocatalytic degradation of methylene blue (MB) dye under UV-VIS and visible light by combining the advantages of macroporous inverse opals of improved light harvesting, surface area, and mass transport with the high adsorption capacity, reactivity, and charge separation of GO nanocolloids [32]. In the case of UV-VIS irradiation, nanoGO deposition on the titania inverse opals led to a marked improvement of the MB photocatalytic degradation rate, which was related to the synergy of slow photon amplification upon spectral overlap of the low-energy edge of the incomplete TiO_2 inverse opal PBG (stop band) with the dye electronic absorption, enhanced MB adsorption due to the electrostatic interactions of surface oxygen groups and π–π coupling with the dye molecules [33], and interfacial nanoGO–TiO_2 charge transfer [32]. Likewise, a distinct acceleration of dye photodegradation kinetics was observed for the nanoGO–TiO_2 photonic films under visible light, which was related to the combination of increased dye adsorption with the slow photon-intensified dye photocatalysis based on the self-sensitization mechanism.

In this work, postreduction of PBG-engineered graphene oxide–titania photonic films was explored in order to elucidate the interplay of interfacial electron transfer with pollutant adsorption via GO's surface functional groups and thus provide a means to further improve the photocatalytic performance of nanoGO–TiO_2 photonic crystals. To this end, well-ordered TiO_2 inverse opal films fabricated by the coassembly technique with tuned stop band for the optimal slow photon-assisted MB photodegradation were functionalized by GO nanosheets and reduced by thermal annealing in He at different temperatures (200 and 500 °C). Thermal reduction of nanoGO is expected to increase the sp^2/sp^3 ratio of carbon atoms, leading to a decrease in the energy gap and an increase in the conductivity of rGO nanosheets [34,35]. Comparative performance evaluation of the pristine and reduced nanoGO–TiO_2 inverse opals was performed on MB dye degradation under UV-VIS and visible light as well as UV-VIS photodegradation of colorless salicylic acid (SA), an emerging water pollutant rarely investigated by photonic crystal photocatalysts [4]. Despite the decrease in the GO amount and the reduction in surface oxygen groups on the post-treated photonic films, a clear enhancement of the photocatalytic activity was identified, indicating that optimization of charge transfer between reduced GO and the nanocrystalline TiO_2 inverse opal walls may effectively alleviate electron-hole recombination and promote their photocatalytic performance.

2. Materials and Methods

2.1. Materials

Monodisperse polystyrene (PS) spheres with diameter of 425 nm were purchased from Microparticles GmbH (Berlin, Germany) in the form of colloidal dispersion of 5% solids (w/v) in deionized (DI) water (2.3% CV, SD = 0.01 μm). The specific sphere diameter was selected because the corresponding nanoGO–TiO_2 inverse opals presented the optimal photocatalytic activity on MB degradation in comparison to the ones synthesized by colloidal PS spheres of other diameters (220, 350, and 510 nm) [32]. Titanium(IV) bis(ammonium lactato)dihydroxide (TiBALDH) 50 wt.% aqueous solution (388165, Sigma-Aldrich, St. Louis, MO, USA), graphene oxide aqueous nanocolloidal dispersion consisting of single-layer GO sheets down to a few nanometers in lateral width, 2 mg/mL in H_2O, 42.0%–52.0% C in dry basis (795534, Sigma-Aldrich, St. Louis, MO, USA), and Hellmanex™ III were obtained from Sigma-Aldrich. All other reagents were of analytical or ACS reagent grade: ethanol (EtOH, 32221, Honeywell Riedel-de Haën, puriss. p.a., absolute, ≥99.8%), methanol (MeOH 179337, Sigma-Aldrich, ACS reagent, ≥99.8%), acetone (32201, Honeywell Riedel-de Haën, ACS reagent, ≥99.5%), hydrochloric acid (HCL, 30721 Fluka, ACS reagent, fuming, ≥37%).

2.2. Inverse Opal Fabrication and Surface Modification

Titania inverse opal films were fabricated via the evaporative coassembly of the PS colloidal spheres with the hydrolyzed TiBALDH titania precursor [36]. This synthesis route was selected over the more common successive deposition method, which involves liquid infiltration and subsequent removal of the polymeric opals, as it produces high-quality inverse opal films with large photonic domains [32]. In particular, cleaned glass substrates (S8902 Sigma-Aldrich) by Hellmanex™ III andultrasound acetone–EtOH washing) were nearly vertically suspended into glass vials, each containing 8 ml of 0.2 wt.% PS dilute sphere suspension and 0.168 mL of titania precursor, synthesized by stirring of 1.23 mL TiBALDH solution, 1.5 mL HCl 0.1 Mand 2.85 mL EtOH for 1 h. The vials were placed in a heating oven at 55 °C until the solvent fully evaporated over 3 days, leading to the deposition of a thin film on the glass substrates. The obtained films were calcined at 500 °C in air for 2 h, resulting in the removal of the polymer matrix and crystallization of titania in the inverse opal structure. The fabricated TiO_2 photonic crystals were designated as PC. The surface functionalization of the inverse opals was performed by dipping the films for 24 h in the nanocolloidal GO dispersion, whose pH had been stabilized at 10 by periodically adding 1–2 drops of NaOH aqueous solution (1 M) and intermediate 10 min stirring. The modified photonic films were denoted as GOnano-PC. The reduction was carried out thermally by calcinating the GOnano-PC films for 2 h under He flow at 200 and 500 °C, producing the rGO-modified photonic crystals named as rGOnano(200)-PC and rGOnano(500)-PC, respectively.

2.3. Material Characterization

The morphology and phase composition of the inverse opals was studied using scanning electron microscopy (SEM, Quanta Inspect, FEI, Eindhoven, The Netherlands) coupled with energy-dispersive X-ray spectroscopy (EDXDX4, EDAX, Mahwah, USA, and transmission electron microscopy (TEM, CM20, FEI, Eindhoven, The Netherlands) augmented by energy-filtered TEM (GIF 200, Gatan,

Pleasanton, CA, USA). The composition of the films was analyzed by X-ray photoelectron spectroscopy (XPS, SPECS, Berlin, Germany). Core-level photoemission spectra were collected with a PHOIBOS 100 (SPECS, Berlin, Germany) hemispherical analyzer at a pass energy of 15 eV utilizing Mg Kα radiation at 1253.6 eV. The binding energy scale was calibrated using the position of both Au $4f_{7/2}$ and $Ag3_{d5/2}$ peaks at 84 and 368.3 eV, respectively, measured on clean gold and silver foils. Gaussian–Lorentzian shapes (Voigt functions) were used for deconvolution of the recorded spectra after standard Shirley background subtraction. The structural properties of the films were investigated by micro-Raman spectroscopy (inVia Reflex, Renishaw, UK) with 514.5 nm excitation from an Ar^+ ion laser. The laser beam was focused on the sample by means of a ×50 (NA = 0.75) objective under low power density (0.05 mW/μm^2) to avoid local heating. The optical properties of the photonic films were determined by diffuse and specular reflectance UV-vis spectroscopy (Cary 60 UV-Vis, Agilent, USA) equipped with a fiber optic diffuse reflectance accessory (Barrelino) and a 15° specular reflectance one (PIKE, UV-VIS 15Spec), using a Halon standard and a UV Al mirror for baseline measurements, respectively. Charge transfer was investigated by photoluminescence (PL) measurements performed by lock-in amplification techniques. The excitation beam was generated by a light-emitting diode at 275 nm and modulated by a mechanical chopper (SR540, Stanford Research Systems, Sunnyvale, USA). The modulated PL signal was analyzed by a 1/4 monochromator (77200, Oriel, Irvine, USA) and detected by a Si photodiode (FDS1010, Thorlabs GmbH, Munich, Germany) using a benchtop photodiode amplifier (PDA200C, Thorlabs GmbH, Munich, Germany) and a lock-in amplifier (PAR 126, Princeton Applied Research, Oak Ridge, TN, USA).

2.4. Photocatalytic Performance

The photocatalytic activity of the inverse opals was evaluated on the aqueous phase degradation of methylene blue [3,7-bis(dimethylamino)phenazathionium chloride] (MB, A18174, Alfa Aesar) and salicylic acid (SA, 247588, Sigma-Aldrich) under UV-VIS and visible light. The photonic films (~1.5 cm^2) were placed horizontally at the bottom of beakers containing aqueous MB (4 mL, 3 μM) or SA (3 mL, 25 μM) solutions, where they were left for 60 min under dark conditions to reach adsorption–desorption equilibrium under stirring [32,37]. To further enhance the adsorption of salicylic acid on the titania films, the pH of the SA solution had been stabilized at 3 using dilute HCL solution [37]. The illumination source was a 150 W Xe lamp (6255, ORIEL GmbH, Darmstadt, Germany), and UV-VIS irradiation was selected by the combination of a long-pass edge filter with cut-on wavelength of 305 nm (20CGA-305, Newport) and a heat reflective mirror (Newport 20CLVS-3 CoolView™, T_{avg} = 85% at 332–807 nm, R_{avg} = 95% at 840–1500 nm, 20CLVS-3 CoolView™, Newport, Irvine, USA). Visible-light irradiation was selected by an additional long-pass edge filter with cut-on at 400 nm (20CGA-400, Newport, Irvine, USA). The horizontal Xe beam was directed on the film surface via a UV-enhanced Al mirror (Newport ValuMax 20D520AL.2, λ/10, R_{avg} > 90% at 250–600 nm, ValuMax 20D520AL.2, Newport, Irvine, USA). The power density of the incident beam reaching the film surface was 2.8 mW/cm^2 in the case of MB degradation and nearly 1 sun (96 mW/cm^2) for the SA in the UV-VIS spectral range. At given time intervals, a small (0.5 mL) aliquot of the MB/SA solution was withdrawn and quantitatively analyzed using a 10 mm path length quartz micro cell (105B-QS, 500 μL, HELMA Analytics) in the Cary 60 spectrophotometer. Subsequently, the analyzed aliquot was poured back into the reacting solution and the illumination continued. The photocatalytic experiments were performed in triplicate, and standard errors were calculated for the mean kinetic constants. The stability of the photocatalytic films was tested by recycling three times the SA degradation test for the same film, with intermediate cleaning of SA residues on the used film by 1 h UV-VIS illumination in 4 mL deionized water.

3. Results and Discussion

3.1. Morphology, Structural, and Optical Properties

The morphology of the TiO_2 photonic crystals was examined by SEM and TEM. Figure 1a–c displays representative SEM images of the rGOnano(200)-PC inverse opal featuring a well-ordered, periodic macroporous structure corresponding to the (111) planes of an fcc lattice. The spherical macropores were hexagonally arranged and were well interconnected through smaller pores with diameter of ca. 50–90 nm. The average diameter of the macropores was 245(10) nm, a value much smaller than that of the original PS colloidal spheres, namely 425 nm, confirming the persistent shrinkage of metal oxide inverse opals by means of the amorphous-to-crystalline phase transition and the inorganic matrix volume decrease after calcination [38]. According to cross-section SEM, shown in Figure 1c, the film thickness was around 4.5 μm. Furthermore, it was observed that surface modification of the TiO_2 inverse opals by nanoGO as well as their subsequent thermal reduction at 200 and 500 °C left both the morphology and the size of the macropores pores intact, identical to those of the pristine photonic films [32].

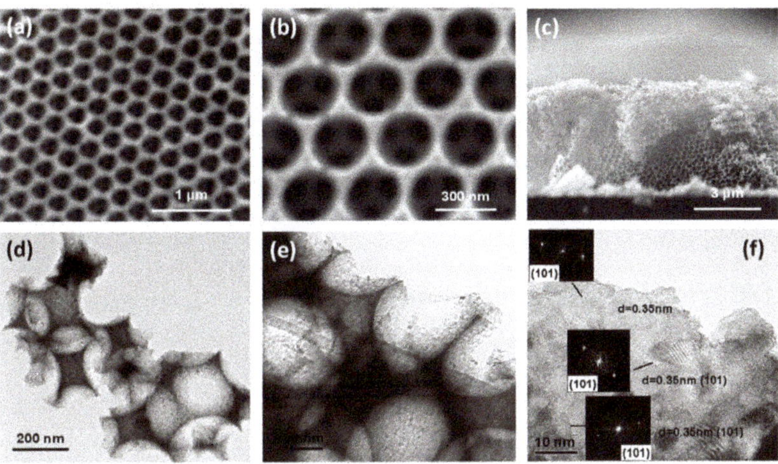

Figure 1. (**a**)–(**c**) Top view and cross-section SEM and (**d**)–(**f**) TEM images of the rGOnano(200)-PC inverse opal at different magnifications. The insets in the high-resolution TEM (HR-TEM) image (**f**) show the fast Fourier transform (FFT) patterns of the indicated areas.

TEM images at successively higher magnifications, displayed in Figure 1d–f, show that the walls of the inverse opal skeleton were mesoporous, consisting of ca. 10 nm titania nanoparticles with the most common 0.35 nm *d*-spacing arising from the (101) planes of the anatase TiO_2 phase. To verify the deposition of graphene oxide and reduced graphene oxide sheets on the inverse opals, elemental mapping was performed by energy-filtered (EF) TEM. Figure 2a–h show a bright-field TEM image of the inverse opal skeleton and the corresponding C, O, and Ti EF-TEM elemental maps for the GOnano-PC and rGOnano(200)-PC films, respectively. The presence of carbon, besides Ti and O, on the nanocrystalline TiO_2 walls of a void macropore could be identified in both cases, verifying the successful surface modification of the titania photonic films with GO nanosheets.

Due to their reduced lateral size, GO nanosheets could be fully incorporated in the inverse opal macropores, justifying the absence of any noticeable effects on the periodicity and macropore diameter of the photonic crystals. On the other hand, surface functionalization of the TiO_2 inverse opals by regular, micrometer-sized GO sheets resulted in extended coverage of the nanocrystalline TiO_2 walls by GO sheets over several macropores [32]. The extensive GO coverage of the TiO_2 skeleton can impede the pore accessibility to the diffusing pollutant molecules and thus impose an adverse effect to the photocatalytic process, similar to the partial surface clogging observed for benchmark mesoporous P25 films subjected to the same functionalization treatment [32].

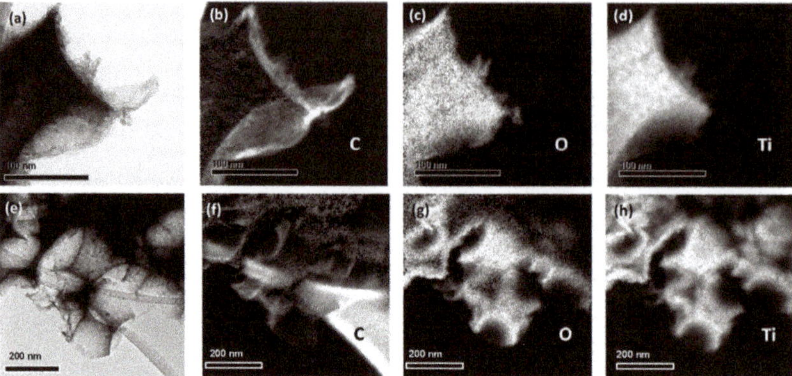

Figure 2. Bright-field TEM image and the corresponding C, O, and Ti energy-filtered TEM (EF-TEM) elemental maps of (**a**)–(**d**) GOnano-PC and (**e**)–(**h**) rGOnano(200)-PC.

The structural properties of the TiO_2 photonic crystals and their functionalization with GO were further investigated by Raman spectroscopy. Figure 3a compares the Raman spectra of the pristine and modified photonic films at 514.5 nm. All inverse opals exhibited the characteristic bands of anatase TiO_2 at approximately 148 (E_g), 398 (B_{1g}), 520 (A_{1g} + B_{1g}), and 643 (E_g) cm^{-1}. No traces of polystyrene, rutile, or brookite could be detected, confirming the removal of the polymer matrix as well as the crystallization of titania in the anatase phase after thermal treatment [39]. Furthermore, appreciable broadening and shift of the anatase Raman peaks was observed, especially for the low-frequency E_g mode, which shifted to 148 cm^{-1} and broadened to full-width at half-maximum of 18 cm^{-1}, indicative of the formation of anatase nanocrystals with size ≤10 nm [40], in close agreement with the direct imaging of the nanocrystalline inverse opal walls by HR-TEM (Figure 1).

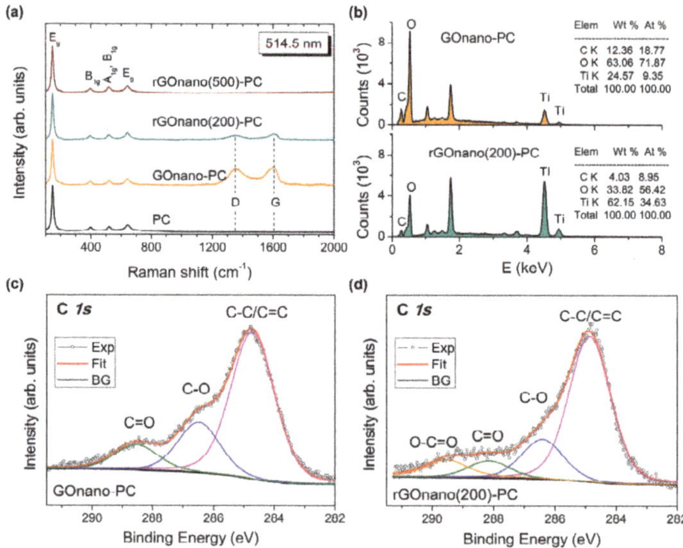

Figure 3. (**a**) Raman spectra of the pristine and modified inverse opals. (**b**) Energy-dispersive X-ray spectroscopy (EDX) spectra and C, O, Ti elemental analysis for GOnano-PC and reduced rGOnano(200)-PC. C *1s* X-ray photoelectron spectroscopy (XPS) spectra for the (**c**) GOnano-PC and (**d**) rGOnano(200)-PC films.

In addition to the anatase Raman bands, the GOnano-PC film presented the characteristic G-mode of graphene oxide, stemming from the stretching of sp^2 carbon atoms, and the highly dispersive D-band, activated by defects in GO in the frequency range of 1000–2000 cm^{-1}. The G-band displayed a distinct asymmetric lineshape with spectral weight being shifted at higher frequency [24], related to the diverse contributions of alternating single–double carbon bonds [41]. Reduction at 200 °C led to a significant decrease in the intensity of the G- and D-modes relative to that of anatase by nearly 65%, suggesting that a considerable amount of GO is removed from the film when it is thermally treated [30]. This behavior could be clearly identified in the Raman spectrum of rGOnano(500)-PC, where the G- and D-peaks could be hardly detected, indicating that the film almost recovered its pristine state after annealing at 500 °C. In addition, the intensity ratio I_D/I_G determined from the integrated areas of the D- and G-bands, which is the characteristic probe of the sp^2 domain size as well as the defect density in graphitic materials [42,43], decreased from the value of 1.60 for GOnano-PC to 1.54 for rGOnano(200)-PC films. The partial restoration of sp^2 bonds and the decrease of defects in the GO nanosheets could be accordingly inferred [44–46], considering that GO thermal reduction does not cause a large expansion of the average graphitic domain size [47]. The loss of GO sheets was further supported by EDX, as shown in Figure 3b, where a nearly 70% decrease in the C content was observed for rGOnano(200)-PC with respect to the GOnano-PC. The surface composition of the nanoGO–TiO_2 inverse opals was investigated by XPS. Figure 3c,d compares the deconvoluted XPS C1s spectra of the GOnano-PC and rGOnano(200)-PC films. Both spectra were highly functionalized with oxygenated groups such as epoxy/hydroxyl groups (C–O at ~286.5 eV), carbonyl group (C=O at ~288.3 eV), and carboxyl group (O–C=O at ~290 eV) [24,47]. As expected, increased amounts of oxygenated groups were present in GOnano-PC compared to the thermally reduced film. It was found out that, for the nanoGO-PC, the ratio between the strongest C *1s* peak, corresponding to the signal arising from C atoms in the C–C/C=C configurations and the other components corresponding to the oxygenated groups, was 1.5, indicating a considerable oxidation degree. After thermal reduction, this ratio reached a value of 2.1, indicative of the loss of surface oxygen groups and the enhancement of the sp^2 carbon bonds [47].

The decrease in the GO amount on the TiO_2 inverse opals upon thermal reduction was further evidenced in the diffuse reflectance (DR%) spectra, as shown in Figure 4. Besides the anatase band gap edge below 400 nm, all films presented a broad DR% peak at approximately 580 nm, close to the low energy (red) edge of the stop band identified at 525 nm in the pristine PC by the 15° incidence specular (R%) spectra. It should be noted that the DR% intensity considerably exceeded the specular one due to the presence of disorder with respect to the relatively coarse beam size (1.5 mm^2) that probes film areas comprising several domains of variable thickness and surface flatness [32]. Diffuse reflectance was more intense at higher wavelengths than the stop band predicted by the R% spectra because slow photons at the "red" stop band edge, which are localized at the high dielectric medium, i.e., the titania skeleton, experience a longer optical path and are thus more likely to be scattered. Surface modification of the titania photonic films with nanoGO as well as the post-thermal treatment hardly affected the position of the broad DR% photonic peak, reflecting the weak effect of the GO nanosheets on the inverse opal macropores and periodicity, in agreement with SEM analysis (Figure 1). However, a distinct decrease in the initial DR% was observed for GOnano-PC, arising from the broadband absorption of the GO nanosheets [30]. On the other hand, the DR% reflectance was gradually restored after post-treatment at 200 and 500 °C, corroborating the corresponding GO losses on the thermally reduced photonic films.

The stop band spectral position can be approximated by modified Bragg's law for first-order diffraction from the (111) planes [4]:

$$\lambda = 2d_{111}\sqrt{n_{eff}^2 - \sin^2\theta}$$

where λ is the stop band wavelength; $d_{111} = \sqrt{2/3}D$ is the interplanar spacing of the (111) planes, with D being the macropore diameter; and n_{eff} is the volume-weighted average of the void spheres' refractive index n_{void} and titania n_{TiO_2} occupying the inverse opal skeleton, defined by $n_{eff}^2 = n_{void}^2 f + n_{TiO_2}^2 (1-f)$, with f being the filling fraction ($f = 0.74$ for the ideal *fcc* lattice) and θ being the angle between the incident beam and the plane normal. Using $\theta = 15°$ together with the measured stop band position and macropore diameter for $n_{TiO_2} = 2.55$ and $n_{air} = 1.0$, the n_{eff} value of 1.34 and titania skeleton filling fractions $(1-f)$ of 0.14 were determined in air, in perfect agreement with previous results [32]. Moreover, using the obtained TiO_2 filling fraction and $n_{H_2O} = 1.33$, the stop band position of 626 nm was calculated for the inverse opal films in aqueous medium, where the photocatalytic reaction takes place.

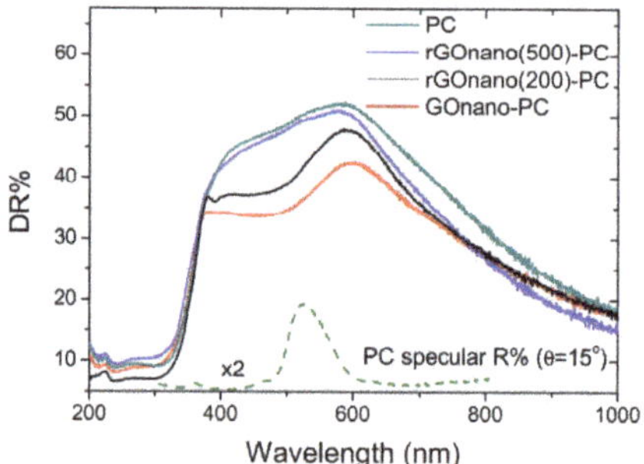

Figure 4. UV-VIS diffuse reflectance (DR%) spectra of the pristine and modified photonic crystals. The dashed line shows the specular R% spectrum of the pristine PC film at 15° incident angle.

3.2. *Photocatalysis*

Figure 5 displays the adsorption of methylene blue on the different inverse opals under dark conditions. A marked rise of MB dark adsorption was observed after nanoGO functionalization, which can be attributed to the electrostatic interactions between the negatively charged oxygenated groups of GO and the positively charged amino groups of the cationic MB dye. These interactions can be further assisted through the π–π coupling of GO itinerant sp^2 electrons with delocalized electrons in the aromatic rings of the MB molecules [33]. Dark adsorption lessened with thermal reduction, reflecting the decrease in surface oxygen groups as well as the GO amount on the photonic films, as inferred from the Raman, EDX, XPS, and UV-VIS results. In fact, the MB adsorption of the rGOnano(500)-PC film was only slightly higher than that of the pristine one, indicating the defunctionalization of the sample after thermal annealing at 500 °C. The small difference in MB adsorption between the pristine and the rGOnano(500)-PC photonic film can be related to residual rGO that could not be detected by Raman spectroscopy, but it enhanced the adsorption slightly.

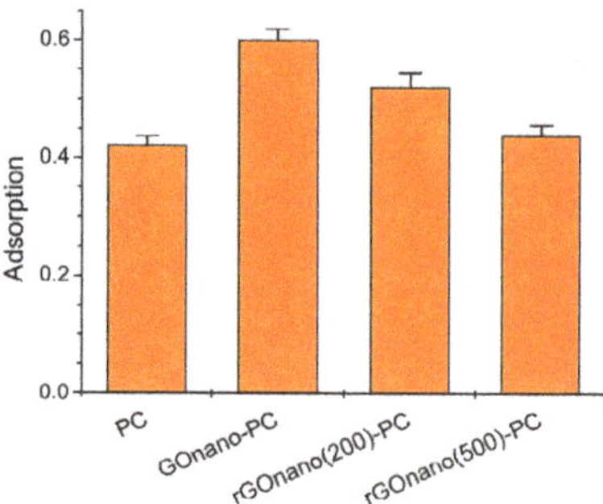

Figure 5. Methylene blue (MB) adsorption on the pristine, GO- and rGO-modified TiO_2 photonic crystals under dark conditions.

Figure 6a,b presents the MB photodegradation kinetics for the pristine, GOnano and rGOnano photonic crystals under UV-VIS and VIS irradiation, respectively. In all cases, the $\ln(C/C_0) = f(t)$ plots were linear, indicating that the MB photodegradation followed first-order kinetics. Figure 6c summarizes the apparent rate constants $k_{UV\text{-}VIS}$ and k_{VIS} calculated from the slopes of the corresponding plots. It was evident that the GO-functionalized film presented the highest apparent rate under both UV-VIS and VIS illumination, implying that deposition of GO nanosheets on the titania films can boost their photocatalytic performance. This behavior can be attributed not only to the higher MB adsorption but also to the enhanced charge separation induced by GO as well as the slow photon amplification produced by the specific photonic crystal [32]. In particular, photogenerated electrons in TiO_2 by UV-VIS light can be scavenged and shuttled in the GO nanosheets, leading to the decrease in electron-hole recombination. As a result, MB molecules could be degraded not only by hydroxyl radicals formed by valence band holes of TiO_2 but also via intermediate peroxide or other radicals formed in the GO sheets [18]. Nevertheless, photocatalytic experiments in the presence of 0.01 M methanol (MeOH) as hydroxyl radical scavenger [48] showed roughly 50% and 30% reduction of the corresponding $k_{UV\text{-}VIS}$ and k_{VIS} constants, respectively, as shown in Figure 3 for the rGOnano(200)-PC, verifying the dominant role of •OH radicals in MB degradation, especially under UV-VIS light [49]. In addition, the PC film fabricated by the 425 nm PS spheres presented the optimal photocatalytic performance due to slow photon effects [32]. In the present case, the red edge of the 626 nm stop band for the PC film in water is also very close to the MB electronic absorption at 664 nm, where multiple scattering due to "red" slow photons localized in the titania skeleton occur, leading to the acceleration of MB photodegradation kinetics. Furthermore, GO can promote the self-sensitized photocatalytic oxidation of MB molecules, which is the main degradation mechanism under visible light, where TiO_2-photogenerated carriers are absent due to the wide anatase band gap. In this case, the strong MB adsorption on the GO nanosheets is crucial for the interfacial electron transfer and the enhanced photodegradation as the electrons of the adsorbed MB molecules excited by visible light are scavenged by GO–TiO_2, forming the oxidation radicals that degrade the pollutant.

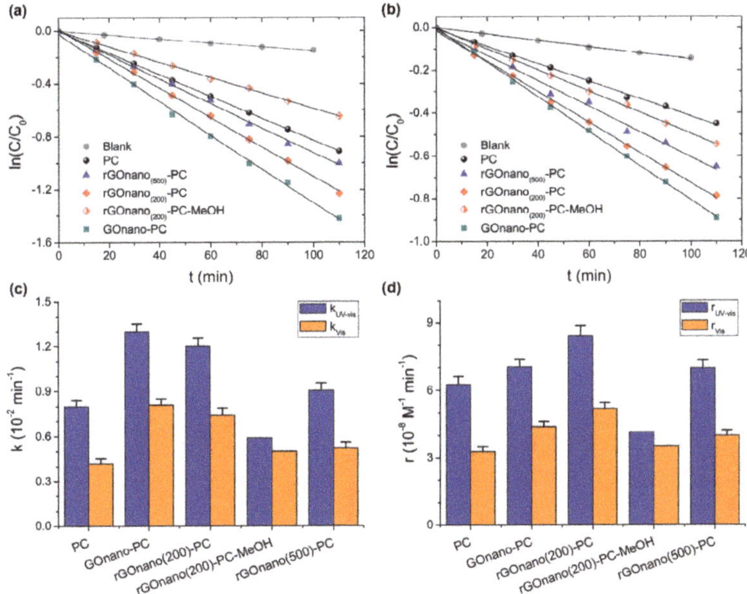

Figure 6. MB photodegradation kinetics of the inverse opals under (**a**) UV-VIS and (**b**) VIS irradiation. (**c**) Apparent rate constants k and (**d**) reaction rates $r = kC_0$ under UV-VIS and visible light.

Nevertheless, it can be noticed that the initial concentration C_0 of the MB solution in the photocatalytic experiments was not the same for all samples because of the different degree of MB adsorption on each film (Figure 5). To take the C_0 variation into account, the reaction rate r was estimated, which, for low (<mM) MB concentrations, is proportional to C_0, according to the Langmuir–Hinshelwood (L–H) model, $r = kC_0$ [49]. Figure 6d compares the reaction rates for all inverse opals under UV-VIS and visible illumination. In both cases, the maximum r value was observed for the rGOnano(200)-PC and not for the GOnano-PC film despite the high apparent constant k of the latter, indicating that the photocatalytic activity of the GO-modified sample improved after reduction at 200 °C. This behavior would further suggest enhanced charge transfer and transport on the rGO nanosheets due to their higher conductivity and favorable work function [23,35]. It appears that, in the case of the rGOnano(200)-PC film, the enhancement of electron transfer/transport is the definitive factor in its photocatalytic efficiency as MB adsorption was lower than that for the GO-functionalized sample. On the other hand, the reaction rate of GOnano-PC decreased when the film was thermally reduced at 500 °C, with its value being almost the same as that for the pristine one. This behavior complies favorably with previous results on the fabrication of highly efficient GO–TiO_2 powder nanocomposites for the photocatalytic decomposition of different water pollutants [16,24,50], where a mild thermal reduction post-treatment at 200 °C in N_2 atmosphere was found to enhance the optimal assembly and photocatalytic activity of TiO_2 nanoparticles on GO sheets. Although direct comparison of the films' photocatalytic performance with literature values is rather difficult due to differences in dye concentration, film size and mass, and irradiation density, the present rates are comparable or even higher than the $r \sim 10^{-8}$ M^{-1} min^{-1} that can be derived from the UV-VIS MB degradation k values reported in a thorough study of TiO_2 inverse opals (5 ppm MB with 2 mg and 5 cm^2 area of 2.5–3 μm thick photonic films under simulated solar light) [51] and most importantly exceed those of the benchmark mesoporous P25 films under the same conditions [32].

To explore charge separation effects on the GO- and rGO-functionalized TiO_2 photonic crystals, the photodegradation of salicylic acid under UV-VIS irradiation was subsequently investigated. Salicylic acid is a colorless water pollutant that absorbs in the UV range, as shown in Figure 7a, away from the stop band of the photonic films (626 nm in water), excluding the contribution of slow photon effects that are dominant in the MB dye degradation.

Moreover, its adsorption on the titania films, although enhanced at pH = 3 due to the SA chemisorption on the TiO_2 surface [37,52], is relatively weaker compared to that of MB molecules with minimal effects on the corresponding reaction rates [53]. Comparative SA photodegradation experiments for the PC, GOnano-PC and rGOnano(200)-PC films were accordingly exploited as an indirect means to assess the efficiency of charge separation for GO- and rGO-functionalized films. SA degradation followed first-order kinetics, as shown in Figure 7b. The corresponding plot for rGOnano(500)-PC was similar to that of the pristine one and thus not included in the graph. The reaction rates r, depicted in Figure 7c, revealed an improvement of the photocatalytic activity for GOnano-PC, which was further increased after thermal reduction, despite the loss of GO nanosheets and surface oxygen groups. In this case, the improved photocatalytic activity for rGOnano(200)-PC can be mainly attributed to the improved charge separation arising from the enhanced electron transfer from TiO_2 to the rGO nanosheets due to their lower work function and higher conductivity [23,34,47]. SA photodegradation by rGOnano(200)-PC in the presence of MeOH as $^{\bullet}OH$ scavenger revealed a weak effect on the reaction rate, corroborating the major role of direct SA oxidation by TiO_2 valence band holes [37]. Moreover, consecutive SA photodegradation tests for the optimal rGOnano(200)-PC photocatalyst, shown in Figure 7d, corroborated the film stability (roughly 6% reduction of the SA degradation was observed after three photocatalytic cycles), similar to the excellent stability of the GOnano–TiO_2 inverse opals on MB degradation [32].

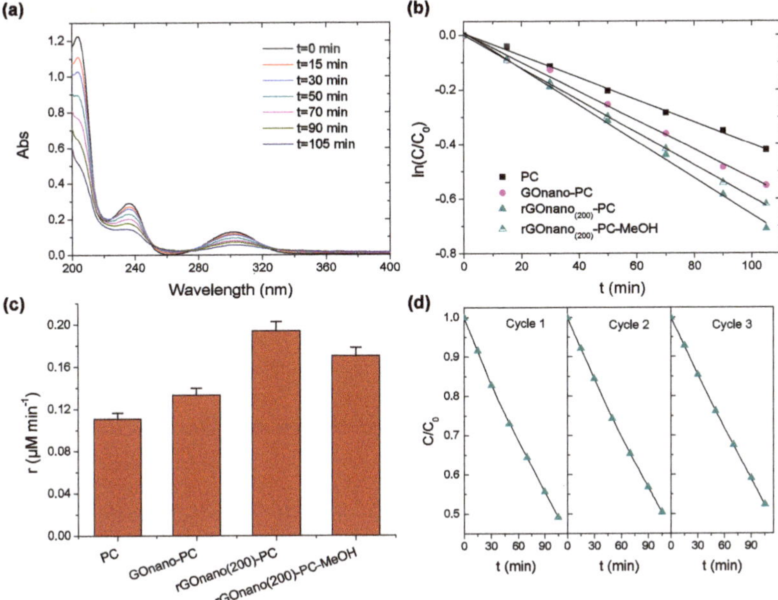

Figure 7. (**a**) Salicylic acid (SA) absorbance variation for the rGOnano(200)-PC film under UV–VIS light. Photodegradation (**b**) kinetics and (**c**) reaction rates of SA for the PC, GOnano-PC and rGOnano(200)-PC inverse opals under UV-VIS irradiation. (**d**) SA photodegradation kinetics after three successive photocatalytic tests using the rGOnano(200)-PC film.

The charge separation on the rGO-modified photonic films was further supported by PL spectroscopy (Figure 8). The PL spectra of the pristine film exhibited a broad band at 380 nm, whose intensity was drastically reduced after the surface deposition of GO, verifying the decrease in electron-hole recombination and corroborating the interfacial transfer of UV photogenerated electrons from TiO_2 to the GO nanosheets. Postreduction of GO at 200 °C led to a further decrease in the PL intensity, reflecting the promotion of electron transfer. On the other hand, PL intensity was augmented after reduction at 500 °C due to the defunctionalization of the film. The PL intensity of the rGOnano(500)-PC, however, did not reach that of the pristine photonic crystal, probably owing to residual graphene oxide on the sample even after its calcination.

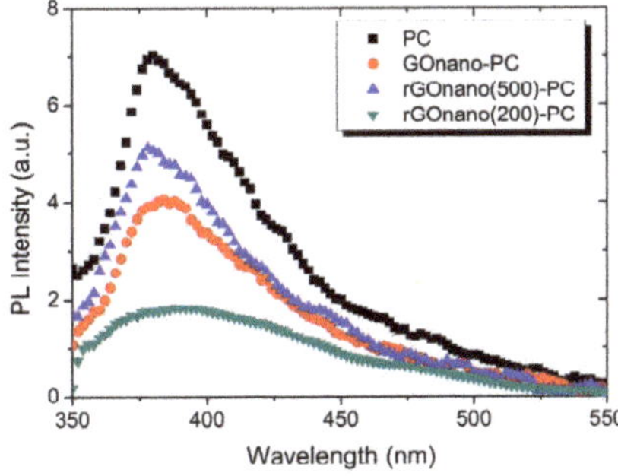

Figure 8. Photoluminescence (PL) spectra of the pristine, GO-functionalized and rGO-functionalized photonic crystals.

4. Conclusions

TiO_2 inverse opals, fabricated via the coassembly of polystyrene colloidal spheres with the water soluble TiBALDH precursor, were surface-functionalized with GO nanosheets and, subsequently, thermally reduced at 200 and 500 °C. Neither GO surface modification nor postreduction affected the highly ordered macroporous structure of the TiO_2 photonic films, leaving the photonic stop band positions intact. Raman, EDX, and XPS spectroscopies corroborated the successful GO functionalization of the TiO_2 inverse opals and disclosed that, upon post-thermal treatment, the amount of nanoGO was moderated on the modified films along with the partial recovery of sp^2 domains. Aqueous phase photodegradation of the MB dye under UV-VIS and visible light showed that thermal reduction of the GO–TiO_2 photonic films at 200 °C, in synergy with slow photon amplification, improved the MB photocatalytic degradation rate despite the loss of oxygen functional groups and GO nanosheets, indicative of enhanced charge separation due to the lower work function and higher conductivity of the rGO nanosheets. The intensification of interfacial charge transfer was further supported by both photocatalytic experiments under UV-VIS light using salicylic acid as emerging water pollutant and PL spectroscopy. In that case, the SA photocatalytic degradation was drastically increased on the post-treated rGO–TiO_2 inverse opals despite the absence of photonic effects. This confirms that further optimization of the amount of reduced GO nanosheets and fine-tuning of their interfacial coupling with the TiO_2 nanoparticles on the nanocrystalline walls of the macroporous inverse opal films is a promising route to alleviate electron-hole recombination and boost their photocatalytic performance.

Author Contributions: Conceptualization, V.L.; Formal analysis, A.D. (Angeliki Diamantopoulou), E.S., D.T. and V.L.; Investigation, A.D. (Angeliki Diamantopoulou), E.S., S.G., D.T., A.D. (Athanasios Dimoulas), S.G. and N.B.; Methodology, A.D. (Angeliki Diamantopoulou) and V.L.; Resources, S.G., N.B., A.D. (Athanasios Dimoulas) and V.L.; Supervision, V.L.; Writing—original draft, A.D. (Angeliki Diamantopoulou); Writing—review & editing, A.D. (Angeliki Diamantopoulou) and V.L.

Funding: This research was funded by the Research Projects For Excellence IKY/SIEMENS, "Enhanced light harvesting by innovative photonic structures of titanium dioxide and graphene oxide for applications in photocatalytic environmental cleaning and solar-to-electric energy conversion". We acknowledge the support for part of this work by the project MIS 5002772, implemented under the Action "Reinforcement of the Research and Innovation Infrastructure", funded by the Operational Programme "Competitiveness, Entrepreneurship and Innovation" (NSRF 2014-2020) and cofinanced by Greece and the European Union (European Regional Development Fund).

Conflicts of Interest: The authors declare no conflict of interest.

References

1. Schneider, J.; Matsuoka, M.; Takeuchi, M.; Zhang, J.; Horiuchi, Y.; Anpo, M.; Bahnemann, D.W. Understanding TiO_2 photocatalysis: mechanisms and materials. *Chem. Rev.* **2014**, *114*, 9919–9986. [CrossRef] [PubMed]
2. Emeline, A.V.; Zhang, X.; Jin, M.; Murakami, T.; Fujishima, A. Application of a "black body" like reactor for measurements of quantum yields of photochemical reactions in heterogeneous systems. *J. Phys. Chem. B.* **2006**, *110*, 7409–7413. [CrossRef] [PubMed]
3. Banerjee, S.; Pillai, S.C.; Falaras, P.; O'shea, K.E.; Byrne, J.A.; Dionysiou, D.D. New insights into the mechanism of visible light photocatalysis. *J. Phys. Chem. Lett.* **2014**, *5*, 2543–2554. [CrossRef] [PubMed]
4. Likodimos, V. Photonic crystal-assisted visible light activated TiO_2 photocatalysis. *Appl. Catal. B Environ.* **2018**, *230*, 269–303. [CrossRef]
5. Vogel, N.; Retsch, M.; Fustin, C.A.; del Campo, A.; Jonas, U. Advances in colloidal assembly: The design of structure and hierarchy in two and three dimensions. *Chem. Rev.* **2015**, *115*, 6265–6311. [CrossRef] [PubMed]
6. Phillips, K.R.; England, G.T.; Sunny, S.; Shirman, E.; Shirman, T.; Vogel, N.; Aizenberg, J. A colloidoscope of colloid-based porous materials and their uses. *Chem. Soc. Rev.* **2016**, *45*, 281–322. [CrossRef]
7. Chen, J.I.L.; von Freymann, G.; Choi, S.Y.; Kitaev, V.; Ozin, G.A. Amplified photochemistry with slow photons. *Adv. Mater.* **2006**, *18*, 1915–1919. [CrossRef]
8. Curti, M.; Schneider, J.; Bahnemann, D.W.; Mendive, C.B. Inverse opal photonic crystals as a strategy to improve photocatalysis: underexplored questions. *J. Phys. Chem. Lett.* **2015**, *6*, 903–3910. [CrossRef]

9. Liu, J.; Zhao, H.; Wu, M.; Van der Schueren, B.; Li, Y.; Deparis, O.; Ye, J.; Ozin, G.A.; Hasan, T.; Su, B.L. Slow photons for photocatalysis and photovoltaics. *Adv. Mater.* **2017**, *29*, 1605349. [CrossRef]
10. Lim, S.Y.; Law, C.S.; Markovic, M.; Kirby, J.K.; Abell, A.D.; Santos, A. Engineering the slow photon effect in photoactive nanoporous anodic alumina gradient-index filters for photocatalysis. *ACS. Appl. Energy Mater.* **2019**, *22*, 1169–1184. [CrossRef]
11. Dervin, S.; Dionysiou, D.D.; Pillai, S.C. 2D nanostructures for water purification: graphene and beyond. *Nanoscale* **2016**, *8*, 15115–15131. [CrossRef] [PubMed]
12. Faraldos, M.; Bahamonde, A. Environmental applications of titania-graphene photocatalysts. *Catal. Today.* **2017**, *285*, 13–28. [CrossRef]
13. Li, X.; Shen, R.; Ma, S.; Chen, X.; Xie, J. Graphene-based heterojunction photocatalysts. *Appl. Surf. Sci.* **2018**, *430*, 53–107. [CrossRef]
14. Novoselov, K.S.; Geim, A.K.; Morozov, S.V.; Jiang, D.; Katsnelson, M.I.; Grigorieva, I.V.; Dubonos, S.V.; Firsov, A.A. Two-dimensional gas of massless Dirac fermions in graphene. *Nature (London)* **2005**, *438*, 197–200. [CrossRef] [PubMed]
15. Dimiev, A.M.; Tour, J.M. Mechanism of Graphene Oxide Formation. *ACS Nano.* **2014**, *8*, 3060–3068. [CrossRef] [PubMed]
16. Pastrana-Martínez, L.M.; Morales-Torres, S.; Likodimos, V.; Figueiredo, J.L.; Faria, J.L.; Falaras, P.; Silva, A.M.T. Advanced nanostructured photocatalysts based on reduced graphene oxide–TiO_2 composites for degradation of diphenhydramine pharmaceutical and methyl orange dye. *Appl. Catal. B Environ.* **2012**, *123–124*, 241–256. [CrossRef]
17. Morawski, A.W.; Kusiak-Nejman, E.; Wanag, A.; Kapica-Kozar, J.; Wróbel, R.J.; Ohtani, B.; Aksienionek, M.; Lipińska, L. Photocatalytic degradation of acetic acid in the presence of visible light-active TiO_2-reduced graphene oxide photocatalysts. *Catal. Today* **2017**, *280*, 108–113. [CrossRef]
18. Giovannetti, R.; Rommozzi, E.; Zannotti, M.; D'Amato, C.A. Recent advances in graphene based TiO_2 nanocomposites ($GTiO_2Ns$) for photocatalytic degradation of synthetic dyes. *Catalysts* **2017**, *7*, 305. [CrossRef]
19. Dickinson, W.W.; Kumar, H.V.; Adamson, D.H.; Schniepp, H.C. High-throughput optical thickness and size characterization of 2D materials. *Nanoscale* **2018**, *10*, 14441–14447. [CrossRef]
20. Kumar, H.V.; Huang, K.Y.S.; Ward, S.P.; Adamson, D.H. Altering and investigating the surfactant properties of graphene oxide. *J. Colloid Interface Sci.* **2017**, *493*, 365–370. [CrossRef]
21. Contreras Ortiz, S.N.; Cabanzo, R.; Mejía-Ospino, E. Crude oil/water emulsion separation using graphene oxide and amine-modified graphene oxide particles. *Fuel* **2019**, *240*, 162–168. [CrossRef]
22. Abutalib, M.M. Insights into the structural, optical, thermal, dielectric, and electrical properties of PMMA/PANI loaded with graphene oxide nanoparticles. *Physica B.* **2019**, *552*, 19–29. [CrossRef]
23. Yeh, T.F.; Cihlar, J.; Chang, C.Y.; Cheng, C.; Teng, H.S. Roles of graphene oxide in photocatalytic water splitting. *Mater. Today.* **2013**, *16*, 78–84. [CrossRef]
24. Pastrana-Martínez, L.M.; Morales-Torres, S.; Likodimos, V.; Falaras, P.; Figueiredo, J.L.; Faria, J.L.; Silva, A.M.T. Role of oxygen functionalities on the synthesis of photocatalytically active graphene-TiO_2 composites. *Appl. Catal. B Environ.* **2014**, *158–159*, 329–340. [CrossRef]
25. Minella, M.; Sordello, F.; Minero, C. Photocatalytic process in TiO_2/graphene hybrid materials. Evidence of charge separation by electron transfer from reduced graphene oxide to TiO_2. *Catal. Today.* **2017**, *281*, 29–37. [CrossRef]
26. Tolosana-Moranchel, A.; Casas, J.A.; Bahamonde, A.; Pascual, L.; Granone, L.I.; Schneider, J.; Dillert, R.; Bahnemann, D.W. Nature and photoreactivity of TiO_2-rGO nanocomposites in aqueous suspensions under UV-A irradiation. *Appl. Catal. B Environ.* **2019**, *241*, 375–384. [CrossRef]
27. Chen, C.; Cai, W.; Long, M.; Zhou, B.; Wu, Y.; Wu, D.; Feng, Y. Synthesis of visible-light responsive graphene oxide/TiO_2 composites with p/n heterojunction. *ACS Nano* **2010**, *4*, 6425–6432. [CrossRef] [PubMed]
28. Loh, K.P.; Bao, Q.; Eda, G.; Chhowalla, M. Graphene oxide as a chemically tunable platform for optical applications. *Nat. Chem.* **2010**, *2*, 1015–1024. [CrossRef]
29. Ge, M.Z.; Li, S.H.; Huang, J.Y.; Zhang, K.Q.; Al-Deyab, S.S.; Lai, Y.K. TiO_2 nanotube arrays loaded with reduced graphene oxide films: Facile hybridization and promising photocatalytic application. *J. Mater. Chem. A* **2015**, *3*, 3491–3499. [CrossRef]
30. Luo, J.; Cote, L.J.; Tung, V.C.; Tan, A.T.L.; Goins, P.E.; Wu, J.; Huang, J. Graphene oxide nanocolloids. *J. Am. Chem. Soc.* **2010**, *132*, 17667–17669. [CrossRef]

31. Kim, H.I.; Moon, G.H.; Monllor-Satoca, D.; Park, Y.; Choi, W. Solar photoconversion using graphene/ TiO_2 composites: nanographene shell on TiO_2 core versus TiO_2 nanoparticles on graphene sheet. *J. Phys. Chem. C.* **2012**, *116*, 1535–1543. [CrossRef]
32. Diamantopoulou, A.; Sakellis, E.; Romanos, G.E.; Gardelis, S.; Ioannidis, N.; Boukos, N.; Falaras, P.; Likodimos, V. Titania photonic crystal photocatalysts functionalized by graphene oxide nanocolloids. *Appl. Catal. B Environ.* **2019**, *240*, 277–290. [CrossRef]
33. Kim, H.; Kang, S.O.; Park, S.; Park, H.S. Adsorption isotherms and kinetics of cationic and anionic dyes on three-dimensional reduced graphene oxide macrostructure. *J. Ind. Eng. Chem.* **2015**, *21*, 1191–1196. [CrossRef]
34. Acik, M.; Lee, G.; Mattevi, C.; Pirkle, A.; Wallace, R.M.; Chhowalla, M.; Cho, K.; Chabal, Y. The role of oxygen during thermal reduction of graphene oxide studied by infrared absorption spectroscopy. *J. Phys. Chem. C* **2011**, *115*, 19761–19781.
35. Kumar, P.V.; Bernardi, M.; Grossman, J.C. The impact of functionalization on the stability, work function, and photoluminescence of reduced graphene oxide. *ACS Nano* **2013**, *7*, 1638–1645. [CrossRef] [PubMed]
36. Hatton, B.; Mishchenko, L.; Davis, S.; Sandhage, K.H.; Aizenberg, J. Assembly of large-area, highly ordered, crack-free inverse opal films. *PNAS* **2010**, *107*, 10354–10359. [CrossRef]
37. Tunesi, S.; Anderson, M. Influence of Chemisorption on the photodecomposition of salicylic acid and related compounds using suspended TiO_2 ceramic membranes. *J. Phys. Chem.* **1991**, *95*, 3399–3405. [CrossRef]
38. Phillips, K.R.; Shirman, T.; Shirman, E.; Shneidman, A.V.; Kay, T.M.; Aizenberg, J. Nanocrystalline precursors for the co-assembly of crack-free metal oxide inverse opals. *Adv. Mater.* **2018**, *30*, 1706329. [CrossRef]
39. Toumazatou, A.; Arfanis, M.K.; Pantazopoulos, P.A.; Kontos, A.G.; Falaras, P.; Stefanou, N.; Likodimos, V. Slow-photon enhancement of dye sensitized TiO_2 photocatalysis. *Mater. Lett.* **2017**, *197*, 123–126. [CrossRef]
40. Balaji, S.; Djaoued, Y.; Robichaud, J. Phonon confinement studies in nanocrystalline anatase-TiO_2 thin films by micro Raman spectroscopy. *J. Raman Spectrosc.* **2006**, *37*, 1416–1422. [CrossRef]
41. Kudin, K.N.; Ozbas, B.; Schniepp, H.C.; Prud'homme, R.K.; Aksay, I.A.; Car, R. Raman spectra of graphite oxide and functionalized graphene sheets. *Nano Lett.* **2007**, *8*, 36–41. [CrossRef]
42. Pimenta, M.A.; Dresselhaus, G.; Dresselhaus, M.S.; Cancado, L.G.; Jorio, A.; Saito, R. Studying disorder in graphite-based systems by Raman spectroscopy. *Phys. Chem. Chem. Phys.* **2007**, *9*, 1276–1291. [CrossRef]
43. Romanos, G.E.; Likodimos, V.; Marques, R.R.N.; Steriotis, T.A.; Papageorgiou, S.K.; Faria, J.L.; Figueiredo, J.L.; Silva, A.M.T.; Falaras, P. Controlling and quantifying oxygen functionalities on hydrothermally and thermally treated single-wall carbon nanotubes. *J. Phys. Chem. C* **2011**, *115*, 8534–8546. [CrossRef]
44. López-Díaz, D.; López Holgado, M.; García-Fierro, J.L.; Velázquez, M.M. Evolution of the Raman spectrum with the chemical composition of graphene oxide. *J. Phys. Chem. C* **2017**, *121*, 20489–20497. [CrossRef]
45. Muzyka, R.; Drewniak, S.; Pustelny, T.; Chrubasik, M.; Gryglewicz, G. Characterization of graphite oxide and reduced graphene oxide obtained from different graphite precursors and oxidized by different methods using Raman spectroscopy. *Materials* **2018**, *11*, 1050. [CrossRef]
46. Kumar, H.V.; Woltornist, S.J.; Adamson, D.H. Fractionation and characterization of graphene oxide by oxidation extent through emulsion stabilization. *Carbon.* **2016**, *98*, 491–495. [CrossRef]
47. Mattevi, C.; Eda, G.; Agnoli, S.; Miller, S.; Mkhoyan, K.A.; Celik, O.; Mastrogiovanni, D.; Granozzi, G.; Carfunkel, E.; Chhowalla, M. Evolution of electrical, chemical, and structural properties of transparent and conducting chemically derived graphene thin films. *Adv. Funct. Mater.* **2009**, *19*, 2577–2583. [CrossRef]
48. Pelaez, M.; Falaras, P.; Likodimos, V.; Shea, K.O.; Armah, A.; Dunlop, P.S.; Byrne, J.A.; Dionysiou, D.D. Use of selected scavengers for the determination of NF-reactive oxygen species during the degradation of microcystin-LR under visible light irradiation. *J. Mol. Catal. A Chem.* **2016**, *425*, 183–189. [CrossRef]
49. Houas, A.; Lachheb, H.; Ksibi, M.; Elaloui, E.; Guillard, C.; Hermann, J.M. Photocatalytic degradation pathway of methylene blue in water. *Appl. Catal. B Environ.* **2001**, *31*, 145–157. [CrossRef]
50. Pastrana-Martínez, L.M.; Morales-Torres, S.; Kontos, A.G.; Moustakas, N.G.; Faria, J.L.; Doña-Rodríguez, J.M.; Falaras, P.; Silva, A.M.T. TiO_2, surface modified TiO_2 and graphene oxide-TiO_2 photocatalysts for degradation of water pollutants under near-UV/Vis and visible light. *Chem. Eng. J.* **2013**, *224*, 17–23. [CrossRef]
51. Zheng, X.Z.; Meng, S.G.; Chen, J.; Wang, J.J.; Xian, J.J.; Shao, Y.; Fu, X.Z.; Li, D.Z. Titanium dioxide photonic crystals with enhanced photocatalytic activity: Matching photonic band gaps of TiO_2 to the absorption peaks of dyes. *J. Phys. Chem. C* **2013**, *117*, 21263–21273. [CrossRef]

52. Regazzoni, A.E.; Mandelbaum, P.; Matsuyoshi, M.; Schiller, S.; Bilmes, S.A.; Blesa, M.A. Adsorption and photooxidation of salicylic acid on titanium dioxide: A surface complexation description. *Langmuir* **1998**, *14*, 868–874. [CrossRef]
53. Arfanis, M.K.; Adamou, P.; Moutakas, N.G.; Theodoros, M.T.; Kontos, A.G.; Falaras, P. Photocatalytic degradation of salicylic acid and caffeine emerging contaminants using titania nanotubes. *Chem. Eng. J.* **2017**, *310*, 525–536. [CrossRef]

Article

Incidence Dependency of Photonic Crystal Substrate and Its Application on Solar Energy Conversion: Ag_2S Sensitized WO_3 in FTO Photonic Crystal Film

Xi Ke [1], Mengmeng Yang [2], Weizhe Wang [1], Dongxiang Luo [2,*] and Menglong Zhang [1,*]

1 Institute of Semiconductors, South China Normal University, Guangzhou 510631, China
2 School of Materials and Energy, Guangdong University of Technology, Guangzhou 510006, China
* Correspondence: luodx@gdut.edu.cn (D.L.); mlzhang@scnu.edu.cn (M.Z.)

Received: 7 July 2019; Accepted: 8 August 2019; Published: 11 August 2019

Abstract: In addition to the most common applications of macroporous film: Supplying a large surface area, PC-FTO (macroporous fluorine-doped tin oxide with photonic crystal structure) can be employed as a template to control the morphologies of WO_3 for exposing a more active facet, and enhance the overall photo-electron conversion efficiency for the embedded photoactive materials under changing illumination incidence through refracting and scattering. The optical features of PC-FTO film was demonstrated by DRUVS (diffuse reflectance UV-vis spectra). Plate-like WO_3 were directly synthesized inside the PC-FTO film as a control group photoanode, Ag_2S quantum dots were subsequently decorated on WO_3 to tune the light absorption range. The impact of photonic crystal film on the photoactivity of Ag_2S/WO_3 was demonstrated by using the photoelectrochemical current density as a function of the incidence of the simulated light source.

Keywords: photocatalytic materials; photonic crystals; nanocomposites

---

1. Introduction

Photocatalysis for applications on a photon-electricity or photon-chemical bond have been paid much attention due to the energy crisis and environmental pollution [1]. Photoelectrochemical devices were therefore developed to perform solar energy conversion including photocurrent generation [2], pollutant degradation [3], and H_2 or CO evolution [4,5]. In consideration of the photocatalysis mechanism, improving the performance of a solar energy conversion device is based on several critical factors, such as the enhancement of light harvesting, efficient utilization of photons, rapid charge carrier separation and transportation. To this end, many efforts including elemental doping [6], dye sensitization [7], adding co-catalysts [8], making composites [9], plasma enhancement [10] and nanostructuring the photoactive materials [11,12] have been made, leading to improved solar energy conversion efficiency being achieved. Though, in addition to the above attempts, a more convenient strategy is correlating the nature of the sun and materials. To be more specific, the sun can be considered as a light source of changing angle relative to a certain location of the Earth, due to the changing solar elevation angle. Thus, the solar energy conversion devices receive photons of varied incidence and density, typically, the photon density (mW cm^{-2}) reaches its zenith at noon (ca. 90° solar elevation angle). In this case, the photonic crystals (PCs) could take advantage of this nature of the sun for more effective utilization of solar energy, due to its capability for manipulating the photon migration depending on their incidence.

PCs are long-range ordered arrays materials and possess periodic modulation of refractive index (RI) within an optical wavelength. The formation of photonic stop band (PSB) of a certain PCs structural material is attributed to the differences between the RI of the PCs material and the RI of its external media (including gas, liquid or solid), and the periodicity. The concept of photonic effect within the

PSB range in PCs was first proposed by John [13] and Yablonovitch [14], independently. After that, the photonic properties of PCs were further explored by theoretical calculations [15–17]. That is, within a particular wavelength range that satisfies Bragg-Snell's law [18], the PSB inhibits the propagation of photons from a certain direction, reduces the group speed of photons (slow photons) and induces multiple scattering [19]. The position of PSB in wavelength is commonly tunable, as is demonstrated by Bragg-Snell's law, it is reliant on the periodic order, the RI of PCs material and external media, filling factors and incidence of illumination. Whereas for a given PCs based solar energy conversion device in certain environment, filling factor, periodic order and RI should be considered as constants, and thus the incidence of illumination would be the only tunable parameter for the manipulation of PSB positions. In addition, different from tuning the RI, the filling factor or periodic order, the changing of incidence would not affect the other properties of PCs (including geometric surface area, electric conductivity and interfaces between PCs and embedded photoactive materials), which makes it very convenient to study the PSB impact on photoactivity of embedded photoactive materials.

TCOs (transparent conductive oxides) including ITO (indium-doped tin oxide), ATO (antimony-doped tin oxide) and FTO (fluorine-doped tin oxide) have been intensively applied in functional windows, solar cells and displays [2,20]. For a solar energy conversion device, TCOs are an ideal material for the synthesis of PCs due to their transparency in visible range and good electrical conductivity. For instance, in previous studies, powder or film PCs have been employed to supply enlarged surface areas available to support the embedded photoactive materials including g-C_3N_4 [21], CdS [22], TiO_2 [23], and halide perovskite photocatalyst [24]. Although the studies of the photonic effect of PSB towards photoactivity of embedded photoactive material are limited. Of direct relevance to this work, CdS embedded WO_3 PCs powders were fabricated [25], the manipulation of PSB was obtained through varying the periodic order (pore size) of WO_3 PCs powder. Enhancement of the hydrogen evolution rate is achieved in case of absorption range of CdS overlapping with the PSB of WO_3 PCs. However, the varied periodic order of WO_3 PCs would not only change the PSB but also suggest a different surface area and reaction sites for the embedded photoactive materials, which make it less convenient to study the photonic effect on photocatalysis. In our recent report [26], g-C_3N_4 as the photoactive material was embedded in an FTO PCs film, overlapped the PSB of FTO PCs film and absorption range of g-C_3N_4 was achieved by tuning the position of the PSB through controlling incidences. An alternative approach to fit the PSB and absorption range of photoactive materials is tuning the absorption of photoactive materials by adding dyes.

Inorganic semiconductors such as CdS, CdSe, PbS, MnTe, and Ag_2S, or organic dyes have been exploited as sensitizers for light-absorption enhancement in dye-sensitized solar cells (DSSCs) or quantum dot-sensitized solar cells (QDSCs) [3,9]. Among them, Ag_2S is a convenient candidate compound for applications in photoanodes. The energy band gap of Ag_2S is ca. 1.1 eV [27], corresponding to a broad absorption range of visible light and near-IR regions. In addition, Ag_2S reveals a promising absorption coefficient of ca. 104 cm^{-1} [28]. In the previous study, Ag_2S quantum dots were prepared on TiO_2 of various architectures (including nanotubes and nanorods) [29] for the formation of DSSCs, and also being synthesized on ZnO and SnO_2 for the preparation of QDSCs [30,31]. The adding of Ag_2S exhibited impressive light absorbability for the electrodes and thus led to improved photoactivity. However, in addition to these wide bandgap semiconductors (TiO_2, ZnO and SnO_2), WO_3 exhibits a narrower band gap which can be photoactivated by visible light, and the band structure allows an efficient photoelectron injection when composited with Ag_2S [32].

In this work, WO_3 was exploited as the photoactive materials, plate-like WO_3 were directly synthesized in FTO PCs film via a solvothermal method. Ag_2S was then loaded on WO_3 via SILAR (successive ion layer adsorption and reaction) method. As control groups, WO_3 and Ag_2S sensitized WO_3 were also prepared on p-FTO (planar FTO glass). DRUVS (diffuse reflectance UV-vis spectra) and PEC (photoelectrochemistry) were correlated to analyze the photonic effect on the photocatalytic performance of Ag_2S sensitized WO_3.

2. Materials and Methods

H_2SO_4 (≥95%), H_2O_2 (30 vol.%), FTO glass slide (11 Ω sq^{-1}), monodispersed polystyrene (2.5 wt.% aqueous suspension, 450 nm in diameter), $SnCl_4{\cdot}5H_2O$ (99.99%), NH_4F (99.99%), $Na_2S{\cdot}9H_2O$ (99.5%), $AgNO_3$ (99.8%), WCl_6 (99.95%) and methanol were purchased from Sigma-Aldrich (Saint Louis, MO, USA) and used as received.

2.1. Synthesis of PC-FTO Films

The PC-FTO film was prepared through a modified well-established soft-template method [33]. A planar FTO glass was stood vertically in a glass sample vial (10 mL volume) containing 5 mL of PS aqueous suspension at 60 °C for 16 h to obtain the PS film template. The FTO precursor solution was prepared from $SnCl_4{\cdot}5H_2O$ (1.4 g, 4 mmol) sonicated in ethanol (20 mL) until dissolved, next, the saturated NH_4F solution (0.24 g, 2 mmol) was added, and the resulting mixture was further sonicated until optically clear and colorless. The PS film was pre-soaked in ethanol for 30 min before being stood vertically and submerged in FTO precursor solution for another 30 min under a vacuum (0.1 Pa). The wet slide was then removed from the glass vial and transferred to a furnace oven for calcination at 450 °C for 2 h with a heat ramp rate of 1 °C min^{-1} in the air to burn off the PS spheres.

2.2. In-Situ Synthesis of WO_3 Platelets in PC-FTO Films

The WO_3 precursor solution was prepared using WCl_6 (1.0 g, 6 mmol) sonicated in methanol for ca. 15 min (40 mL) until dissolved. A PC-FTO substrate was put in an autoclave (25 mL volume) filled with 15 mL WO_3 precursor solution. Subsequently, the autoclave was transferred to a furnace oven at 100 °C for 6 h. The sample was then transferred from the autoclave to a furnace oven for a heat treatment at 475 °C for 2 h with a heat ramp rate of 1 °C min^{-1} in the air.

2.3. Sensitizing of WO_3 with Ag_2S Quantum Dots via SILAR Method

The Ag_2S precursor was prepared from $AgNO_3$ (50 mM) and Na_2S (50 mM) aqueous solutions. Typically, a WO_3@PC-FTO film was soaked in $AgNO_3$ solution for 30 s before being dried by nitrogen stream. The film was then soaked in Na_2S solution for another 30 s. Subsequently, the film was rinsed with DI water and dried with nitrogen stream. This process was repeated several times to get Ag_2S/WO_3@mac-FTO photoanodes sensitized with the varied amounts of Ag_2S dots.

2.4. Characterization

Samples were stuck to an aluminium stage by sticky carbon tape, and then the SEM images were achieved by a Hitachi S-4800 field emission scanning electron microscope (Tokyo, Japan). For the preparation of TEM samples, films were scraped off from the FTO substrates and carefully ground. The samples were then transferred to methanol for 15 min sonication. One drop of the suspension was added to 3 mm porous carbon-coated copper grids and allowed to dry under air. JEOL 2011 transmission electron microscopes (Tokyo, Japan) were exploited to collect the TEM images with 200 kV accelerating voltage. An attached kit of an EDAX Phoenix X-ray spectrometer (Mahwah, NJ, USA) incorporated to the TEM was employed to perform energy dispersive analysis of X-rays (EDX) mapping. An Ocean Optics HR2000+ high-resolution spectrometer (Edinburgh, UK) was incorporated with a DH-2000-BAL lamp with a light wavelength of 200 nm to 1100 nm (deuterium/helium). An R400-7-UV-Vis transmission probe (Ocean Optics, Edinburgh, UK) was used to record the diffuse reflectance UV-vis spectra using deuterium/helium lamb (200 nm–1100 nm). Spectra were collected using Spectra Suite software (Version 2.7) of 10 s integration time, 30 boxcar smoothing width, and 10 scans to average. Wide angle PXRD patterns were obtained by a Lynx eye incorporated detector Bruker-AXS D8 Advance instrument (Billerica, MA, USA), with Cu Kα (1.54Å) radiation, the slit on the source was 1 mm and the detector slit was 2.5 mm. PXRD data were scanned from 10° to 70° 2θ, with 0.02° step size and a scan speed of 0.1 s per step.

PEC measurements were carried out using a standard 3 electrode setup. The Ag/AgCl (3 M KCl internal solution) was exploited as the reference electrode and a platinum sheet (10×10 mm^2) was used as the counter electrode. The samples were used as a working electrode, the connection was achieved using copper tape on the top of the electrode and the bottom 10 mm of the electrode was immersed into the electrolyte solution. The PEC cell contained a quartz window allowing the illumination of simulated solar light. A 150 W Xe lamp with the (irradiance of ca. 100 mW cm^{-2}) was exploited as the simulated light source. NaOH (1 M) with pH at 13.6 was prepared as the aqueous electrolyte solution using the Millipore system (≥18 MΩ cm^{-3}, Burlington, MA, USA) filter water. All potentials can be referenced to the reversible hydrogen electrode (RHE) using the following equation: $E_{ref}(Ag/AgCl)$ = 0.0210 V vs. NHE at 25 °C.

$$E(\text{vs. RHE}) = E(\text{vs. Ag/AgCl}) + E_{ref}(\text{Ag/AgCl}) + 0.0591\text{V} \times \text{pH}$$

3. Result and Discussion

3.1. Geometrical Properties

The geometrical properties of the as-prepared PC-FTO films are initially characterized by scanning electron microscope (SEM), Figure 1a exhibits an inverse opal structural (face-centered cubic, FCC) film with the pore size of ca. 330 ± 30 nm and a smooth FTO skeleton. This PC-FTO film was fabricated on FTO glass via a well-established soft template method [34]. Figure 1b presents the edge of the as-prepared PC-FTO film, which was approximately nine layers corresponding to ca. 2 μm in thickness, indicating a large surface area available for the embedded materials, in comparison to that of planar analogs. The long-range ordered FCC array and the periodic RI (between FTO skeleton and air pore) suggests intensive photonic effects as expected by Bragg-Snell's law. The hexagonal arrangement of the air pore corresponds to the (111) plane of an FCC structure, which is the predominant plane for the formation of a PSB from a PCs film. After the coating of WO_3, homogeneous plate-like materials could be observed on the skeleton of PC-FTO (Figure 1c and Figure S1), without blocking the pores. The WO_3@PC-FTO exhibits a relatively rougher surface in comparison to the bare PC-FTO (Figure 1a).

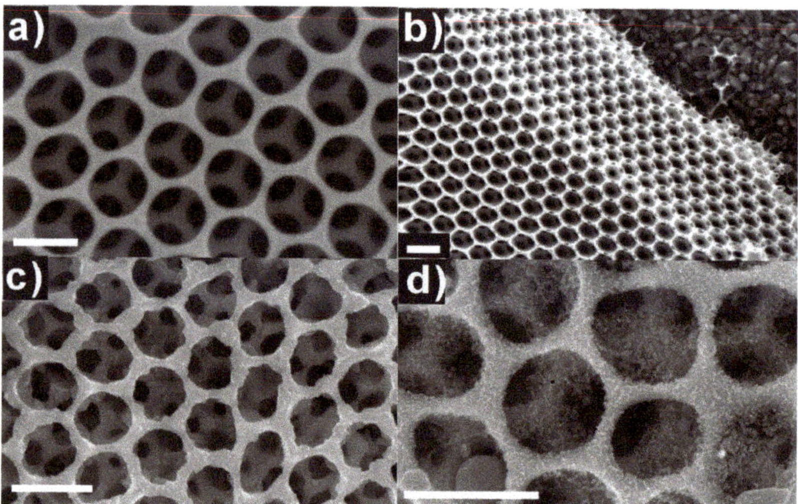

Figure 1. (**a**) SEM image of macroporous fluorine-doped tin oxide with photonic crystal structure (PC-FTO) viewed perpendicularly; (**b**) SEM image of the edge of PC-FTO film; (**c**) SEM image of WO_3@PC-FTO; (**d**) SEM image of Ag_2S/WO_3@PC-FTO, scale bar = 500 nm.

It should be noted that the amount of the embedded photoactive materials in a PC film should be carefully controlled because the excessive coating will reduce the periodicity of PCs, on the other hand, the insufficient coating may lead to weak photoresponse. Furthermore, the geometrical shape of the WO_3 in a PC-FTO film can also be compared to that which is synthesized on a planar FTO substrate using the same method (Figure 2). Within PC-FTO films, the feature size of the materials can be restrained to a certain scale by the nanostructural substrate due to the complex geometrical surface, similar feature size control effects can also be found in Fe_2O_3 embedded SiO_2 [35] or SnO_2 [12] in previous reports. Figure 1d and Figure S2 reveal the WO_3@PC-FTO film after the decoration of Ag_2S, the Ag_2S decorated samples led to rougher surface due to the relatively small crystal size of Ag_2S.

Figure 2. SEM images of WO_3 synthesized on planar FTO glass. (**a**) SEM of WO_3 of low magnification; (**b**) SEM of WO_3 of high magnification.

3.2. Optical Properties

Since this work was focused on the impact of photonic crystal film on the photocatalytic performance of Ag_2S/WO_3, the photonic effects and the optical properties of Ag_2S/WO_3 were characterized by diffuse reflectance UV-vis spectra (DRUVS) and UV-vis absorption spectra, respectively. The PSB position (expressed as the reflectance maximum) of PC-FTO film was recorded at the different incidence of illumination with respect to the surface normal (Figure 3a) through DRUVS. For comparison, the PSB position for (hkl) plane of a PC-FTO was also estimated by mathematic calculation from a combination of Bragg's and Snell's laws as the following expresses:

$$\lambda = \frac{2d_{hkl}}{m}\left[\varphi n^2 + (1-\varphi)n_0^2 - sin^2\theta\right]^{\frac{1}{2}}$$

$$d_{hkl} = \frac{D\sqrt{2}}{\left[h^2 + k^2 + l^2\right]^{\frac{1}{2}}}$$

where φ is the filling factor of PC-FTO film (assumed to be 0.26 for FCC structure), n and n_0 are the RI of FTO (ca. 1.61) and the external media, respectively. D is the periodic order of PC-FTO film (ca. 340 nm). Since the calculation of a PSB position using Bragg-Snell's law is based on an ideal FCC structural model, thus a comparison between the calculated and experimental incident dependence of PSB can be exploited to evaluate the optical quality of the as-prepared PC-FTO film. As shown in Table S1, the experimental PSB is in good agreement with the Bragg diffraction from (111) sets of planes, in an incidence range from 15° to 45°. The PC-FTO exhibits varied PSB positions (from 585 nm to 635 nm) in the DRUVS (Figure 3b) according to the incidence of illumination. The PSB shift can also be confirmed by eye when holding the PC-FTO film at a certain angle relative to the lamp, the colors (attributed to the diffuse reflectance of PCs) varied from red to violet as shown in Figure 3c, indicating a wide tunable range of PSB positions. It should be noted that the DRUVS can hardly collect the reflected photons when there is a wide angle between sample and probe, but the photographs in Figure 3c could be evidence for the wide tunable range of the PSB.

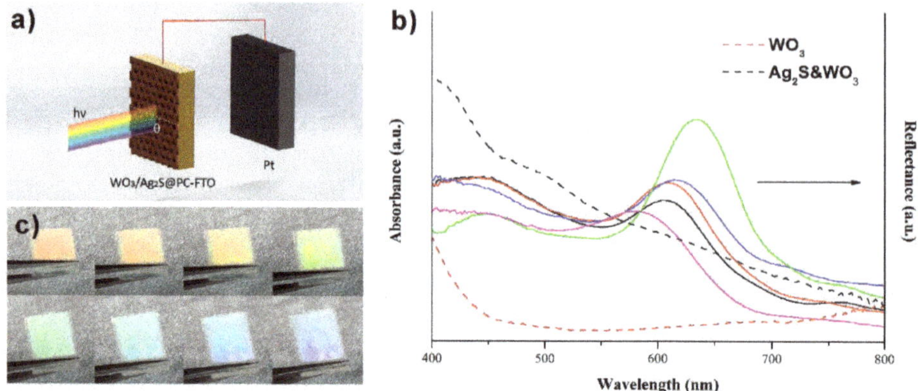

Figure 3. (**a**) Schematic diagram of the varied incidence illumination on a PC-FTO based electrode. (**b**) diffuse reflectance UV-vis spectra (DRUVS) (solid line) of photonic stop band (PSB) positions under different incidence and UV-vis absorption spectra (dash line) of WO_3 and $Ag_2S/WO3$. (**c**) digital photographs of one PC-FTO film under different incidences.

3.3. *Qualitative Analysis*

Figure 4a and Figure S4 reveal the high-resolution transmission electron microscopy (HR-TEM) images of the as-prepared Ag_2S/WO_3 on PC-FTO, the embedded WO_3 platelets were attached by Ag_2S (few nanometer in size), in which the lattice fringes corresponded to the (200) plane of Ag_2S (JCPDS 04-0774) and the (002) crystal plane of WO_3 (JCPDS 43-1034), respectively. The elemental mapping of Ag_2S/WO_3@PC-FTO was achieved by energy dispersive X-ray spectroscopy (EDX), from which the skeleton of PC-FTO, WO_3 platelets and Ag_2S dots are clearly presented in Figure 4b and Figure S3. The elemental ratio of the as-prepared Ag_2S/WO_3@PC-FTO sample was presented in Table S2. The powder X-ray diffraction (PXRD) was also employed for the characterization of an as-prepared Ag_2S/WO_3@PC-FTO, from which the diffraction patterns were consistent with the HR-TEM results. The loading amounts of Ag_2S via SILAR led to no obvious difference in the PXRD patterns of samples but the slightly increased diffraction patterns from Ag_2S (Figure 5).

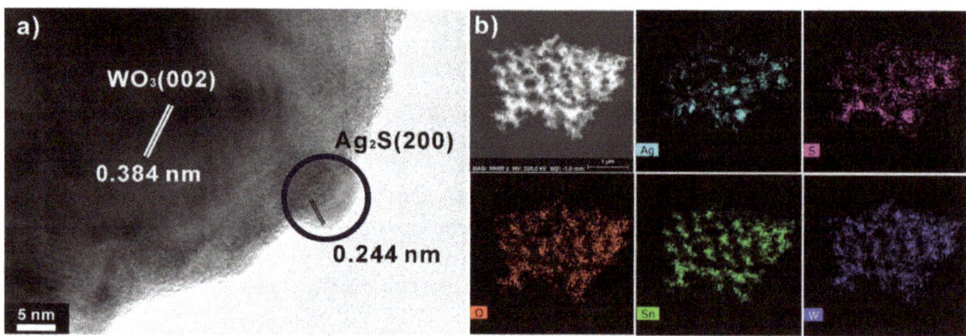

Figure 4. (**a**) high-resolution transmission electron microscopy (HR-TEM) image of Ag_2S/WO_3@PC-FTO photoanode; (**b**) TEM-EDX elemental mapping of Ag_2S/WO_3@PC-FTO.

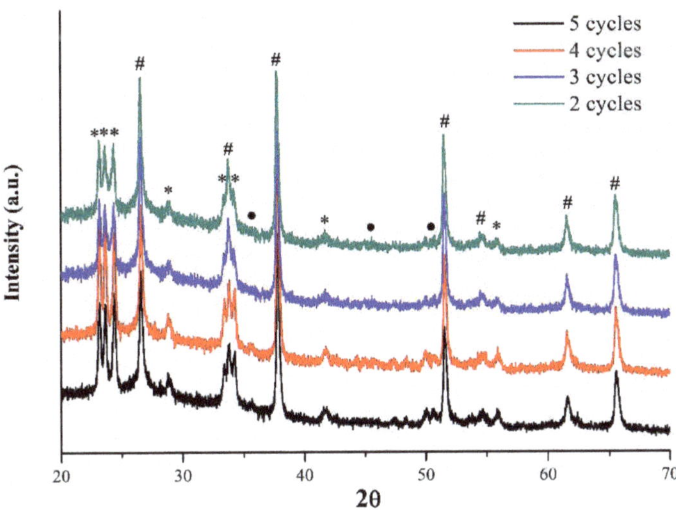

Figure 5. Powder X-ray diffraction (PXRD) of Ag_2S/WO_3@PC-FTO electrode with different loading of Ag_2S, * = WO_3, # = SnO_2 and • = Ag_2S.

3.4. Photocatalytic Properties

Photoelectrochemistry (PEC) was exploited to evaluate the photoactivity of Ag_2S/WO_3@PC-FTO photoanode, as a function of incidence. The setup of a PEC cell was built up with a standard three electrode system in 1 M NaOH aqueous electrolyte, in which the sample, Pt wire, and Ag/AgCl were employed as a working electrode, counter electrode and reference electrode, respectively. The incidence of illumination was obtained by rotating the working electrode to a certain degree relative to the Xe lamp (as shown in the schematic diagram in Figure 3a). Since the rotating of electrode will lead to a reduction of received photon density (mW cm^{-2}) by electrode, the photo density received by electrode at a certain angle was obtained by rotating a detector (shell removed) of photometer at the same angle, thus a correction (photocurrent divided by photon density received by photometer at certain angle) was conducted to evaluate the photocurrent generation under constant irradiance (1 sun, 100 mW cm^{-2}).

Ag_2S/WO_3@PC-FTO electrodes of different SILAR cycles performed varied photocatalytic activities when the surface normal was illuminated, in which the 3 SILAR cycles performed the maximum photocurrent density of ca. 1.54 mA cm^{-2} at 0 V vs. Ag/AgCl (Figure S5) and good stability (Figure S7) in a linear sweep voltammogram (LSV), and thus this electrode was selected for the study of incidence.

For comparison, the Ag_2S/WO_3@planar-FTO photoanode and WO_3@PC-FTO photoanode were also prepared using the same method. Due to the absence of PCs substrate, the current density generated by Ag_2S/WO_3@planar-FTO is ca. 32% (0.52 mA cm^{-2} vs. 1.54 mA cm^{-2}) in comparison to Ag_2S/WO_3@PC-FTO at surface normal illumination, this is due to the lack of surface area available for the photoactive materials. The current density of Ag_2S/WO_3@planar-FTO was reduced 61.5% along with the incidence from 0° to 75° (Figure 6a), while the current density per sun shows no obvious variation, suggesting that the electrode could generate stable current density relative to 1 sun. In terms of the WO_3@PC-FTO electrode, the current density reduced 47.6% from 0° to 75° illumination (Figure 6b). Interestingly, an obvious enhanced current density per sun was obtained at 75° and 60° illumination, where the edge of the PSB (60°) and PSB (75°) overlapped with the absorption range of WO_3. Though for the Ag_2S/WO_3@PC-FTO electrode (Figure 6c), due the narrow band gap of Ag_2S (1.1 eV, Scheme 1) [36] in visible, the PSB of PC-FTO can always overlap with the absorption range of Ag_2S/WO_3, and thus reducing the enhancement of current density per sun when rotating the electrode, but it still performed a slower reduction (54.5%) of current density with respect to that on planar-FTO.

In theory, the PCs could suppress the recombination of charge carriers in PSB or its edge [37], and this was proved in some previous experimental reports, for instance, a 40% longer lifetime of PL decay was observed when the emission frequency of Tb^{3+} is close to the PSB in a PC-SiO_2 system [38]. With respect for a photocatalysis system, the external media is typically aqueous (RI $\geq$ 1.33), and thus the RI differences between PCs materials and external media is smaller than PCs in the air (RI $\approx$ 1.00), which may reduce the photonic effects from PSB (Figure S6). For example, the g-C_3N_4 exhibited no excited state lifetime change when its absorption range overlaps with the PSB [8]. In our case, although the absorption of Ag_2S/WO_3 always overlapped with the PSB, we also saw no obvious PL decay lifetime enhancement. Even though, the scattering and refracting will still improve the utilization of light, when the incidence of light changes. The observed slower reduction of current density and the improved current density per sun at certain incidence suggests the realistic value of PCs.

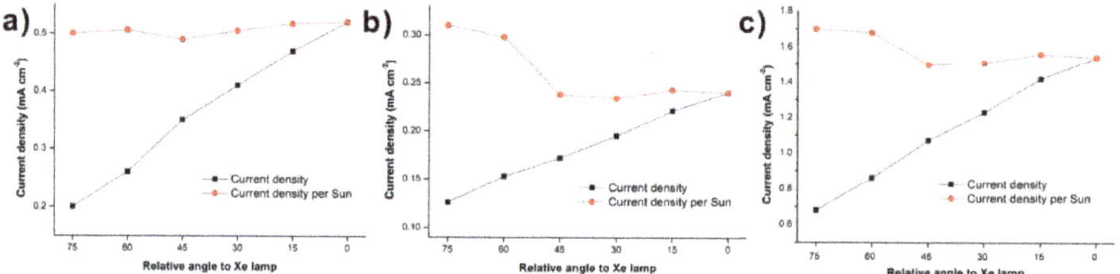

Figure 6. Incidence dependence of photoelectrochemistry (PEC) current density and current density relative to 1 sun (ca. 100 mW cm^{-2}) for (**a**) Ag_2S/WO_3@planar-FTO; (**b**) WO_3@PC-FTO and (**c**) Ag_2S/WO_3@PC-FTO at 0 V vis $V_{Ag/AgCl}$, the irradiance to the surface normal is adjusted to 1 sun.

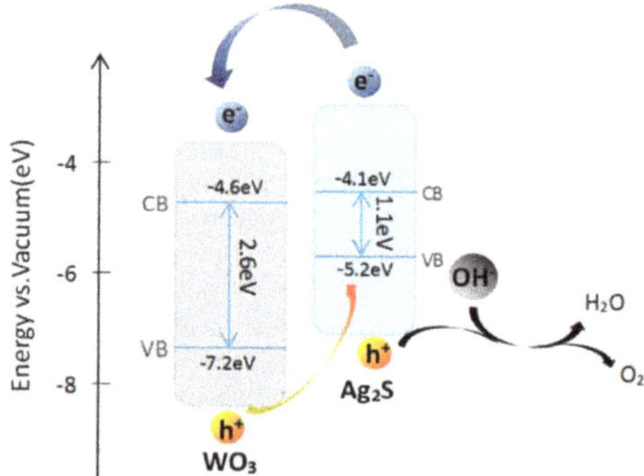

Scheme 1. Mechanism of Ag_2S decorated WO_3 under illumination.

4. Conclusions

Ag_2S/WO_3 and WO_3 were synthesized in a PC-FTO film to demonstrate the incidence of light as a function of photocurrent density. In addition to the enhanced surface area, the embedded photoactive materials in PC-FTO perform enhanced PEC at certain incidences in comparison to that on planar analogs. Due to the multiple scattering and refracting nature of PCs, photons could be collected more effectively at certain incidence ranges (in comparison to planar analogs), resulting in enhanced photocurrent generation when integrating the product from all incidences. In consideration of the nature of sun (a changing angle light source), the results suggest that the PCs based substrate would support more efficient utilization of solar energy for the embedded photoactive materials.

Supplementary Materials: The following are available online at http://www.mdpi.com/1996-1944/12/16/2558/s1, Figure S1: SEM image of a bare WO_3@mac-FTO photoanode; Figure S2: SEM images of mac-FTO film coated with only Ag_2S quantum dots; Figure S3: EDX elemental mapping of Ag_2S/WO_3@mac-FTO with 3 SILAR cycles; Figure S4: TEM of Ag_2S/WO_3@PC-FTO in different magnification; Figure S5: (a) UV-vis absorption and (b) LSV of Ag_2S/WO_3@PC-FTO with different SILAR cycles; Figure S6: PSB spectra of PC-FTO in NaOH electrolyte; Figure S7: Stability test of Ag_2S/WO_3@PC-FTO electrode under illumination from surface normal for 5 min; Table S1: Calculated and experimental PSB of PC-FTO film from (111) plane; Table S2: Elemental ratios of Ag_2S/WO_3@PC-FTO with 3 SILAR cycles.

Author Contributions: Conceptualization, D.L.; data curation, W.W.; formal analysis, M.Y.; funding acquisition, D.L.; methodology, X.K.; project administration, M.Z.; supervision, M.Z.; writing—original draft, X.K.; writing—review and editing, M.Z.

Funding: This research was funded by the National Natural Science Foundation of China (Grant No. 61704034), the Key Platforms and Research Projects of Department of Education of Guangdong Province (Grant No. 2016KTSCX034), and the Natural Science Foundation of Guangdong Province (Grant No. 2018A0303130199 and 2017A030313632).

Conflicts of Interest: The authors declare no conflict of interest.

References

1. Zhou, M.; Hou, Z.; Zhang, L.; Liu, Y.; Gao, Q.; Chen, X. n/n junctioned g-C3N4for enhanced photocatalytic H2generation, Sustain. *Energy Fuels* **2017**, *1*, 317–323.
2. Zhang, M.; Robert, W.; Mitchell, H.; Huang, R.E. Douthwaite, Ordered multilayer films of hollow sphere aluminium-doped zinc oxide for photoelectrochemical solar energy conversion. *J. Mater. Chem. A* **2017**, *5*, 22193–22198. [CrossRef]
3. Liu, H.; Wang, X.; Wu, D. Tailoring of bifunctional microencapsulated phase change materials with CdS/SiO_2 double-layered shell for solar photocatalysis and solar thermal energy storage. *Appl. Therm. Eng.* **2018**, *134*, 603–614. [CrossRef]
4. Pesci, F.M.; Sokolikova, M.S.; Grotta, C.; Sherrell, P.C.; Reale, F.; Sharda, K.; Ni, N.; Palczynski, P.; Mattevi, C. MoS2/WS2 Heterojunction for Photoelectrochemical Water Oxidation. *ACS Catal.* **2017**, *7*, 4990–4998. [CrossRef]
5. Krbal, M.; Prikryl, J.; Zazpe, R.; Sopha, H.; Macak, J.M. CdS-coated TiO_2 nanotube layers: downscaling tube diameter towards efficient heterostructured photoelectrochemical conversion. *Nanoscale* **2017**, *9*, 7755–7759. [CrossRef] [PubMed]
6. Ye, K.-H.; Wang, Z.; Gu, J.; Xiao, S.; Yuan, Y.; Zhu, Y.; Zhang, Y.; Mai, W.; Yang, S. Carbon quantum dots as a visible light sensitizer to significantly increase the solar water splitting performance of bismuth vanadate photoanodes. *Energy Environ. Sci.* **2017**, *10*, 772–779. [CrossRef]
7. Marandi, M.; Goudarzi, Z.; Moradi, L. Synthesis of randomly directed inclined TiO_2 nanorods on the nanocrystalline TiO_2 layers and their optimized application in dye sensitized solar cells. *J. Alloy. Compd.* **2017**, *711*, 603–610. [CrossRef]
8. Qiao, S.; Mitchell, R.W.; Coulson, B.; Jowett, D.V.; Johnson, B.R.; Brydson, R.; Isaacs, M.; Lee, A.F.; Douthwaite, R.E. Douthwaite, Pore confinement effects and stabilization of carbon nitride oligomers in macroporous silica for photocatalytic hydrogen production. *Carbon* **2016**, *106*, 320–329. [CrossRef]
9. Shuang, S.; Lv, R.; Cui, X.; Xie, Z.; Zheng, J.; Zhang, Z. Efficient photocatalysis with graphene oxide/Ag/Ag_2S–TiO_2 nanocomposites under visible light irradiation. *RSC Adv.* **2018**, *8*, 5784–5791. [CrossRef]
10. Zheng, Z.; Xie, W.; Huang, B.; Dai, Y. Plasmon-Enhanced Solar Water Splitting on Metal-Semiconductor Photocatalysts. *Chemistry* **2018**, *24*, 18322–18333. [CrossRef]
11. Nayak, A.K.; Sohn, Y.; Pradhan, D. Facile Green Synthesis of $WO_3 \cdot H_2O$ Nanoplates and WO_3 Nanowires with Enhanced Photoelectrochemical Performance. *Cryst. Growth Des.* **2017**, *17*, 4949–4957. [CrossRef]
12. Xiao, Y.; Chen, M.; Zhang, M. Multiple layered macroporous SnO_2 film for applications to photoelectrochemistry and morphology control of iron oxide nanocrystals. *J. Power Sources* **2018**, *402*, 62–67. [CrossRef]
13. John, S. Strong localization of photons in certain disordered dielectric superlattices. *Phys. Rev. Lett.* **1987**, *58*, 2486–2489. [CrossRef] [PubMed]

14. Yablonovitch, E. Inhibited spontaneous emission in solid-state physics and electronics. *Phys. Rev. Lett.* **1987**, *58*, 2059–2062. [CrossRef] [PubMed]
15. Zhang, H.-F.; Liu, S.-B. Analyzing the photonic band gaps in two-dimensional plasma photonic crystals with fractal Sierpinski gasket structure based on the Monte Carlo method. *AIP Adv.* **2016**, *6*, 085116. [CrossRef]
16. Zhang, H.-F. Investigations on the two-dimensional aperiodic plasma photonic crystals with fractal Fibonacci sequence. *AIP Adv.* **2017**, *7*, 075102. [CrossRef]
17. Zhang, H.-F.; Chen, Y.-Q. The properties of two-dimensional fractal plasma photonic crystals with Thue-Morse sequence. *Phys. Plasmas* **2017**, *24*, 042116. [CrossRef]
18. Zhang, L.; Lin, C.Y.; Valev, V.K.; Reisner, E.; Steiner, U.; Baumberg, J.J. Plasmonic enhancement in $BiVO_4$ photonic crystals for efficient water splitting. *Small* **2014**, *10*, 3970–3978. [CrossRef] [PubMed]
19. Mitchell, R.; Brydson, R.; Douthwaite, R.E. Enhancement of hydrogen production using photoactive nanoparticles on a photochemically inert photonic macroporous support. *Phys. Chem. Chem. Phys. PCCP* **2015**, *17*, 493–499. [CrossRef]
20. Xu, Y.F.; Rao, H.S.; Chen, B.X.; Lin, Y.; Chen, H.Y.; Kuang, D.B.; Su, C.Y. Achieving Highly Efficient Photoelectrochemical Water Oxidation with a $TiCl_4$ Treated 3D Antimony-Doped SnO_2 Macropore/Branched alpha-Fe_2O_3 Nanorod Heterojunction Photoanode. *Adv. Sci.* **2015**, *2*, 1500049. [CrossRef]
21. Zhu, B.; Zhang, L.; Cheng, B.; Yu, J. First-principle calculation study of tri- s -triazine-based g-C 3 N 4: A review. *Appl. Catal. B Environ.* **2018**, *224*, 983–999. [CrossRef]
22. Cho, K.-H.; Sung, Y.-M. The formation of Z-scheme CdS/CdO nanorods on FTO substrates: The shell thickness effects on the flat band potentials. *Nano Energy* **2017**, *36*, 176–185. [CrossRef]
23. Chen, J.I.L.; von Freymann, G.; Choi, S.Y.; Kitaev, V.; Ozin, G.A. Amplified Photochemistry with Slow Photons. *Adv. Mater.* **2006**, *18*, 1915–1919. [CrossRef]
24. Schünemann, S.; Tüysüz, H. An Inverse Opal Structured Halide Perovskite Photocatalyst. *Eur. J. Inorg. Chem.* **2018**, *2018*, 2350–2355. [CrossRef]
25. Cui, X.; Wang, Y.; Jiang, G.; Zhao, Z.; Xu, C.; Wei, Y.; Duan, A.; Liu, J.; Gao, J. A photonic crystal-based CdS–Au–WO_3 heterostructure for efficient visible-light photocatalytic hydrogen and oxygen evolution. *RSC Adv.* **2014**, *4*, 15689–15694. [CrossRef]
26. Xiao, Y.; Chen, M.; Zhang, M. Improved utilization of sun light through the incidence dependence of photonic stop band: g-C3N4 embedded FTO photonic crystal film. *ChemPhotoChem* **2018**, *3*, 101–106. [CrossRef]
27. Lin, S.; Feng, Y.; Wen, X.; Harada, T.; Kee, T.W.; Huang, S.; Shrestha, S.; Conibeer, G. Observation of Hot Carriers Existing in Ag_2S Nanoparticles and Its Implication on Solar Cell Application. *J. Phys. Chem. C* **2016**, *120*, 10199–10205. [CrossRef]
28. Tubtimtae, A.; Wu, K.-L.; Tung, H.-Y.; Lee, M.-W.; Wang, G.J. Ag_2S quantum dot-sensitized solar cells. *Electrochem. Commun.* **2010**, *12*, 1158–1160. [CrossRef]
29. Park, K.-H.; Dhayal, M. Simultaneous growth of rutile TiO_2 as 1D/3D nanorod/nanoflower on FTO in one-step process enhances electrochemical response of photoanode in DSSC. *Electrochem. Commun.* **2014**, *49*, 47–50. [CrossRef]
30. Wu, J.-J.; Chang, R.-C.; Chen, D.-W.; Wu, C.-T. Visible to near-infrared light harvesting in Ag_2S nanoparticles/ZnO nanowire array photoanodes. *Nanoscale* **2012**, *4*, 1368. [CrossRef]
31. Shen, H.; Jiao, X.; Oron, D.; Li, J.; Lin, H. Efficient electron injection in non-toxic silver sulfide (Ag_2S) sensitized solar cells. *J. Power Sources* **2013**, *240*, 8–13. [CrossRef]
32. Tubtimtae, A.; Cheng, K.-Y.; Lee, M.-W. Ag_2S quantum dot-sensitized WO_3 photoelectrodes for solar cells. *J. Solid State Electrochem.* **2014**, *18*, 1627–1633. [CrossRef]
33. Wang, W.; Dong, J.; Ye, X.; Li, Y.; Ma, Y.; Qi, L. Heterostructured TiO_2 Nanorod@Nanobowl Arrays for Efficient Photoelectrochemical Water Splitting. *Small* **2016**, *12*, 1469–1478. [CrossRef] [PubMed]
34. Xie, H.; Li, Y.; Jin, S.; Han, J.; Zhao, X. Facile Fabrication of 3D-Ordered Macroporous Nanocrystalline Iron Oxide Films with Highly Efficient Visible Light Induced Photocatalytic Activity. *J. Phys. Chem. C* **2010**, *114*, 9706–9712. [CrossRef]
35. Brillet, J.; Gratzel, M.; Sivula, K. Decoupling feature size and functionality in solution-processed, porous hematite electrodes for solar water splitting. *Nano Lett.* **2010**, *10*, 4155–4160. [CrossRef] [PubMed]

36. Hou, W.; Xiao, Y.; Han, G.; Zhang, Y.; Chang, Y. Titanium dioxide/zinc indium sulfide hetero-junction: An efficient photoanode for the dye-sensitized solar cell. *J. Power Sources* **2016**, *328*, 578–585. [CrossRef]
37. Sakoda, K. Enhanced light amplification due to group-velocity anomaly peculiar to two- and three-dimensional photonic crystals. *Opt. Express* **1999**, *4*, 167–176. [CrossRef]
38. Aloshyna, M.; Sivakumar, S.; Venkataramanan, M.; Brolo, A.G.; van Veggel, F.C.J.M. Significant Suppression of Spontaneous Emission in SiO_2 Photonic Crystals Made with Tb3+-Doped LaF3Nanoparticles. *J. Phys. Chem. C* **2007**, *111*, 4047–4051. [CrossRef]

Article

Effective Photocatalytic Activity of Sulfate-Modified $BiVO_4$ for the Decomposition of Methylene Blue Under LED Visible Light

Vinh Huu Nguyen [1,2], Quynh Thi Phuong Bui [3,*], Dai-Viet N. Vo [1], Kwon Taek Lim [4], Long Giang Bach [2], Sy Trung Do [5], Tuyen Van Nguyen [5], Van-Dat Doan [6], Thanh-Danh Nguyen [7] and Trinh Duy Nguyen [1,2,*]

1. Center of Excellence for Green Energy and Environmental Nanomaterials (CE@GrEEN), Nguyen Tat Thanh University, Ho Chi Minh 755414, Vietnam
2. NTT Hi-Tech Institute, Nguyen Tat Thanh University, Ho Chi Minh 755414, Vietnam
3. Faculty of Chemical Technology, Ho Chi Minh City University of Food Industry, Ho Chi Minh 705800, Vietnam
4. Department of Display Engineering, Pukyong National University, Busan 608-737, Korea
5. Institute of Chemistry, Vietnam Academy of Science and Techology, Hanoi 10072, Vietnam
6. Faculty of Chemical Engineering, Industrial University of Ho Chi Minh city, Ho Chi Minh 700000, Vietnam
7. Institute of Research and Development, Duy Tan University, Da Nang City 550000, Vietnam
8. Institute of Chemical Technology, Vietnam Academy of Science and Technology, Ho Chi Minh 700000, Vietnam

* Correspondence: quynhbtp@hufi.edu.vn (Q.T.P.B.); ndtrinh@ntt.edu.vn (T.D.N.)

Received: 21 June 2019; Accepted: 12 August 2019; Published: 22 August 2019

Abstract: In this study, we investigated sulfate-modified $BiVO_4$ with the high photocatalytic activity synthesized by a sol-gel method in the presence of thiourea, followed by the annealing process at different temperatures. Its structure was characterized by thermal gravimetric analysis (TGA), powder X-ray diffraction (XRD), Raman spectroscopy, scanning electron microscopy/energy-dispersive X-ray spectroscopy (SEM/EDS), X-ray photoelectron spectroscopy (XPS), and ultraviolet-visible diffuse reflectance spectroscopy (UV-Vis DRS). The $BiVO_4$ synthesized in the presence of thiourea and calcined at 600 °C (T-BVO-600) exhibited the highest photocatalytic degradation efficiency of methylene blue (MB) in water; 98.53% MB removal was achieved within 240 min. The reaction mechanisms that affect MB photocatalytic degradation on the T-BVO-600 were investigated via an indirect chemical probe method, using chemical agents to capture the active species produced during the early stages of photocatalysis, including 1,4-benzoquinone (scavenger for O_2^-), ethylenediaminetetraacetic acid disodium salt (scavenger for h^+), and *tert*-butanol (scavenger for $HO^\bullet$). The results show that holes (h^+) and hydroxyl radicals ($HO^\bullet$) are the dominant species of MB decomposition. Photoluminescence (PL) measurement results of terephthalic acid solutions in the presence of $BiVO_4$ samples and $BiVO_4$ powders confirm the involvement of hydroxyl radicals and the separation efficiency of electron-hole pairs in MB photocatalytic degradation. Besides, the T-BVO-600 exhibits good recyclability for MB removal, achieving a removal rate of above 83% after five cycles. The T-BVO-600 has the features of high efficiency and good recyclability for MB photocatalytic degradation. These results provide new insight into the purpose of improving the photocatalytic activity of $BiVO_4$ catalyst.

Keywords: sulfate-modified $BiVO_4$; methylene blue; LED visible light; photodecomposition

---

1. Introduction

Bismuth vanadate ($BiVO_4$) has recently been extensively studied by researchers around the world and has been used as a new catalyst in the photocatalytic field because of the economic advantage of synthetic materials, low toxicity, excellent chemical stability, and narrow bandgap (about 2.4 eV for monoclinic scheelite $BiVO_4$) [1]. Researchers have discovered that $BiVO_4$ offers outstanding photocatalytic performance in water splitting and oxidation of toxic organic compounds [2–6]. The photoinduced charge carrier formation was highly efficient due to low bandgap energy properties. However, recombination of excess electrons and holes are extensive due to its poorly charged transfer characteristics and weakly adsorbed surface, limiting the photocatalytic activity of $BiVO_4$ [7]. In order to improve the separation efficiency of the photogenerated electron-hole pairs to the catalytic surface for high-photodynamic catalysis, the researchers proposed several measures, such as (1) control of crystal structure, crystal form, and crystal surface [8,9], (2) formation of p-n bonds and the establishment of an internal electromagnetic interaction region extending from n-type semiconductor ($BiVO_4$) to p-type semiconductor materials [10], and (3) formation of the monoclinic-tetragonal structure of $BiVO_4$ [11]. Such approaches mainly involve improving the photocatalytic activity of $BiVO_4$, which is enhanced through either material synthesis with crystal form control or doping with nonmetal elements which have been proven to be efficient and promising research directions recently.

According to previously published studies, the concentration of the reactants and the solution medium (such as the pH, the effect of anions) had significant implications for the crystalline form of $BiVO_4$ in solution. For example, small changes in the affecting factors (such as pH, temperature, and reactants) will alter the growth of $BiVO_4$ crystals. As a result, crystalline forms such as nanoplates [12], micro bar [13], elliptic structure [14], and various crystal forms are formed [15]. Therefore, the addition of a surfactant to the reaction solution to control the crystal growth process could facilitate the designing of the ideal catalyst. For this purpose, recently, researchers have been using urea for the synthesis of $BiVO_4$. Urea can control the precipitation of cation by slowly forming hydroxide ions in solution through hydrolysis [9,16]. Thus, the slow hydrolysis of urea leads to a gradual increase in the pH of the reaction solution and provides a special solution for controlling the crystal growth process. Also, the doping of nonmetal elements, such as S [17], C [18], N [19], and P [20], into the structure of $BiVO_4$ photocatalyst also enhances the photocatalytic efficiency of this material. Similar to urea, thiourea ($CS(NH_2)_2$) also hydrolyzes to form NH_3, which participates in the pH adjustment of the reaction solution, thus contributing to the process of controlling the crystal growth of the material. Also, previous studies have used thiourea as the S source to modify $BiVO_4$ to enhance the photocatalytic activity of the material [4,17,21]. The previous studies have used thiourea or Na_2S as the S source to modify $BiVO_4$ to enhance the photocatalytic activity of the material. However, S-doped $BiVO_4$ was synthesized by hydrothermal method, and surfactants were introduced in the synthesis process. In some of these studies, these surfactants may have a bad effect on the environment. Therefore, easier methods to synthesize S-doped $BiVO_4$ are necessary.

Hence, we report on the synthesis of sulfate-modified $BiVO_4$ by a sol-gel method using thiourea as the reducing agent to control simultaneous crystal morphology as well as the S source for material modification. At the same time, we investigated the effect of heating temperature on the structure of $BiVO_4$ as well as on the photocatalytic activity of the material in the decomposition of organic compounds using visible light.

2. Experimental

2.1. Materials

Ammonium metavanadate (NH_4VO_3, ≥98%), bismuth(III) nitrate pentahydrate ($Bi(NO_3)_3{\cdot}5H_2O$, ≥98.0%), *tert*-butanol (TBA, $(CH_3)_3COH$, ≥99.5%), and 1,4-benzoquinone (BQ, $C_6H_4O_2$) were purchased from Sigma-Aldrich. Thiourea (CH_4N_2S, 99.8%) was purchased from Prolabo (France). Nitric acid (HNO_3, 65–68%), ethanol (CH_3CH_2OH, 99.7%), ethylenediaminetetraacetic acid disodium salt (EDTA-2Na), and methylene blue (MB, 99%) were obtained from Xilong Chemical Co., Ltd. (Shantou, China).

2.2. Fabrication of Sulfate-Modified Bismuth Vanadate ($BiVO_4$)

We synthesized the sulfate-modified $BiVO_4$ using the sol-gel combustion method with a coupling of the sol-gel process and the annealing process. Firstly, 20 mmol of $Bi(NO_3)_3{\cdot}5H_2O$ was dissolved in 200 mL of HNO_3 (2M) and stirred for about 30 min to form a clear solution (solution A). At the same time, 20 mmol of NH_4VO_3 was dissolved in 200 mL of water and stirred for about 180 min at 70 °C to form a uniform transparent yellow solution (solution B). After adding solution B to solution A drop by drop, a dark yellow solution was obtained. Then, 62.5 mmol of thiourea was added to the mixture. The obtained mixture was vigorously stirred for 30 min before being heated at 85 °C to evaporate the water under continuous stirring overnight. Finally, the obtained powder was finely ground and calcined at different temperatures (400–700 °C) for 3 h with a heating rate of 5 °C/min in the air. The obtained samples were denoted as T-BVO-400, T-BVO-500, T-BVO-600, and T-BVO-700, corresponding to the annealing temperature of 400, 500, 600, and 700 °C, respectively. For comparison, $BiVO_4$ was synthesized with the absence of thiourea and calcined at 600 °C (denoted as BVO-600).

2.3. Characterization

Thermal gravimetric analysis (TGA) was conducted on a TGA Q500 V20.10 Build 36 under air condition with a heating rate of 5 °C/min from room temperature to 800 °C. X-ray diffraction (XRD) patterns were recorded in a D8 Advance Bruker powder diffractometer (Bruker, Billerica, MA, USA) with a Cu Kα excitation source at a scan rate of 0.030°/s in the 2-theta range of 5–80°. The surface morphologies and particle size of $BiVO_4$ samples were observed by scanning electron microscope (SEM, JEOL JSM 7401F, Peabody, MA, USA). X-ray photoelectron spectroscopy (XPS) was recorded on Thermo VG Multilab 2000 (Thermo VG Scientific, Waltham, MA, USA). Fourier transform infrared (FT-IR) spectra were recorded on an EQUINOX 55 spectrometer (Bruker, Billerica, MA, USA). Raman spectroscopy was carried out on the HORIBA Jobin Yvon spectrometer (Horiba Scientific, Kyoto, Japan) with a laser beam of 633 nm in the wavenumber of 100–1000 cm^{-1}. The optical absorption characteristics of the photocatalysts were determined by ultraviolet-visible (UV-Vis) diffuse reflectance spectroscopy (UV-Vis DRS, Shimazu UV-2450, Kyoto, Japan) in the range of 300–900 cm^{-1}. Photoluminescence (PL) measurements were recorded using an F-4500 Spectro-fluorometer (Hitachi, Chiyoda, Japan).

2.4. Photocatalytic Activity Test

The photocatalytic activities of the $BiVO_4$ samples were evaluated by the photodegradation of methylene blue (MB) in a 250 mL double-layer interbed glass beaker photocatalytic reactor under visible light irradiation by six daylight Cree® Xlamp® XM-L2 LEDs (Cree, Inc., Durham, NC, USA) with max power of 10 W and max light output of 1052 lumen). In each run, a mixture consisting of MB aqueous solution (15 ppm, 100 mL) and the given catalyst (100 mg) was magnetically stirred in the dark for 1 h to reach the adsorption-desorption equilibrium of the dye on the catalyst surface. After this time, the LEDs light source was switched on. Five mL of the suspension was withdrawn at the same intervals and immediately centrifuged to separate photocatalyst particles at 7000 rpm for 15 min. The MB concentration was monitored by measuring the absorption intensity at its maximum absorbance wavelength of $\lambda = 664$ nm using a UV-visible spectrophotometer (Model Evolution 60S, Thermo Fisher Scientific, Massachusetts, MA, USA) in a 1 cm path length spectrometric quartz cell.

2.5. Active Species Trapping Experiments

The active species generated during the early stages of photocatalytic processes such as O_2^-, h^+, and $HO^\bullet$ are responsible for MB degradation and are determined by an indirect chemical probe method. Chemical agents, namely BQ, EDTA-2Na, and TBA were agents that capture O_2^-, h^+, and $HO^\bullet$, respectively. The experimental procedure was similar to the above photocatalytic experiments except that the chemical agents are added before the beginning of the photocatalytic experiment. The concentration of agents added was 0.3 M except for BQ. The concentration of BQ was 1.0×10^{-3} M because higher concentrations of BQ might hinder the determination of MB concentration by UV-Vis absorption spectra.

2.6. *Analysis of Hydroxyl Radical (HO•)*

The generation of hydroxyl radicals (HO•) in the LED/sulfate-modified $BiVO_4$ system was detected by the PL technique using terephthalic acid (TA) as the sensor molecule. The formation of 2-hydroxyterephthalic acid (HTA) from the reaction of TA with HO• radicals exhibits strong photoluminescence. The process was similar to the above photocatalytic activity test with the replacement of MB solution by TA solution containing TA 0.5 mM and NaOH 2 mM. A fluorescence spectrophotometer analyzed the clear solution at an excitation wavelength of 315 nm after 240 min of irradiation.

3. Results and Discussion

3.1. *Characterization of Sulfate-Modified $BiVO_4$ and Pure $BiVO_4$*

The crystal structure of sulfate-modified $BiVO_4$ and pure $BiVO_4$ was confirmed by X-ray diffraction. Figure 1A displays the XRD pattern of the $BiVO_4$ samples that were synthesized with/without the presence of thiourea and calcined at 600 °C (T-BVO-600 and BVO-600 samples). For the BVO-600 sample, the diffraction peaks on the XRD pattern aligned with the monoclinic scheelite phase of $BiVO_4$ (m-s $BiVO_4$, JCPDS no. 01-075-1867) with weakly diffracted peaks at $2\theta = 15.5°$, strongly diffracted peaks at $2\theta = 28.9°$, and the splitting of peaks at $2\theta = 18.5°$, 35°, and 47°. However, with the presence of thiourea, the XRD pattern of the T-BVO-600 sample not only shows the characteristic diffraction peaks in the monoclinic scheelite structure, but also exhibits weak diffraction peak characteristics for bismuth oxide sulfate ($Bi_{34.67}O_{36}(SO_4)_{16}$, JCPDS no. 00-041-0689), indicating that the presence of thiourea did not alter the phase structure of m-s $BiVO_4$ crystal. In addition, there was no observable change in the position of the (121) and (040) planes of both T-BVO-600 and BVO-600 samples, implying that S could not be doped into the lattice of $BiVO_4$ crystal. The negligible differences between lattice parameters of T-BVO-600 and BVO-600, which are shown in Table 1, also confirm that no significant change in the crystal phase of $BiVO_4$ occurred. These analyses indicated the possible loading of S to the $BiVO_4$ surface as sulfate instead of doping into the lattice of $BiVO_4$.

To investigate the presence of functional groups and bonds in the material structure, the material was analyzed by FT-IR, as shown in Figure 1B. In general, both T-BVO-600 and BVO-600 samples have characteristic vibration peaks for m-s $BiVO_4$, which is consistent with the FT-IR results for the $BiVO_4$ material in the previous study. These peaks are the stretching vibration (δ) and bending vibration mode (υ) respectively, of the O–H bond of the water molecules adsorbed onto the surface of the material [22], the bending vibration of the Bi–O bond [23], the stretching vibration of the VO_4^{3-} group [22,23], and the asymmetric stretching vibration of V=O [24]. Also, the presence of the SO_4^{2-} group on the T-BVO-600 sample was detected through FT-IR spectra at wavenumbers of 1110 and 1016 cm^{-1}, corresponding to the asymmetric and symmetric stretching vibrations of S=O bonds [25].

Table 1. Lattice parameters and the V–O bond lengths at 829 cm^{-1} for the Bismuth vanadate ($BiVO_4$) prepared by the different annealing temperature.

Sample No.	Preparation Condition (°C)	Cell Parameters [a]					Crystallite Size [a] (nm)	Bond Length [b] (Å)
		a (Å)	b (Å)	c (Å)	β (Å)	V_{cell} ($Å^3$)		V–O
T-BVO-85	85	–	–	–	–	–	–	–
T-BVO-400	400	5.136	5.094	11.686	90.257	305.277	36.890	1.7002 ± 0.0005
T-BVO-500	500	5.174	5.101	11.664	90.222	307.809	34.420	1.6961 ± 0.0002
T-BVO-600	600	5.186	5.093	11.669	90.174	308.195	33.700	1.6940 ± 0.0002
T-BVO-700	700	5.192	5.094	11.667	90.198	308.576	32.740	1.6937 ± 0.0006
BVO-600	600 (without thiourea)	5.192	5.095	11.664	90.206	308.501	33.190	1.6935 ± 0.0006

[a] Data obtained by XRD data; [b] Data obtained by Raman data.

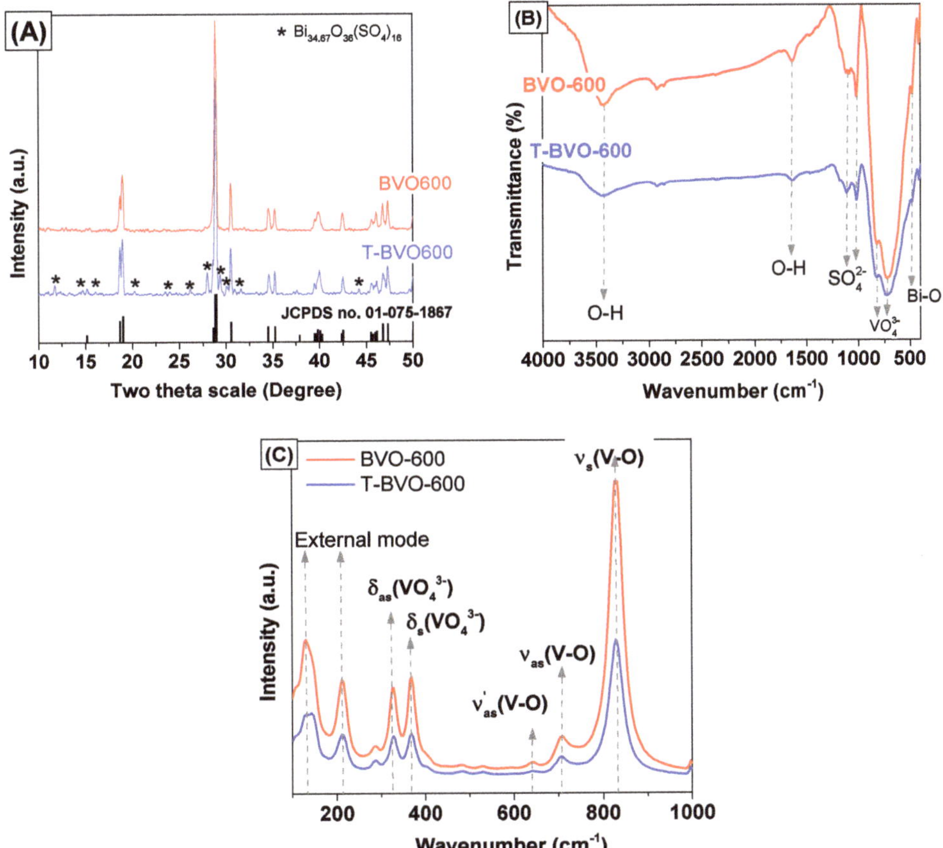

Figure 1. The X-ray diffraction (XRD) patterns (**A**), Fourier transform infrared (FT-IR) spectra (**B**), and Raman spectra (**C**) of T-BVO-600 and BVO-600.

Figure 1C displays the Raman spectra of sulfate-modified $BiVO_4$ and pure $BiVO_4$. As shown in Figure 1C, both samples have characteristic vibration peaks for m-s $BiVO_4$, consisting of (i) the stretching vibration (δ_s) and asymmetric vibration mode (δ_{as}) of V–O at 829 and 709 cm^{-1} respectively, (ii) the symmetrical bending vibrations (υ_s) and asymmetric bending vibrations (υ_{as}) of the VO_4^{3-} group at 368 and 327 cm^{-1} respectively, and (iii) the external modes (rotation/translation) in $BiVO_4$ at 212 and 129 cm^{-1} [26]. The V–O bond length of sulfate-modified $BiVO_4$ and pure $BiVO_4$ samples can be calculated via the empirical expression [27,28]:

$$\nu = 21349 \cdot e^{-19176 \cdot R} \tag{1}$$

where, ν is the stretching vibration frequency of V–O (cm^{-1}) and R is the V–O bond length (Å). The R values of T-BVO-600 and BVO-600 samples are 1.694 and 1.692 Å, respectively. This result further confirmed the monoclinic scheelite phase of $BiVO_4$.

To provide more structure information, the chemical states of T-BVO-600 and BVO-600 samples were determined by XPS. As shown in Figure 2A, the chemical composition of the two samples mainly consists of Bi, V, C, and O elements. The presence of S element in the T-BVO-600 sample indicated that S is deposited on the $BiVO_4$ surface. The high-resolution XPS spectrum of the T-BVO-600 showed a broad peak of S 2 s at 233.9 eV (Figure 2B), assigned to S^{+6} in the SO_4^{2-} group [29,30]. Figure 2C shows the Bi 4f spectrum containing two strongly symmetric peaks Bi $4f_{7/2}$ and Bi $4f_{5/2}$. They are 158.97 eV and 164.36 eV for the T-BVO-600 sample, and 158.52 eV and 163.88 eV for the BVO-600 sample respectively, which are characteristics of Bi^{3+} [31,32]. The shoulder of the Bi $4f_{7/2}$ (159.41 eV) and Bi $4f_{5/2}$ (164.66 eV) spectra for the BVO-600 sample could be ascribed to the presence of Bi^{5+} oxidation state in $BiVO_4$ [33,34]. The shoulder of the Bi $4f_{7/2}$ (156.55 eV) and Bi $4f_{5/2}$ (162.34 eV) spectra for the T-BVO-600 sample could be ascribed to the presence of Bi^{2+} oxidation state in BiO [35,36]. The asymmetric V $2p_{3/2}$ signals can be decomposed into two subpeaks at 514.85 eV and 516.95 eV for T-BVO-600, and 514.9 and 516.51 eV for BVO-600 (Figure 2C), which is assigned for V^{4+} and V^{5+} species, respectively [31]. The surface molar of V^{4+}/V^{5+} for T-BVO-600 (0.502) was higher than that of the BVO-600 (0.027) sample, which confirmed that the T-BVO-600 sample was oxygen-deficient [19]. The BE positions of O 1s at 530.10 eV in the BVO-600 sample and 530.9 eV in the T-BVO-600 sample (Figure 2D) are assigned to O^{2-} [17]. In the case of BVO-600, this peak can be decomposed into two subpeaks at 529.76 eV and 531.69 eV (Figure 2D), which are assigned to the lattice oxygen (O_{latt}) in crystalline $BiVO_4$ and the adsorbed oxygen (O_{ads}) on the $BiVO_4$ surface, respectively. Similarly, the O 1s peak of T-BVO-600 can also be decomposed into two subpeaks at 530.25 eV and 532.57 eV (Figure 2D). However, the slight shifting of the peak indexed to the adsorbed oxygen (O_{ads}) toward a higher BE for the T-BVO-600 (Figure 2D) could be due to the presence of sulfate on the $BiVO_4$ surface. The surface molar of O_{ads}/O_{latt} for T-BVO-600 (0.354) was lower than that of the BVO-600 sample (0.436). The low number of O_{ads} species play an important role in the photocatalytic performance of T-BVO-600 (to be seen in investigating the mechanisms of dye degradation section). The XPS results, together with XRD and FT-IR, confirm the presence of bismuth oxide sulfate ($Bi_{34.67}O_{36}(SO_4)_{16}$) in the T-BVO-600 structure.

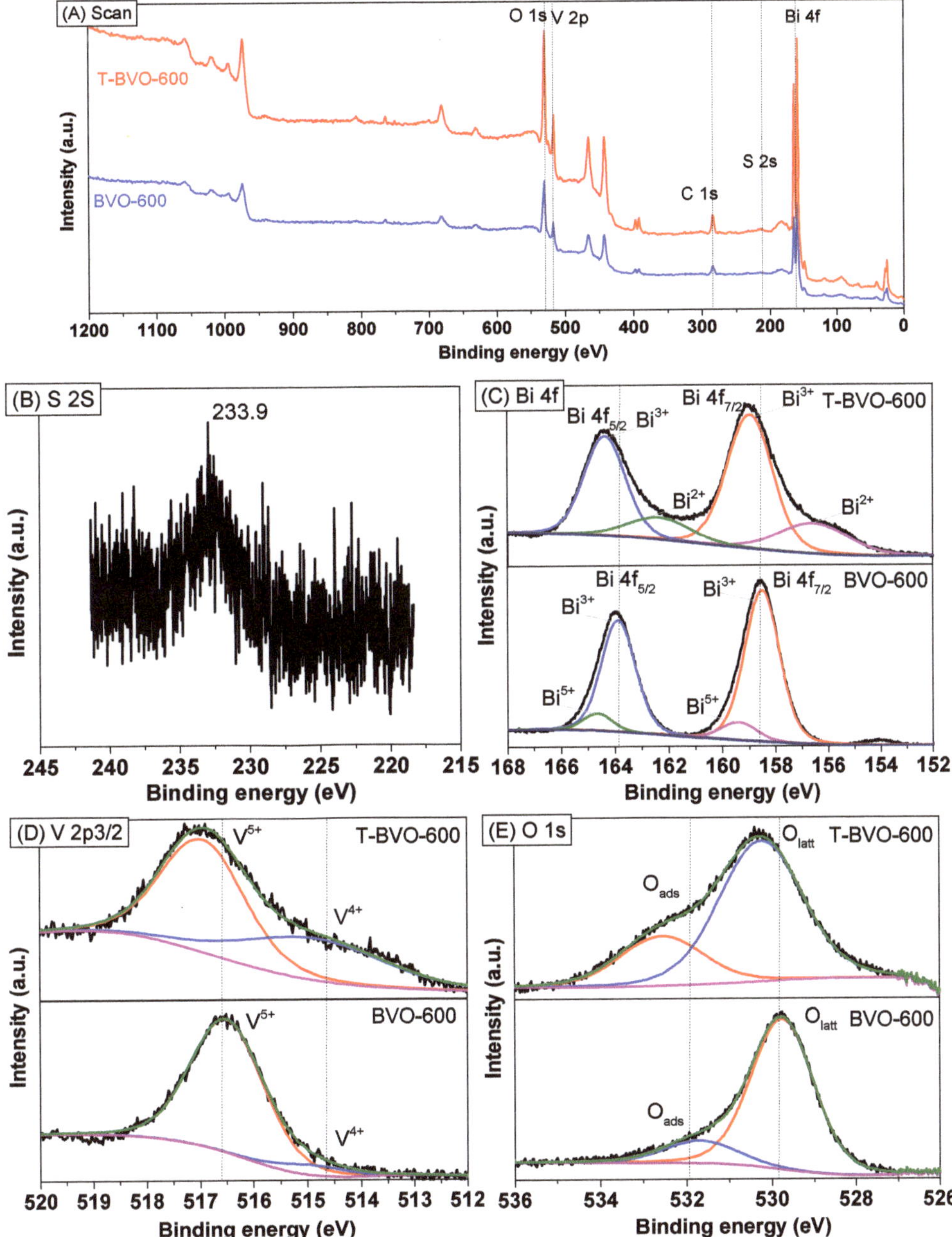

Figure 2. Full scan (**A**), S 2s (**B**), Bi 4f (**C**), V 2p3/2 (**D**), and O 1s (**E**) X-ray photoelectron spectroscopy (XPS) spectra of T-BVO-600 and BVO-600.

Crystal morphology, particle size, and particle distribution of the material were observed through SEM images. The SEM images of the samples were synthesized under different conditions, as shown in Figure 3A,B. The obtained shape and crystal size of the materials were very different when synthesizing under different conditions. For the $BiVO_4$ sample which was synthesized without thiourea and calcined at 600 °C, the crystal morphology has a granular shape and is approximately 1 μm in size. For the $BiVO_4$ synthesized using thiourea, the forming material has a granular crystalline form with insignificant granular boundaries, and particles are deposited into large plates with openings formed between the particles. In addition, the presence of S element in the T-BVO-600 sample was also confirmed by FE-SEM/EDS images (Figure 3C). These results, along with the XPS result (Table 2), revealed that sulfate was successfully deposited into the $BiVO_4$ surface.

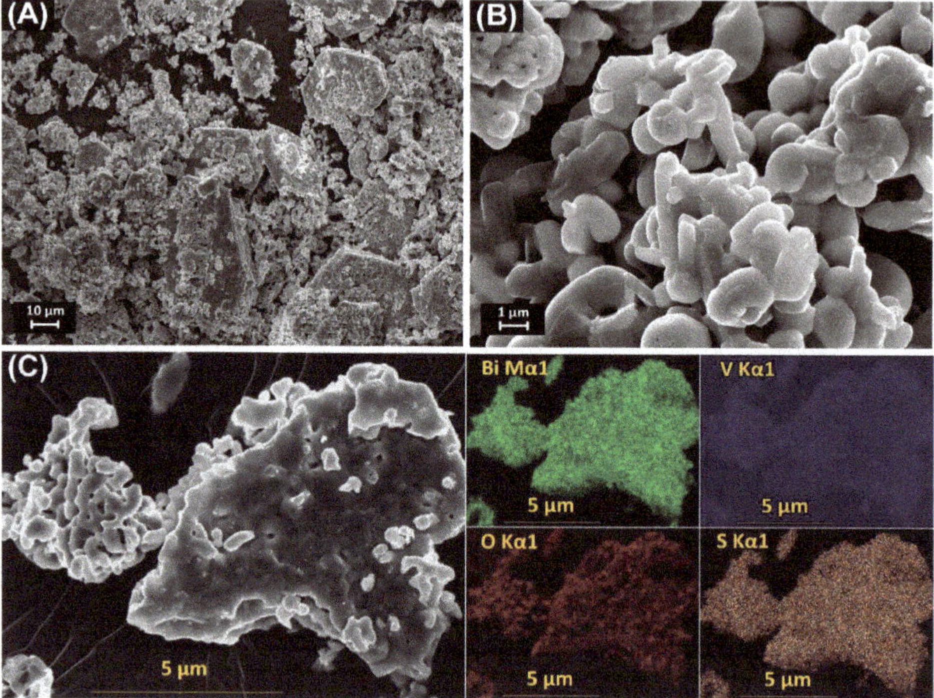

Figure 3. Scanning electron microscopy (SEM) images the $BiVO_4$ samples: T-BVO-600 (**A**) and BVO-600 (**B**) and field emission scanning electron microscopes and energy-dispersive X-ray spectrometer (FESEM/EDS) images of the T-BVO-600 sample with maps of Bi Ma1, V Ka1, OKa1, and S Ka1 (**C**).

Table 2. Surface element compositions of the BVO-600 and T-BVO-600 samples.

Sample No.	Atomic % [a]				
	O 1s	C 1s	Bi 4f	V 2p3	S 2s
BVO-600	46.55	29.25	15.85	8.36	–
T-BVO-600	50.94	28.44	11.67	6.22	2.72

[a] Data obtained by XPS data

The light absorption properties of photocatalytic materials were analyzed using the UV-Vis DRS technique. The results are shown in Figure 4A. All $BiVO_4$ samples showed narrow absorption in the visible light region, which can facilitate the enhancement of photocatalytic property under visible light irradiation. The bandgap energy of $BiVO_4$ samples can be estimated from the $(ah\nu)^2$–$h\nu$ curves (Figure 4B). The E_g value of the $BiVO_4$ samples was determined to be 2.29 eV for T-BVO-600 and 2.23 eV for BVO-600.

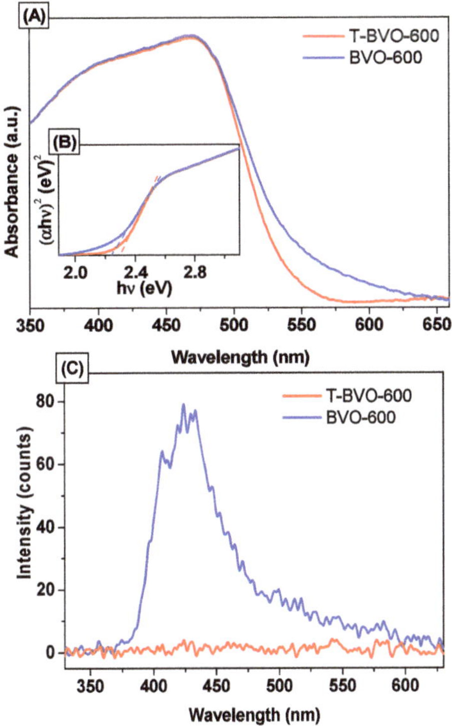

Figure 4. Ultraviolet-visible diffuse reflectance spectroscopy (UV-Vis DRS) spectra (**A**), $(ah\nu)^2$–hν curves (**B**), and photoluminescence (PL) spectra of T-BVO-600 and BVO-600 (**C**).

The photoluminescence spectra of the T-BVO-600 and BVO-600 samples were also used to study the separation efficiency of electron-hole pairs generated by irradiation. The higher intensity of the PL spectrum of the sample indicates that the electron-hole pair recombination takes place more rapidly, thus reducing photocatalytic activity. The PL spectrum of the $BiVO_4$ samples upon excitation at 325 nm is shown in Figure 4C. It can be seen that BVO-600 generates broad emission peaks in the wavelength range of 360–625 nm with a maximum emissivity of 430 nm corresponding to the transfer of charges from Bi to V centers [37]. Whereas the synthetic $BiVO_4$ samples using thiourea do not seem to emit peaks in this wavelength range, indicating that the recombination of electrons and holes in these samples is very low.

3.2. *Effect of the Annealing Temperature*

To investigate the effect of the annealing temperature on the phase structure of $BiVO_4$ crystals, the amorphous $BiVO_4$, which was obtained after heating at 85 °C under continuous stirring overnight (T-BVO-85), was calcined at a different temperature. XRD patterns of the amorphous $BiVO_4$ and $BiVO_4$ samples annealed from 400 °C to 700 °C are shown in Figure 5A. As shown in Figure 5A, the amorphous $BiVO_4$ exhibited a low crystal with the main phase of vanadium oxide (V_3O_5, JCPDS No. 01-071-0039). When T-BVO-85 was annealed at 400 °C, the XRD pattern showed the coexistence of tetragonal scheelite phase of $BiVO_4$ (t-s $BiVO_4$, JCPDS No. 14-0133) and V_2O_5 (JCPDS No. 00-041-1426). Besides, we also observed that the peak at 2θ = 13.26° for T-BVO-85 and T-BVO-400 samples were related to the typical in-planar peak of graphitic carbon nitride (g-C_3N_4) polymers [38,39] which were formed during heating at 85 °C to evaporate the water and the annealing at 400 °C. When further increasing the annealing temperature to 500 °C (T-BVO-500), the XRD pattern of this sample demonstrated a pattern similar to that of m-s $BiVO_4$ with the coexistence of $Bi_2O(SO_4)_2$ (JCPDS No. 01-078-2087). The peak at 13.26° disappeared, which indicated that the organic compounds were completely burned. The T-BVO-600 sample exhibited the coexistence of m-s $BiVO_4$ and $Bi_{34.67}O_{36}(SO_4)_{16}$, which was analyzed above. When further increasing the annealing temperature to 700 °C, pure m-s $BiVO_4$ was formed. The XRD results indicate that the different annealing temperatures would result in $BiVO_4$ materials with different composition and phase.

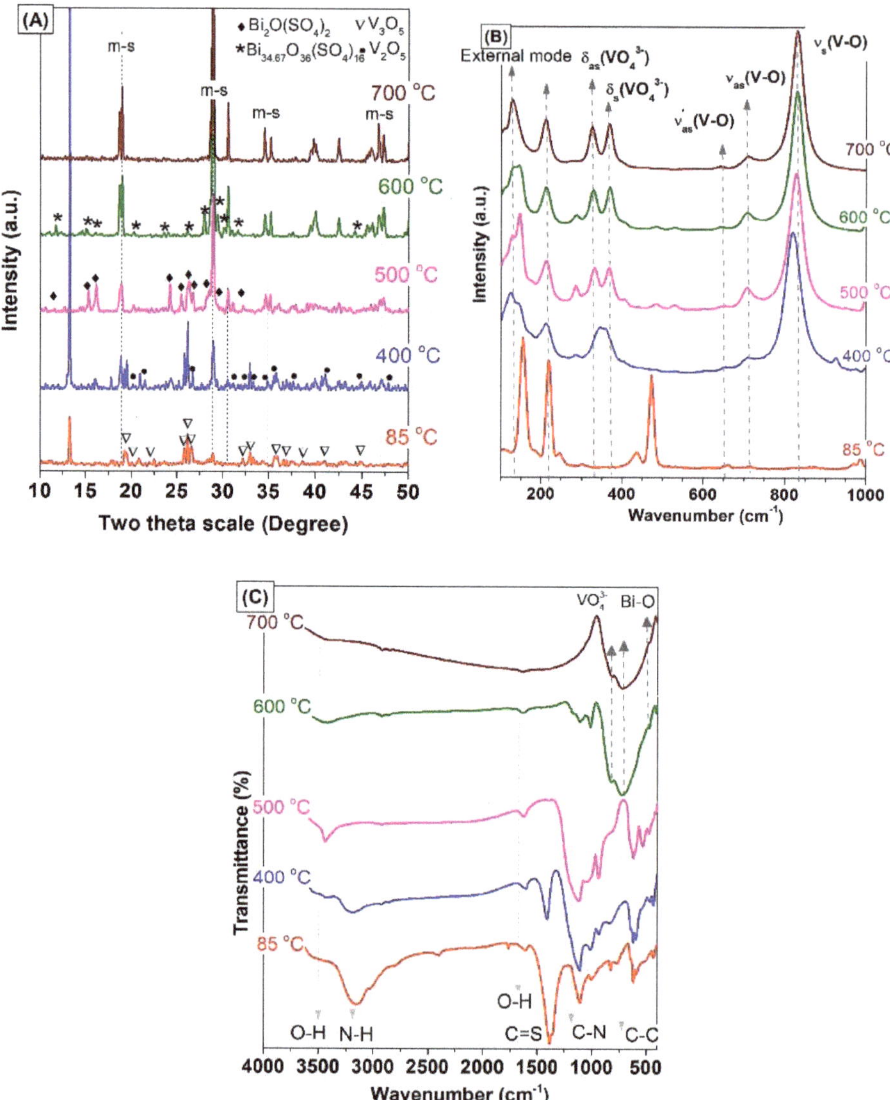

Figure 5. XRD pattern (**A**), Raman spectra (**B**), and FT-IR spectra (**C**) of the BiVO4 amorphous (85 °C) and sulfate-modified $BiVO_4$ annealed at 400, 500, 600, and 700 °C.

The FT-IR spectra of the amorphous $BiVO_4$ and $BiVO_4$ samples annealed from 400 °C to 700 °C were shown in Figure 5B. For the amorphous $BiVO_4$ and $BiVO_4$ samples annealed at 400 °C and 500 °C, absorption bands are located at 620, 730, 806, 940, 1110, 1385, 1406 (and 1598), and 3168 cm^{-1} assigned to the C–C (aromatic), N–H (bending), tri-s-triazine, C–H, C–C, C–H (aromatic), C–N (aromatic), and N–H vibrations respectively, which confirms the formation of g-C_3N_4 [40–43]. The formation of g-C_3N_4 compounds obtained by polycondensation of thiourea at different temperatures is in agreement with previous reports [38,44,45]. The $BiVO_4$ samples annealed at 600 °C and 700 °C have characteristic vibration peaks for m-s $BiVO_4$ and no peaks that could be indexed to the presence of organic compounds.

The Raman results (Figure 5C) show that the $BiVO_4$ samples annealed at 500 °C, 600 °C, and 700 °C have characteristic vibration mode in the m-s $BiVO_4$ while T-BVO-400 has characteristic vibration mode in the t-s $BiVO_4$ [26]. In addition, the R values of the T-BVO-500 and T-BVO-600 samples shown in Table 1 matched with the V–O bond length of m-s $BiVO_4$, while the R values of T-BVO-400 matched with the V–O bond length of m-s $BiVO_4$ [46]. This result further confirmed the significant effect of annealing temperature on the structure phase of $BiVO_4$, which is in agreement with XRD and IR results.

The SEM images of the samples synthesized under different conditions are shown in Figure 6. The obtained shape and crystal size of the materials were very different when synthesized under different conditions. For the $BiVO_4$ synthesized using thiourea, the non-calcined sample has no definite shape. After being heated at 400 °C and 500 °C, the crystals with granular shape are formed and sized less than 1 μm. The particle size becomes more significant with increasing calcining temperature (Figure 6) due to the growth of $BiVO_4$ crystals during high calcining temperature. When the calcining temperature increases to 600 °C, the forming material has a granular crystalline form with insignificant granular boundaries and particles are deposited into large plates with openings formed between the particles. At 700 °C, the material forming the crystal structure is mainly irregular particles, about 2 μm in size. Obviously, heat treatment exerts a significant influence on the morphology and crystallinity of $BiVO_4$ when the $BiVO_4$ was synthesized in the presence of thiourea.

TG analysis was performed to observe the physical and chemical processes that occur when the T-BVO-85 sample was processed at different temperatures as well as the purity of the T-BVO-600 sample. Figure 7 shows the weight loss curve of T-BVO-85 and T-BVO-600. From the TGA curve shown in Figure 7A, the T-BVO-85 has five mass loss processes occurring when the sample is heated from room temperature to 800 °C. The first loss of mass occurs from room temperature to 100 °C with a weight loss of 1.07%, corresponding to the removal of adsorbed water on the surface of the material. The second mass loss process in the range of temperature from 100 °C to 300 °C, related to the decomposition of thiourea, corresponds to approximately 31.00% of the weight loss. The third mass loss in the temperatures range from 300 °C to 470 °C is due to the decomposition of the nitrate salts with a mass loss of about 7.98% [47] The fourth mass loss in the temperature range from 470 °C to 540 °C involves complete oxidation of carbon residue in the sample with mass loss of about 4.81%. The fifth mass loss in the region from 540 °C to 680 °C is the transition between the tetragonal and monoclinic phases, which is accompanied by a weight loss of about 7.61%. The differential thermal analysis (DTA) curve of the T-BVO-85 shows five exothermic peaks, corresponding to five mass loss processes on the TG curve. The sharply exothermic peak at 218.67 °C with significant weight loss in the TG curve is attributed to the decomposition of the thiourea because the decomposition reaction of thiourea is the exothermic reaction. For the T-BVO-600 sample (Figure 7B), only one mass loss occurred during the high-temperature range of 605 °C to 690 °C, corresponding to the phase transition between the tetragonal and monoclinic phases. The weight loss accounts for about 0.68%. In addition, no other mass loss occurred at lower temperatures, indicating that thiourea and nitrate salts were decomposed entirely in the sample when the sample was heated at 600 °C.

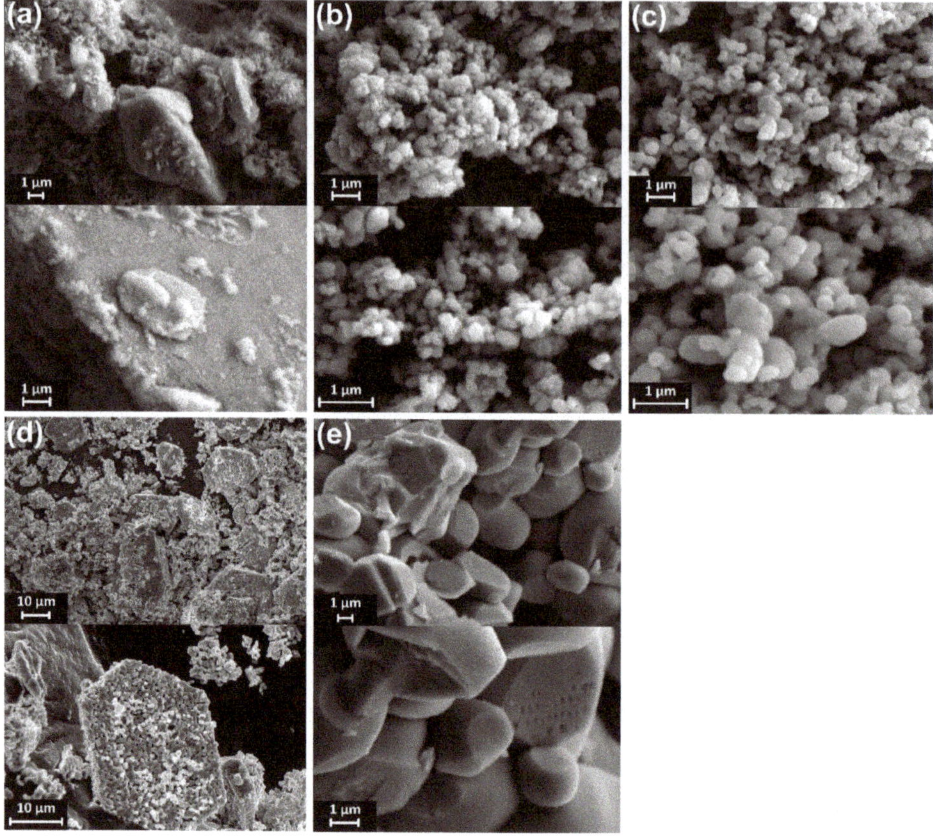

Figure 6. SEM images of the BiVO4 amorphous (**a**) and sulfate-modified $BiVO_4$ annealed at 400 °C (**b**), 500 °C (**c**), 600 °C (**d**), and 700 °C (**e**).

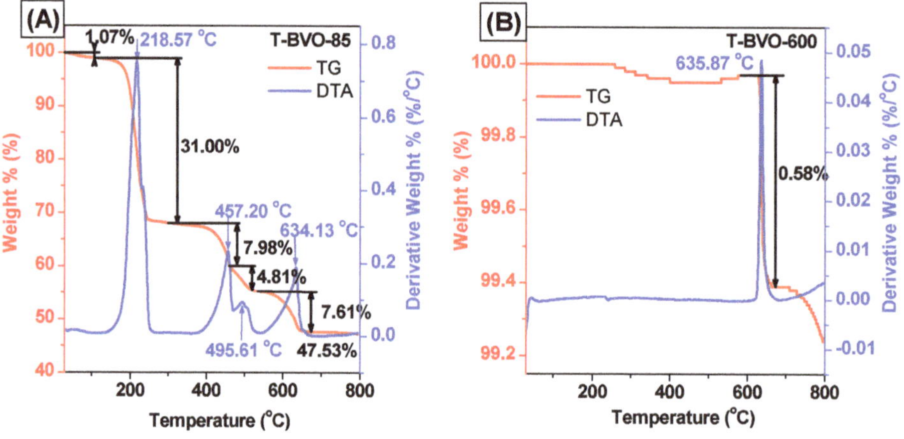

Figure 7. Differential thermal analysis-Thermal gravimetric analysis (DTA-TGA) curves of the $BiVO_4$: (**A**) T-BVO-85 and (**B**) T-BVO-600.

3.3. Photocatalytic Activities

According to the results of the photocatalytic activity shown in Figure 8, it can be seen that on the synthetic $BiVO_4$ samples using thiourea, the calcined samples showed better photocatalytic efficiency than the non-calcined $BiVO_4$ sample (except for the sample calcined at 400 °C). After 240 minutes of irradiation, about 46.02% of MB was removed for the $BiVO_4$ sample without calcining. However, the removal efficiency of MB can reach 87.13%, 98.93%, and 97.57% for the sample calcined at 500, 600, and 700 °C, respectively. The results show that calcining can enhance the photocatalytic activity of the thiourea-based $BiVO_4$ sample, and the calcined sample at 600 °C has the best photocatalytic activity. For the non-thiourea synthesized $BiVO_4$ calcined at 600 °C, low photocatalytic activity, only about 85.54% of MB was removed.

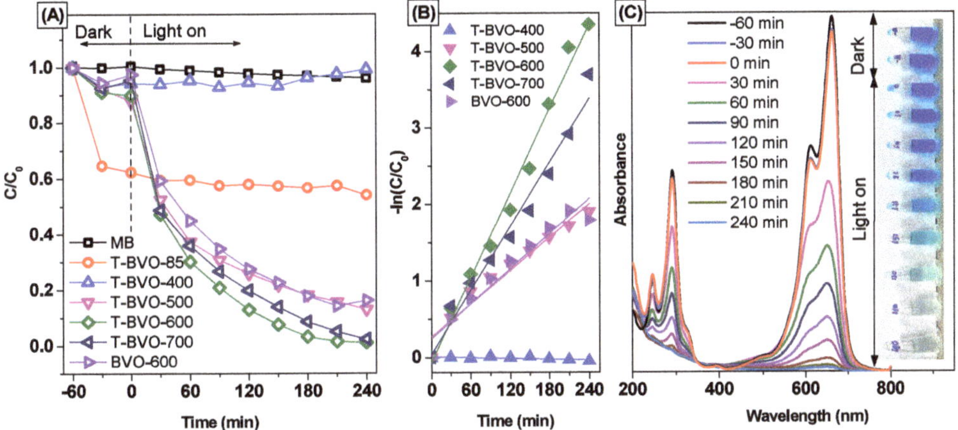

Figure 8. Photocatalytic degradation of methylene blue (MB) over $BiVO_4$ samples (**A**), plots of $\ln(C_o/C)$ versus irradiation time representing the fit using a pseudo-first-order reaction rate (**B**), and UV-vis absorption spectra of MB solution separated from catalyst suspensions during illumination using T-BVO-600, (**C**). Insert displays a digital photo of photodegradation for MB after different illumination times.

The photocatalytic degradation of MB according to the first kinetics [48], as confirmed by the linearity of $\ln(C_0/C_t)$ according to time (t, min) (shown in Figure 8B) and the reaction rate constants of the samples, are listed in Table 3. The results indicate that MB photolysis occurs very slowly with no catalyst. When using $BiVO_4$ catalysts synthesized using thiourea, the photocatalytic activity of the samples increased as the calcining temperature increased, and the photocatalytic activity reaches a maximum when the calcining temperature is 600 °C (T-BVO- 600). The photocatalytic activity on the sulfate-modified $BiVO_4$ increases in the following order: T-BVO-400, T-BVO-500, T-BVO-700, T-BVO-600 with the rate constant (k) respectively, are 1.881×10^{-7} min^{-1}, 7.240×10^{-3} min^{-1}, 13.90×10^{-3} min^{-1}, and 18.37×10^{-3} min^{-1}. The rate constant of BVO-600 is 7.620×10^{-3} min^{-1}, smaller than that of the T-BVO-600 by about 0.548 times.

Table 3. The bandgap energy of the samples and the rate constants (k) values of the samples for MB degradation.

Sample No.	Preparation Condition (°C)	Bandgap [c] (E_g) eV	First Kinetics			
			$K \times 10^{-3}$ (min^{-1})	R^2	$t_{1/2}$ (min)	t_{90} (min)
T-BVO-85	85	–	4.313×10^{-4}	0.824	1.607×10^6	5.339×10^6
T-BVO-400	400	2.230	1.881×10^{-4}	0.506	3.685×10^6	12.241×10^6
T-BVO-500	500	2.112	7.240	0.955	95.738	318.037
T-BVO-600	600	2.288	18.370	0.989	37.733	125.345
T-BVO-700	700	2.277	13.900	0.977	49.867	165.654
BVO-600	600 (without thiourea)	2.246	7.620	0.945	90.964	302.176

[c] Data obtained by UV-Vis-DRS data; [d] Data obtained by the relationship between ln (C_0/C) and irradiation time t (min)

Figure 8C shows the change in the UV-vis absorption spectrum of MB over time in the presence of T-BVO-600. As the lighting time increases, the maximum absorption peak of MB at 664 nm decreases. The decrease in MB concentration was also observed through the dark blue of the MB solution, which began to fade, and the blue color was almost completely lost when the lighting time increased to 240 min. In addition, there is no increase in the absorption peak in the UV region of MB during irradiation, suggesting that most MB has completely decomposed. The high photocatalytic activity of T-BVO-600 indicates that it can be widely used in the treatment of wastewater containing organic dyes.

3.4. *Investigation of the Mechanisms of Dye Degradation*

The photodegradation mechanism of MB by T-BVO-600 and BVO-600 has been investigated via an indirect chemical probe method, using chemical agents to capture the active species produced during the early stages of photocatalysis. During photocatalytic oxidation, organic compounds (especially compounds containing double bonds) are attacked by active species, including holes (h^+), hydroxyl radicals ($HO^•$), and superoxide anion radical (O_2^-). According to previous studies, BQ, EDTA, and TBA were agents that capture O_2^-, h^+, and $HO^•$, respectively [49–51]. As shown in Figure 9A, the MB degradation effect of T-BVO-600 was only slightly reduced by the addition of EDTA and the decomposition efficiency decreased as TBA was added to the photocatalytic system. Meanwhile, the photocatalytic activity of MB degradation did not change significantly when EDTA and TBA were added to the reaction system (Figure 9B). These results indicate that h^+ and $HO^•$ are the major species of MB decomposition under T-BVO-600/visible light system. However, there was a slight increase in the MB degradation effect observed after BQ was added to the T-BVO-600/visible light system. This was also observed in the BVO-600/visible light system. This result is due to the increasing separation efficiency of electron-hole pairs through immediately e^- captured by BQ. The mechanism of the photocatalytic activity under visible light irradiation in the samples used in this study is described in Figure 10. The reaction mechanism can be proposed as follows:

$$
\begin{gathered}
\text{Sulfate-modified } BiVO_4 + h\nu \text{ (visible light)} \rightarrow \text{Sulfate-modified } BiVO_4\ (e^-(CB) + h^+(VB)) \\
H_2O(ads) + h^+(VB) \rightarrow OH^•(ads) + H^+(ads) \\
O_2 + e^-(CB) \rightarrow O_2^-(ads) \\
O_2^-(ads) + H^+ \rightarrow HOO^•(ads) \\
2HOO\ (ads) \rightarrow H_2O_2(ads) + O_2 \\
H_2O_2\ (ads) \rightarrow 2OH^-(ads) \\
MB + OH^- \rightarrow \text{dye intermediates} \rightarrow CO_2 + H_2O
\end{gathered}
\quad (2)
$$

$$MB + h^+(VB) \rightarrow \text{dye intermediates} \rightarrow CO_2 + H_2O$$

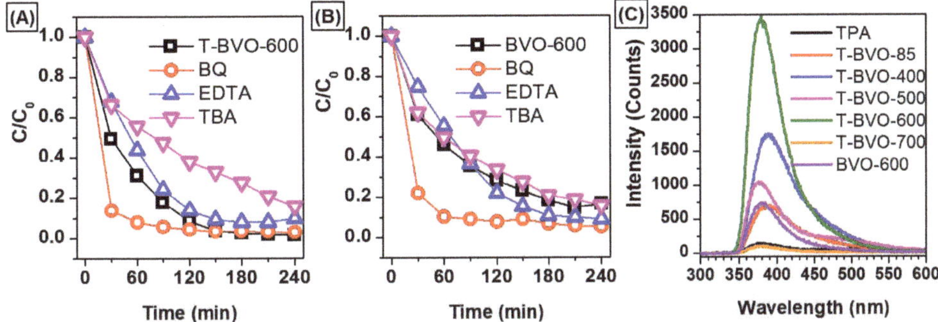

Figure 9. Trapping experiments of photocatalytic degradation of MB over (**A**) T-BVO-600, (**B**) BVO-600 samples (BQ: 1,4-benzoquinone, EDTA: ethylenediaminetetraacetic acid disodium and TBA: *tert*-butanol) and (**C**) PL spectra of terephthalic acid (λ_{ex} = 315 nm) in the presence of $BiVO_4$ samples.

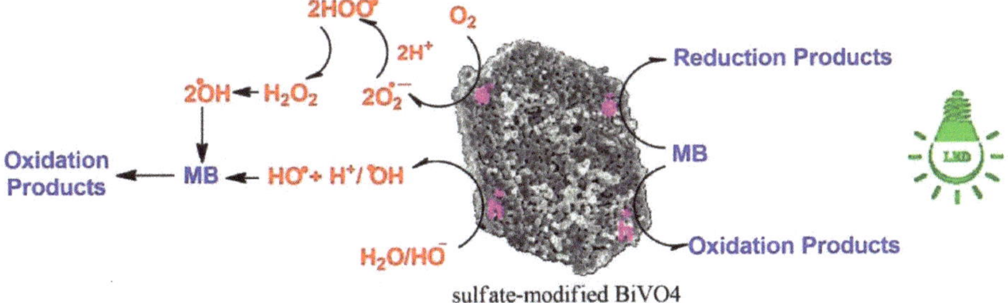

Figure 10. Illustrative representation of direct mechanism for MB photodegradation process.

In addition, the formation of hydroxyl radicals ($HO^\bullet$) on the surface of the $BiVO_4$ catalyst was detected by photoluminescence (PL) technique using terephthalic acid as the sensor molecule. Terephthalic acid immediately reacts with the $HO^\bullet$ radicals to form 2-hydroxyterephthalic acid (HTA) with strong photoluminescence. Figure 9C shows the change in peak intensity of the reaction solution after 240 minutes of irradiation with the presence of $BiVO_4$ samples under different synthesis conditions. In Figure 9C, the fluorescence signal is recorded with very low intensity in the wavelength range from 370 to 600 nm when no catalyst is used. However, in the presence of catalysts, strong fluorescence intensity was observed at 380 nm. The PL signal of the T-BVO-600 sample was higher than those of the other samples, indicating that the formation of $HO^\bullet$ on this sample was highest and correlated with the high photodegradation of MB.

3.5. Reusability and Stability

To be an effective catalyst in practical applications, the reusability of the catalyst is a critical factor. Here, the reusability of the T-BVO-600 was tested five times. At each time, the reaction solution was withdrawn over time. The catalyst was separated by centrifugation and then collected and purified by washing (three times with ethanol and one time with distilled water) for the next experiment. The results are shown in Figure 11A. It can be seen that the photocatalytic activity of the material decreased gradually as the cycle was repeated. The removal efficiencies of MB at each cycle were 98.65%, 97.31%, 95.98%, 90.77%, and 83.11%, respectively. This result is due to the reduction of the catalyst in the purification process since micro-sized T-BVO-600 plates can adhere to the centrifuge tube causing sample loss during washings. Compared with the first cycle, the removal efficiencies of MB decreased slightly at the second and third times, and significantly decreased at the fourth and fifth times. In addition, the crystalline structure of the materials was also tested by XRD (Figure 11B). The XRD pattern of T-BVO-600 after five times of use still exhibits characteristic diffraction peaks as in monoclinic T-BVO-600 at a 2θ angle by 18.5°, 28.9°, 35°, and 47°. Also, the crystalline surface morphology before and after the reaction of T-BVO-600 is shown in Figure 11C. According to the SEM image, there is no clear difference in surface morphology and crystal structure. The results of the above analysis show that the crystal structure, as well as the morphology of the material, does not change after the photocatalytic reaction.

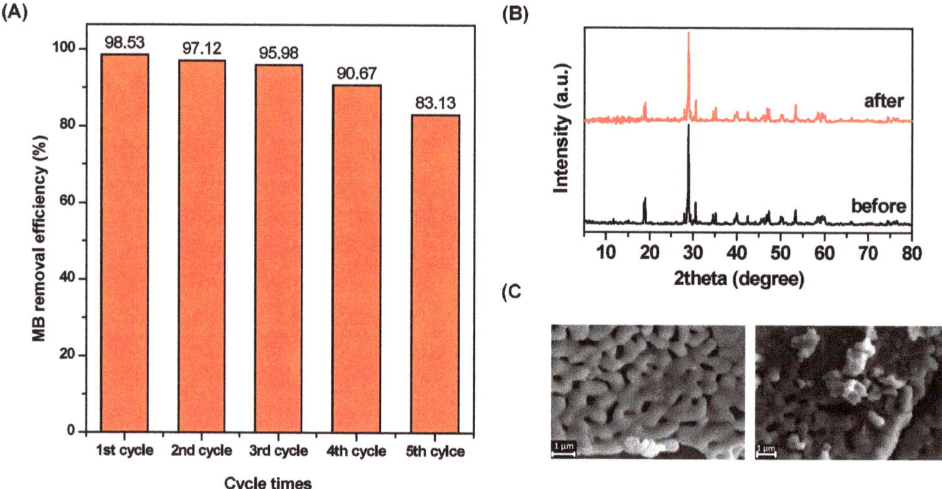

Figure 11. Photo-stability tests over T-BVO-600 sample for the cycling photodegradation of MB (**A**), XRD patterns (**B**), and SEM micrograph (**C**) of T-BVO-600 sample before and after the photo-stability tests.

4. Conclusions

The sulfate-modified $BiVO_4$ photocatalytic material with the high photocatalytic degradation efficiency of MB was successfully synthesized by a sol-gel method. The results indicate that the heat treatment exerted an important influence on the crystal phase, morphology, and crystallinity of $BiVO_4$ when the $BiVO_4$ was synthesized in the presence of thiourea. The thiourea also significantly affected the control of crystal formation and crystal phase of $BiVO_4$ with and without the presence of thiourea and calcined at 600 °C. The as-prepared T-BVO-600 exhibited the highest degradation of MB, in which 98.53% removal of MB was achieved within 240 min. The T-BVO-600 exhibited good recyclability for MB removal, removal of MB was above 83% after five cycles. The T-BVO-600 with the features of high efficiency and good recycling ability is a promising photocatalyst for water purification.

Author Contributions: Methodology, S.T.D. and T.-D.N.; formal analysis, L.G.B.; data curation, Q.T.P.B. and L.G.B.; writing—original draft preparation, V.H.N. and V.-D.D.; writing—review and editing, D.-V.N.V., K.T.L., L.G.B., T.V.N. and T.D.N.; visualization, K.T.L.; supervision, T.-D.N., Q.T.P.B. and T.D.N.

Funding: This research is funded by Vietnam National Foundation for Science and Technology Development (NAFOSTED) under grant number 104.05-2017.315.

Conflicts of Interest: The authors declare no conflict of interest.

References

1. Tokunaga, S.; Kato, H.; Kudo, A. Selective preparation of monoclinic and tetragonal BiVO4 with scheelite structure and their photocatalytic properties. *Chem. Mater.* **2001**, *13*, 4624–4628. [CrossRef]
2. Kudo, A.; Omori, K.; Kato, H. A novel aqueous process for preparation of crystal form-controlled and highly crystalline BiVO4 powder from layered vanadates at room temperature and its photocatalytic and photophysical properties. *J. Am. Chem. Soc.* **1999**, *121*, 11459–11467. [CrossRef]
3. Gan, J.; Rajeeva, B.B.; Wu, Z.; Penley, D.; Liang, C.; Tong, Y.; Zheng, Y. Plasmon-enhanced nanoporous BiVO4 photoanodes for efficient photoelectrochemical water oxidation. *Nanotechnology* **2016**, *27*, 235401. [CrossRef] [PubMed]
4. Zhao, Z.; Dai, H.; Deng, J.; Liu, Y.; Au, C.T. Enhanced visible-light photocatalytic activities of porous olive-shaped sulfur-doped BiVO4-supported cobalt oxides. *Solid State Sci.* **2013**, *18*, 98–104. [CrossRef]
5. García-Pérez, U.M.; Sepúlveda-Guzmán, S.; Martínez-De La Cruz, A. Nanostructured BiVO 4 photocatalysts synthesized via a polymer-assisted coprecipitation method and their photocatalytic properties under visible-light irradiation. *Solid State Sci.* **2012**, *14*, 293–298. [CrossRef]
6. Nguyen, D.T.; Hong, S.S. Synthesis of $BiVO_4$ nanoparticles using microwave process and their photocatalytic activity under visible light irradiation. *J. Nanosci. Nanotechnol.* **2017**, *17*, 2690–2694. [CrossRef] [PubMed]
7. Zhong, D.K.; Choi, S.; Gamelin, D.R. Near-complete suppression of surface recombination in solar photoelectrolysis by "co-Pi" catalyst-modified W: BiVO4. *J. Am. Chem. Soc.* **2011**, *133*, 18370–18377. [CrossRef] [PubMed]
8. Li, R.; Zhang, F.; Wang, D.; Yang, J.; Li, M.; Zhu, J.; Zhou, X.; Han, H.; Li, C. Spatial separation of photogenerated electrons and holes among {010} and {110} crystal facets of BiVO4. *Nat. Commun.* **2013**, *4*, 1432. [CrossRef]
9. Zhang, Y.; Gong, H.; Zhang, Y.; Liu, K.; Cao, H.; Yan, H.; Zhu, J. The controllable synthesis of octadecahedral BiVO4 with exposed {111} Facets. *Eur. J. Inorg. Chem.* **2017**, *2017*, 2990–2997. [CrossRef]
10. Cao, J.; Zhou, C.; Lin, H.; Xu, B.; Chen, S. Surface modification of m-BiVO 4 with wide band-gap semiconductor BiOCl to largely improve the visible light induced photocatalytic activity. *Appl. Surf. Sci.* **2013**, *284*, 263–269. [CrossRef]
11. Lopes, O.F.; Carvalho, K.T.G.; Nogueira, A.E.; Avansi, W.; Ribeiro, C. Controlled synthesis of BiVO4 photocatalysts: Evidence of the role of heterojunctions in their catalytic performance driven by visible-light. *Appl. Catal. B Environ.* **2016**, *188*, 87–97. [CrossRef]
12. Han, M.; Sun, T.; Tan, P.Y.; Chen, X.; Tan, O.K.; Tse, M.S. M-BiVO@4γBi2O3 core-shell p-n heterogeneous nanostructure for enhanced visible-light photocatalytic performance. *RSC Adv.* **2013**, *3*, 24964–24970. [CrossRef]
13. Hong, S.J.; Lee, S.; Jang, J.S.; Lee, J.S. Heterojunction BiVO4/WO3 electrodes for enhanced photoactivity of water oxidation. *Energy Environ. Sci.* **2011**, *4*, 1781–1787. [CrossRef]
14. Zhao, W.; Wang, Y.; Yang, Y.; Tang, J.; Yang, Y. Carbon spheres supported visible-light-driven CuO-$BiVO_4$ heterojunction: Preparation, characterization, and photocatalytic properties. *Appl. Catal. B Environ.* **2012**, *115–116*, 90–99. [CrossRef]
15. Nguyen, H.V.; Thuan, T.V.; Do, S.T.; Nguyen, D.T.; Vo, D.V.N.; Bach, L.G. High Photocatalytic Activity of Oliver-Like $BiVO_4$ for Rhodamine B Degradation under Visible Light Irradiation. *Appli. Mech. Mater.* **2018**, *876*, 52–56. [CrossRef]
16. Jiang, H.Q.; Endo, H.; Natori, H.; Nagai, M.; Kobayashi, K. Fabrication and photoactivities of spherical-shaped BiVO4 photocatalysts through solution combustion synthesis method. *J. Eur. Ceram. Soc.* **2008**, *28*, 2955–2962. [CrossRef]
17. Guo, M.; Wang, Y.; He, Q.; Wang, W.; Wang, W.; Fu, Z.; Wang, H. Enhanced photocatalytic activity of S-doped BiVO 4 photocatalysts. *RSC Adv.* **2015**, *5*, 58633–58639. [CrossRef]
18. Yin, C.; Zhu, S.; Chen, Z.; Zhang, W.; Gu, J.; Zhang, D. One step fabrication of C-doped BiVO4 with hierarchical structures for a high-performance photocatalyst under visible light irradiation. *J. Mater. Chem. A* **2013**, *1*, 8367–8378. [CrossRef]
19. Tan, G.; Zhang, L.; Ren, H.; Huang, J.; Yang, W.; Xia, A. Microwave hydrothermal synthesis of N-doped BiVO4 nanoplates with exposed (040) facets and enhanced visible-light photocatalytic properties. *Ceram. Int.* **2014**, *40*, 9541–9547. [CrossRef]

20. Jo, W.J.; Jang, J.W.; Kong, K.J.; Kang, H.J.; Kim, J.Y.; Jun, H.; Parmar, K.P.S.; Lee, J.S. Phosphate doping into monoclinic BiVO 4 for enhanced photoelectrochemical water oxidation activity. *Angew. Chemie Int. Ed.* **2012**, *51*, 3147–3151. [CrossRef]
21. Zhao, Z.; Dai, H.; Deng, J.; Liu, Y.; Au, C.T. Effect of sulfur doping on the photocatalytic performance of BiVO4 under visible light illumination. *Cuihua Xuebao Chinese J. Catal.* **2013**, *34*, 1617–1626. [CrossRef]
22. Geng, Y.; Zhang, P.; Kuang, S. Fabrication and enhanced visible-light photocatalytic activities of $BiVO_4/Bi_2WO_6$ composites. *RSC Adv.* **2014**, *4*, 46054–46059. [CrossRef]
23. Gotić, M.; Musić, S.; Ivanda, M.; Šoufek, M.; Popović, S. Synthesis and characterisation of bismuth(III) vanadate. *J. Mol. Struct.* **2005**, *744–747*, 535–540. [CrossRef]
24. Zulkifili, A.; Fujiki, A.; Kimijima, S. Flower-like BiVO4 Microspheres and Their Visible Light-Driven Photocatalytic Activity. *Appl. Sci.* **2018**, *8*, 216. [CrossRef]
25. Nam, S.H.; Kim, T.K.; Boo, J.H. Physical property and photo-catalytic activity of sulfur doped TiO 2 catalysts responding to visible light. *Catal. Today* **2012**, *185*, 259–262. [CrossRef]
26. Phu, N.D.; Hoang, L.H.; Vu, P.K.; Kong, M.H.; Chen, X.B.; Wen, H.C.; Chou, W.C. Control of crystal phase of BIVO4nanoparticles synthesized by microwave assisted method. *J. Mater. Sci. Mater. Electron.* **2016**, *27*, 6452–6456. [CrossRef]
27. Hardcastle, F.D.; Wachs, I.E. Determination of vanadium-oxygen bond distances and bond orders by Raman spectroscopy. *J. Phys. Chem.* **2005**, *95*, 5031–5041. [CrossRef]
28. Nguyen, D.T.; Hong, S.S. The effect of solvent on the synthesis of BiVO4 using solvothermal method and their photocatalytic activity under visible light irradiation. *Top. Catal.* **2017**, *60*, 782–788. [CrossRef]
29. Chen, G.; Ford, T.E.; Clayton, C.R. Interaction of sulfate-reducing bacteria with molybdenum dissolved from sputter-deposited molybdenum thin films and pure molybdenum powder. *J. Colloid Interface Sci.* **1998**, *204*, 237–246. [CrossRef]
30. Strohmeier, B.R.; Hercules, D.M. Surface spectroscopic characterization of manganese/aluminum oxide catalysts. *J. Phys. Chem.* **2005**, *88*, 4922–4929. [CrossRef]
31. Wang, M.; Niu, C.; Liu, J.; Wang, Q.; Yang, C.; Zheng, H. Effective visible light-active nitrogen and samarium co-doped $BiVO_4$ for the degradation of organic pollutants. *J. Alloys Compd.* **2015**, *648*, 1109–1115. [CrossRef]
32. Zhou, J.; Tian, G.; Chen, Y.; Shi, Y.; Tian, C.; Pan, K.; Fu, H. Growth rate controlled synthesis of hierarchical Bi2S 3/In2S3 core/shell microspheres with enhanced photocatalytic activity. *Sci. Rep.* **2014**, *4*, 4027. [CrossRef] [PubMed]
33. Antony, R.P.; Baikie, T.; Chiam, S.Y.; Ren, Y.; Prabhakar, R.R.; Batabyal, S.K.; Loo, S.C.J.; Barber, J.; Wong, L.H. Catalytic effect of Bi5+ in enhanced solar water splitting of tetragonal BiV0.8Mo0.2O4. *Appl. Catal. A Gen.* **2016**, *526*, 21–27. [CrossRef]
34. Saison, T.; Chemin, N.; Chanéac, C.; Durupthy, O.; Mariey, L.; Maugé, F.; Brezová, V.; Jolivet, J.P. New insights into BiVO $_4$ properties as visible light photocatalyst. *J. Phys. Chem. C* **2015**, *119*, 12967–12977. [CrossRef]
35. Vila, M.; Díaz-Guerra, C.; Lorenz, K.; Piqueras, J.; Alves, E.; Nappini, S.; Magnano, E. Structural and luminescence properties of Eu and Er implanted Bi2O3nanowires for optoelectronic applications. *J. Mater. Chem. C* **2013**, *1*, 7920–7929. [CrossRef]
36. Yousif, A.; Jafer, R.M.; Som, S.; Duvenhage, M.M.; Coetsee, E.; Swart, H.C. Ultra-broadband luminescent from a Bi doped CaO matrix. *RSC Adv.* **2015**, *5*, 54115–54122. [CrossRef]
37. Usai, S.; Obregón, S.; Becerro, A.I.; Colón, G. Monoclinic-tetragonal heterostructured BiVO4 by yttrium doping with improved photocatalytic activity. *J. Phys. Chem. C* **2013**, *117*, 24479–24484. [CrossRef]
38. Zhang, G.; Zhang, J.; Zhang, M.; Wang, X. Polycondensation of thiourea into carbon nitride semiconductors as visible light photocatalysts. *J. Mater. Chem.* **2012**, *22*, 8083–8091. [CrossRef]
39. Hong, J.; Xia, X.; Wang, Y.; Xu, R. Mesoporous carbon nitride with in situ sulfur doping for enhanced photocatalytic hydrogen evolution from water under visible light. *J. Mater. Chem.* **2012**, *22*, 15006–15012. [CrossRef]
40. Jourshabani, M.; Shariatinia, Z.; Badiei, A. Controllable synthesis of mesoporous sulfur-doped carbon nitride materials for enhanced visible light photocatalytic degradation. *Langmuir* **2017**, *33*, 7062–7078. [CrossRef]
41. Wang, Y.T.; Wang, N.; Chen, M.L.; Yang, T.; Wang, J.H. One step preparation of proton-functionalized photoluminescent graphitic carbon nitride and its sensing applications. *RSC Adv.* **2016**, *6*, 98893–98898. [CrossRef]

42. Chen, F.; Yang, Q.; Wang, Y.; Zhao, J.; Wang, D.; Li, X.; Guo, Z.; Wang, H.; Deng, Y.; Niu, C.; et al. Novel ternary heterojunction photcocatalyst of Ag nanoparticles and g-C3N4 nanosheets co-modified BiVO4 for wider spectrum visible-light photocatalytic degradation of refractory pollutant. *Appli. Catalys. B Environ.* **2017**, *205*, 133–147. [CrossRef]
43. Wang, A.; Lee, C.; Bian, H.; Li, Z.; Zhan, Y.; He, J.; Wang, Y.; Lu, J.; Li, Y.Y. Synthesis of g-C3N4/Silica Gels for White-Light-Emitting Devices. *Part. Part. Syst. Charact.* **2017**, *34*, 1600258. [CrossRef]
44. Oh, W.D.; Lok, L.W.; Veksha, A.; Giannis, A.; Lim, T.T. Enhanced photocatalytic degradation of bisphenol A with Ag-decorated S-doped g-C 3 N 4 under solar irradiation: Performance and mechanistic studies. *Chem. Eng. J.* **2018**, *333*, 739–749. [CrossRef]
45. Bellardita, M.; García-López, E.I.; Marcì, G.; Krivtsov, I.; García, J.R.; Palmisano, L. Selective photocatalytic oxidation of aromatic alcohols in water by using P-doped g-C3N4. *Appl. Catal. B Environ.* **2018**, *220*, 222–233. [CrossRef]
46. Sleight, A.W.; Chen, H.Y.; Ferretti, A.; Cox, D.E. Crystal growth and structure of BiVO4. *Mater. Res. Bull.* **1979**, *14*, 1571–1581. [CrossRef]
47. Zhang, L.; Dai, Z.; Zheng, G.; Yao, Z.; Mu, J. Superior visible light photocatalytic performance of reticular $BiVO_4$ synthesized: Via a modified sol-gel method. *RSC Adv.* **2018**, *8*, 10654–10664. [CrossRef]
48. Nguyen, D.T.; Hong, S.S. Synthesis of Metal Ion-Doped TiO 2 Nanoparticles Using Two-Phase Method and Their Photocatalytic Activity Under Visible Light Irradiation. *J. Nanosci. Nanotechnol.* **2015**, *16*, 1911–1915. [CrossRef]
49. Pingmuang, K.; Chen, J.; Kangwansupamonkon, W.; Wallace, G.G.; Phanichphant, S.; Nattestad, A. Composite photocatalysts containing BiVO 4 for degradation of cationic dyes. *Sci. Rep.* **2017**, *7*, 8929. [CrossRef]
50. Huang, H.; Liu, L.; Zhang, Y.; Tian, N. Novel BiIO4/BiVO4 composite photocatalyst with highly improved visible-light-induced photocatalytic performance for rhodamine B degradation and photocurrent generation. *RSC Adv.* **2015**, *5*, 1161–1167. [CrossRef]
51. Liu, Y.; Guo, H.; Zhang, Y.; Tang, W.; Cheng, X.; Liu, H. Activation of peroxymonosulfate by BiVO4 under visible light for degradation of Rhodamine B. *Chem. Phys. Lett.* **2016**, *653*, 101–107. [CrossRef]

Article

Microwave-Assisted Synthesis of High-Energy Faceted TiO_2 Nanocrystals Derived from Exfoliated Porous Metatitanic Acid Nanosheets with Improved Photocatalytic and Photovoltaic Performance

Yi-en Du [1,2,3,*], Xianjun Niu [1], Wanxi Li [1], Jing An [1], Yufang Liu [1], Yongqiang Chen [1,*], Pengfei Wang [4], Xiaojing Yang [2,*] and Qi Feng [3]

1. School of Chemistry & Chemical Engineering, Jinzhong University, Jinzhong 030619, China; xjniu1984@163.com (X.N.); liwanxi1986@163.com (W.L.); ann.jean@163.com (J.A.); sxyclyf@sina.com (Y.L.)
2. Beijing Key Laboratory of Energy Conversion and Storage Materials, College of Chemistry, Beijing Normal University, Beijing 100875, China
3. Department of Advanced Materials Science, Faculty of Engineering, Kagawa University, 2217-20 Hayashi-cho, Takamatsu-shi 761-0396, Japan; feng@eng.kagawa-u.ac.jp
4. State Key Laboratory of Coal Conversion, Institute of Coal Chemistry, Chinese Academy of Sciences, Taiyuan 030001, China; wangpf@sxicc.ac.cn

* Correspondence: duye@jzxy.edu.cn (Y.-e.D.); chenyongqiang@jzxy.edu.cn (Y.C.); yang.xiaojing@bnu.edu.cn (X.Y.)

Received: 23 September 2019; Accepted: 30 October 2019; Published: 4 November 2019

Abstract: A facile one-pot microwave-assisted hydrothermal synthesis of rutile TiO_2 quadrangular prisms with dominant {110} facets, anatase TiO_2 nanorods and square nanoprisms with co-exposed {101}/[111] facets, anatase TiO_2 nanorhombuses with co-exposed {101}/{010} facets, and anatase TiO_2 nanospindles with dominant {010} facets were reported through the use of exfoliated porous metatitanic acid nanosheets as a precursor. The nanostructures and the formation reaction mechanism of the obtained rutile and anatase TiO_2 nanocrystals from the delaminated nanosheets were investigated. The transformation from the exfoliated metatitanic nanosheets with distorted hexagonal cavities to TiO_2 nanocrystals involved a dissolution reaction of the nanosheets, nucleation of the primary $[TiO_6]^{8-}$ monomers, and the growth of rutile-type and anatase-type TiO_2 nuclei during the microwave-assisted hydrothermal reaction. In addition, the photocatalytic activities of the as-prepared anatase nanocrystals were evaluated through the photocatalytic degradation of typical carcinogenic and mutagenic methyl orange (MO) under UV-light irradiation at a normal temperature and pressure. Furthermore, the dye-sensitized solar cell (DSSC) performance of the synthesized anatase TiO_2 nanocrystals with various morphologies and crystal facets was also characterized. The {101}/[111]-faceted pH2.5-T175 nanocrystal showed the highest photocatalytic and photovoltaic performance compared to the other TiO_2 samples, which could be attributed mainly to its minimum particle size and maximum specific surface area.

Keywords: anatase TiO_2 nanocrystals; high-energy facets; photocatalytic activity; photovoltaic performance

1. Introduction

Inorganic nanocrystals with tailored morphologies and specific facets have received much attention in the past decade due to their many intrinsic shape-dependent properties and excellent technological applications in energy and environmental fields [1,2]. As one of the most studied semiconductor metal oxides, titanium dioxide (TiO_2) has been extensively utilized in photovoltaic cells, dye-sensitized soar cells, photocatalysis, the photocatalytic degradation of organic pollutants, Li-ion batteries, etc., due to its good photocatalytic activity, high biological and chemical inertness, long-term chemical and thermal stability, low price, nontoxicity, and excellent degradation capability [3–7]. Compared to the other two crystalline phases (rutile (tetragonal, space group $P4_2/mnm$) and brookite (orthorhombic, space group *Pbca*)) of TiO_2, anatase (tetragonal, space group $I4_1/amd$) is widely accepted for its higher photocatalytic activity and for having the most superior dye-sensitized solar cell performance [8,9]. However, the photocatalytic and photovoltaic performances of anatase TiO_2 nanocrystals still need to be further improved for their practical application and commercialization. To achieve this purpose, anatase TiO_2 nanocrystals with high crystallinity, well-defined morphology, good architecture, a small crystallite size, a large specific surface area, and proper composition are desirable to improve photocatalytic and photovoltaic performance [10,11]. Besides, anatase TiO_2 nanocrystals with well-defined morphologies and tailored high-energy crystal facets for photocatalysis and photoanodes also have been proven to be an effective approach to significantly improve photocatalytic and photovoltaic performance in recent years [12]. Surface scientists have demonstrated that the average surface energies of anatase TiO_2 increase in the order of {111} facet (1.61 J/m^2) > {110} facet (1.09 J/m^2) > {001} facet (0.90 J/m^2) > {010}/{100} facet (0.53 J/m^2) > {101} facet (0.44 J/m^2) [13,14]. Generally speaking, facets with high surface energies diminish rapidly during the crystal growth process for the minimization of the total surface free energy under equilibrium conditions, which results in the total exposed surface of the anatase TiO_2 crystals being primarily controlled by {101} facets with poor reactivity and thermodynamic stability [15]. As a result, the morphology of the anatase TiO_2 crystals is a slightly truncated tetragonal bipyramid with eight equivalent {101} facets and two equivalent {001} facets based on Wulff construction [8]. Therefore, it is imperative to reasonably design and synthesize various morphologies of TiO_2 nanocrystals with exposed high-energy surfaces to enhance photocatalytic activity and optimize dye-sensitized solar cell (DSSC) performance. Important progress was made by Wen and coworkers, who reported the synthesis of nanosized anatase TiO_2 crystallites with their dominant {010} facets exposed [16]. Subsequently, Yang and coworkers reported the synthesis of microsized anatase TiO_2 crystals (with around 47% and 89% {001} facets being exposed) by using TiF_4 as the raw material and hydrogen fluoride (HF) as the capping agent [15,17].Following these breakthroughs, more and more research work has focused on the design and control of the synthesis of anatase TiO_2 crystals with varied percentages of high active facets [2,8,18–20]. However, the previously reported methods for synthesizing anatase TiO_2 crystals with high-energy facets often involve the use of toxic and corrosive HF or other F-containing species, which restricts the practical application of TiO_2 nanocrystals [10,12]. For example, Feng and coworkers reported the synthesis of regular foursquare anatase TiO_2 mesocrystal sheets with dominant {001} facets by using $TiCl_4$ as the titanium source and HF as the capping agent [21]. Illa and coworkers reported on hollow spheres assembled by high-energy {001}-faceted mesoporous cuboid anatase nanocrystals, which showed higher photocatalytic and photoelectrochemical activity than did those of commercial P25 [22]. Khalil and coworkers reported the synthesis of Au–TiO_2 heterostructures with exposed {001} facets, which showed enhanced photocatalytic performance [23]. Ye and coworkers reported the synthesis of anatase TiO_2 crystals with co-exposed {101}/{001} and {010}/{001} facets by using $(NH_4)_2TiF_6$ as the titanium and fluorine species [24]. Xu and coworkers reported the synthesis of square-shaped plate TiO_2 single crystals with well-defined {111} facets that were exposed using both HF and ammonia as the capping reagents [14]. Amoli and coworkers reported the synthesis of anatase TiO_2 nanocubes and nanoparallelepipeds with well-defined {111} facets exposed by using oleylamine as the morphology controlling agent and NH_3/HF as the stabilizing agent [25]. Recently, we also synthesized truncated tetragonal bipyramid anatase TiO_2 nanocrystals with co-exposed {001}, {010}, and {101} facets and tetragonal cuboid anatase TiO_2 nanocrystals with co-exposed [111]- and {101} facets by using the H^+ form of tetratitanate $H_2Ti_4O_9$ as the precursor and HF as the capping agent [26]. Very recently, using porous metatitanic acid H_2TiO_3 as the precursor and HF as the capping agent, cuboid-like anatase TiO_2 nanocrystals with co-exposed {101}/[111] facets and irregular anatase TiO_2 nanocrystals with co-exposed {101}/{010} facets were also synthesized [27].

The microwave-assisted hydrothermal method is unique in its ability to be scaledup without suffering thermal gradient effects, providing a potentially industrially important improvement over convective methods in the synthesis of nanocrystals [28]. The advantage of using the microwave-assisted hydrothermal method over a conventional heating method lies in its homogenous selective heating and scalability, its shorter reaction time, its rapid formation rate, and the higher purity and better crystallinity of the produced products (as the yields obtained are identical for both heating methods) [29–31]. For example, under microwave-assisted solvothermal conditions, Yoon and coworkers reported the synthesis of various polymorphs of TiO_2 nanocrystals with different morphologies by using $TiCl_4$ or $TiCl_3$ as precursors [29]. Nunes and coworkers reported the production and photocatalytic activity of brookite/rutile mixed-phase TiO_2 nanorod spheres and nanorod arrays grown on polyethylene terephthalate substrates, which displayed remarkable degradability performance and reusability under ultraviolet and solar radiation for the degradation of rhodamine B [32].

In the present work, we report a facile synthesis of shape-tunable TiO_2 nanocrystals with dominant {110}, {010}, and [111] facets through a simple one-step fluorine-free microwave-assisted hydrothermal method. The synthetic process demonstrates that the pH value of the exfoliated nanosheet solution plays a key role in morphology evolution and facet exposure under hydrothermal conditions. The possible transformation reaction mechanism of the TiO_2 nanocrystals with various morphologies and exposure facets and the photocatalytic and photovoltaic performance were investigated. Compared to P25 TiO_2 nanocrystals, the pH2.5-T175 nanocrystals with dominant [111] facets showed higher photocatalytic and photovoltaic performance.

2. Materials and Methods

2.1. Preparation of Layered $Li_2TiO_3$and H_2TiO_3

Lithium metatitanate (Li_2TiO_3) was formed through a conventional solid-state reaction [33]. Lithium carbonate (Li_2CO_3, 99.0%) and titanium dioxide (TiO_2, 98.5%) were taken in a molar ratio of 1.05:3 and milled in a horizontal ball mill at 50 rpm for 6 h, and then the mixture was calcined at 850 °C for 24 h in an air atmosphere. Then, 10.0 g of the obtained earthy gray Li_2TiO_3 powder was acid-treated three times with a 1.0 mol·L^{-1} HCl solution (1 L) at room temperature for 24 h under magnetic stirring to remove Li^+ completely from the metatitanate to obtain a white metatitanic acid H_2TiO_3 sample.

2.2. Exfoliation of H_2TiO_3 into a Nanosheet Colloidal Solution

The obtained metatitanic acid H_2TiO_3 sample (5.0 g) was hydrothermally treated at 100 °C for 24 h with magnetic stirring in a 12.5% tetramethylammonium hydroxide solution (50 mL, $(CH_3)_4NOH$, hereafter TMAOH) to intercalate TMA^+ ions into the interlayer of the metatitanic acid H_2TiO_3 and to obtain a TMA^+-intercalated layered H_2TiO_3 sample. The intercalation compound obtained was delaminated into H_2TiO_3 nanosheet by dispersed it in 450 mL of deionized water with stirring at room temperature for 3 days.

2.3. Microwave Hydrothermal Synthesis of TiO_2 Nanocrystals

TiO_2 nanocrystals were prepared through microwave-assisted hydrothermal treatment of the above exfoliated H_2TiO_3 nanosheet colloidal solution (40 mL) at a desired temperature for 2 h. After the microwave hydrothermal process, the resultant sample was filtered, washed multiple times with deionized water, and then dried using a freeze-drying machine. Here, the obtained TiO_2 sample is specified as pHx-Ty, where x and y are the desired pH value of the nanosheet colloidal solution and the desired temperature of the microwave-assisted hydrothermal treatment, respectively.

2.4. Photocatalytic Activity

The photocatalytic activity of the resulting TiO_2 nanocrystals was assessed through the photodecomposition of carcinogenic and mutagenic methyl orange (MO), one of the most stable azo dyes [34,35]. Here, 200 mL of 2.31×10^{-5} mol/L MO solution was added to a 250mL quartz glass beaker containing 75 mg of as-prepared TiO_2 nanocrystals and then sonicated for 5 min for the complete dispersion of the TiO_2 nanocrystals followed by continuous stirring for 60 min in the dark to prepare the suspension solution of TiO_2 nanocrystals. The above suspension solution was placed in the dark for 2 days to ensure that the adsorption–desorption phenomenon reached equilibrium. After that, photocatalytic degradation was carried out by stirring the above suspensions under ultraviolet irradiation using a 175-W low-pressure mercury lamp (λ = 365 nm). The distance between the mercury lamp and the suspension solution was 30 cm. At intervals of 15 min, 4 mL of suspension solution was taken out from the beaker and centrifuged at 2000 rpm for 5 min to separate the TiO_2 nanocrystals, and the concentration of MO supernatant solution was analyzed by using a TU-1901 spectrophotometer (Beijing Purkinje General Instrument Co. Ltd. Beijing, China). The photodegradation efficiency (η) of the as-prepared TiO_2 nanocrystals (as a percentage) was calculated according to the equation [36]:

$$\eta = \frac{c_0 - c_t}{c_t} \times 100\%$$

where c_0 and c_t represent the concentration of the MO suspensionin the absence and presence of light irradiation, respectively. For a comparison, Degussa P25-TiO_2 (~87% anatase and ~13% rutile) was also investigated.

2.5. Fabrication of Photoanodes and Dye-Sensitized Solar Cells (DSSCs)

For the preparation of the nanoporous TiO_2 layers with the prepared TiO_2 nanocrystals and the Degussa P25-TiO_2 nanocrystals, a viscous slurry was prepared through the following procedure. The prepared TiO_2 nanocrystals (0.5 g) were dispersed in ethanol (2.5 g) and then mixed with α-terpineol (2.0 g), ethyl-cellulose 10 (1.4 g of 10 wt% solution in ethanol), and ethyl-cellulose 45 (1.1 g of 10 wt% solution in ethanol). Subsequently, the resulting mixture was sonicated for 5 min and then ball-milled for 3 days. After removing the ethanol with a rotary evaporator, an optical paste containing 18% TiO_2, 9% ethyl-cellulose, and 73% α-terpineol was obtained. A nanoporous TiO_2 photoanode was obtained according to the following steps. Fluorine-doped tin oxide (FTO)-conducting glass substrates were cleaned in the order of neutral cleaner, deionized water, and absolute ethanol by ultrasonication for 5 min. Subsequently, the FTO glass substrates (12.5×25.0 mm^2) were dipped into titanium tetraisopropoxide solution (0.1 mol·L^{-1}) for 1 min and then washed with ionized water and absolute ethanol. After being dried, the prepared FTO glass substrates were placed in a muffle furnace, and the reaction temperature was increased to 480 °C at 10 °C/min and was kept at this temperature for 1 h to coat a dense TiO_2 film on the surface of the FTO glass substrates. Afterwards, the prepared TiO_2 paste was coated on the above FTO glass substrates by using the doctor-blade technique, was dried at 100 °C for 15 min, and was sintered in the above muffle furnace at 315 °C for 15 min. This process was repeated several times until the desired thickness of TiO_2 film was obtained. The above FTO glass substrates coated with the TiO_2 pastes were first sintered at 450 °C for 30 min, then treated with 0.1 mol·L^{-1} titanium tetraisopropoxide solution as described above, and finally sintered at 480 °C for 1 h to fabricate TiO_2 thin-film photoanodes. After being cooled to 80 °C, the TiO_2 photoanodes were immersed in N719 dye solution for 24 h at room temperature. An N719-sensitized TiO_2 photoelectrode (working electrode) was coupled with a platinum-sputtered FTO glass substrate (counter-electrode) with an electrolyte solution between the electrodes to assemble a sandwich-type DSSC. The electrolyte of the DSSC consisted of a solution of 0.6 mol·L^{-1} 1-butyl-3-methylimidazolium iodide, 0.03 mol·L^{-1} I_2, 0.1 mol·L^{-1} guanidinium thiocyanate, and 0.5 mol·L^{-1} tert-butylpyridine in a cosolvent of acetonitrile and valeronitrile at a volume ratio of 85:15, which was injected into the cell through capillary forces.

2.6. Characterization

X-ray diffraction (XRD) patterns of the resulting specimens were prepared by a XRD-6100 powder X-ray diffractometer (SHIMADZU, Kyoto, Japan) with Cu Kα radiation. Field-emission scanning electron microscopy (FESEM, HITACHI, Tokyo, Japan) and a transmission electron microscope (TEM, JEOL, Tokyo, Japan) were used to characterize the morphological and structural properties of the TiO_2 nanocrystal specimens. The specific surface area of the specimens was determined through aBrunauer-Emmett-Teller (BET) analysis (Quantachrome Quadra Win Instruments, Ashland, VA, USA) using N_2 as the adsorbate gas at −196 °C. All of the specimens were degassed at 120 °C for 3 h prior to nitrogen adsorption measurements. Photovoltaic DSSC performance and photocurrent–voltage characteristic curves were measured using a Hokuto-Denko BAS100B electrochemical analyzer (Hokuto, Yamanashi-ken, Osaka, Japan). The effective irradiation area of the DSSC was fixed at 0.25 cm^2.

3. Results and Discussion

3.1. Preparation of the Metatitanic Acid Titanate Nanosheet Colloidal Solution

The XRD pattern of the layered Li_2TiO_3 sample obtained at 850 °C for 24 h in an air atmosphere is shown in Figure 1a. The diffraction peaks recorded at 2θ = 18.59°, 20.40°, 36.10°, 43.78°, 47.86°, 57.78°, 63.46°, and 66.90° can be attributed to the reflections of (002), (020), (−131), (−133), (−204), (006), (−206), and (062) crystal planes of layered Li_2TiO_3, which were almost in accordance with the characteristic diffraction peaks of the monoclinic Li_2TiO_3 crystal (*C2/c* space group, JCPDS No. 33-0831). The interlayer spacing was estimated to be 4.47 Å according to the (002) crystal plane of Li_2TiO_3, and the chemical formula ($Li_{2.06}TiO_{3.0}$) (on the basis of the chemical analyses) was close to the theoretical formula of Li_2TiO_3. The corresponding crystal structure of layered Li_2TiO_3 is shown in Figure 1d. The crystal structure of Li_2TiO_3 could be represented as a cubic close packing of three different layers (Li_6 layers, O_6 layers, and Li_2Ti_4 layers) in the *c* axis direction [37]. In the Li_2Ti_4 layer, 1/3 of the 4e sites were occupied by lithium atoms and 2/3 by titanium atoms, so the formula for a well-ordered Li_2TiO_3 could be written as $Li(Li_{1/3}Ti_{2/3})O_2$ [37–39]. That is, the atomic fraction of titanium in the Li_6 layers and Li_2Ti_4 layers was 75% and 25%, respectively [38]. Lithium extraction in the layered Li_2TiO_3 precursor was performed using 1 $mol{\cdot}L^{-1}$ hydrochloric acid. According to the chemical analyses ($H_{1.993}Li_{0.007}TiO_3$), the amount of lithium extracted was about 99.65%. Figure 1b depicts the XRD pattern of H_2TiO_3. The main peak corresponding to (002) of H_2TiO_3 slightly shifted to a higher 2θ angle compared to the peak in Figure 1a, indicating a decrease in the interlayer (from 4.77 to 4.72 Å) through an exchange of the lithium atoms (radius of ~0.076 nm) for hydrogen ions (radius of ~0.0012 nm) [33]. It can be seen that the XRD pattern in Figure 1b is similar to that in Figure 1a, which suggests that the structure did not change much after lithium extraction. The FESEM image of Li_2TiO_3 shows that the irregular particles with smooth surfaces were nonagglomerated, and the average size of the particles was in the range of 0.5–2.0 μm (Figure 2a). The particle shape of H_2TiO_3 (nonagglomerated) was similar to that of the Li_2TiO_3 precursor, except that the smooth surface of the H_2TiO_3 particles became rough and irregular cracks appeared on the surface (Figure 2b,c), implying that the extraction process hardly destroyed the structure of the precursor even if the Li^+ ions in the Li_2Ti_4 intralayers were extracted. The XRD pattern of the TMA^+-intercalated H_2TiO_3 sample (i.e., the dried TMA^+-exfoliated H_2TiO_3 nanosheets) is shown in Figure 1c. Compared to the diffraction peaks of the H_2TiO_3, a set of new diffraction peaks was observed at ~5.30° (002), ~10.52° (004), and ~15.84° (006) in the TMA^+-intercalated H_2TiO_3 sample. The basal diffraction peak of H_2TiO_3 significantly shifted to a lower 2θ angle (Figure 1b), indicating the expansion of the interlayer (from 4.72 to 16.66 Å) through the intercalation of $(CH_3)_4N^+$. The gallery heights of $[(CH_3)_4N]_2TiO_3$, which were determined by subtracting the $TiO_3{}^{2-}$ layer thicknesses of 4.6 Å, were 12.06 Å (16.66–4.6 Å). Since the height of $(CH_3)_4N^+$ and the diameter of a water molecule are ca. 5.3 and 2.8 Å [40,41], respectively, it was suspected that two molecules of water (2 × 2.8 Å) and a molecule of $(CH_3)_4N^+$ (5.3 Å) were vertically arranged in the interlayer. Overall, the data in Figure 1c shows that $[(CH_3)_4N]_2TiO_3$ formed nanosheet stacks after drying [42].

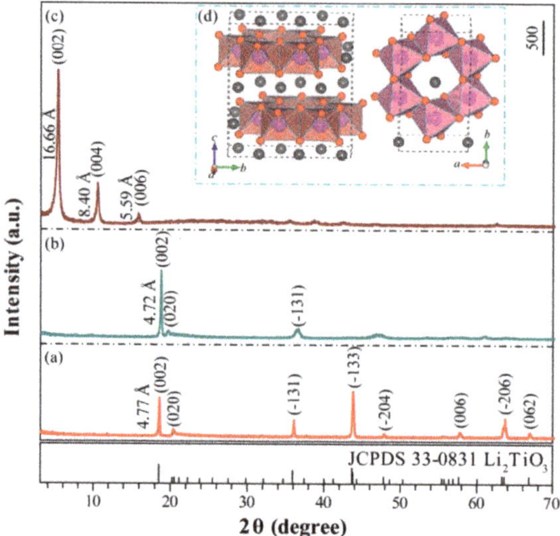

Figure 1. X-ray diffraction (XRD) patterns of (**a**) Li_2TiO_3, (**b**) H_2TiO_3, and (**c**) TMA^+-intercalated H_2TiO_3 samples; and (**d**) structural model of Li_2TiO_3 viewed in the *a* axis and *c* axis directions.

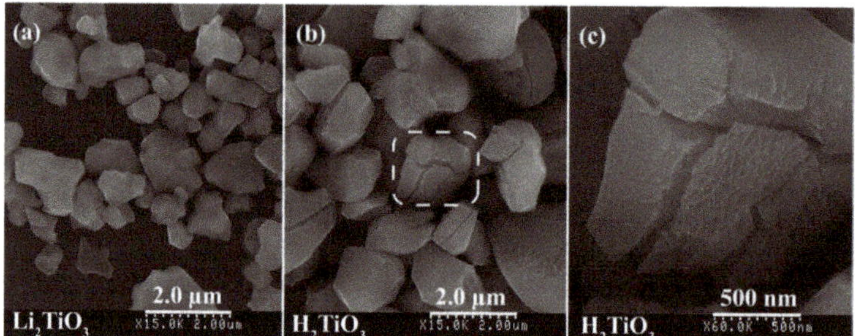

Figure 2. Field-emission (FE)SEM images of (**a**) Li_2TiO_3, and (**b**,**c**) H_2TiO_3.

3.2. Synthesis of TiO_2 Nanocrystals from Metatitanic Acid Nanosheet Colloidal Solution

TiO_2 nanocrystals with different structures and morphologies were rapidly synthesized by microwave-assisted hydrothermal treatment of the prepared metatitanic acid nanosheet colloidal solution at various pH values (0.5–13.5) and temperatures (105–185 °C) for 2 h. The crystallographic structures of the TiO_2 nanocrystals obtained at different pH values and various temperatures were identified by XRD analysis (see Supplementary Materials, Figures S1–S8). The dependence of the TiO_2 nanocrystals on the microwave-assisted hydrothermal reaction conditions is summarized in Figure 3. It can be seen that the rutile phase was formed preferentially under strong acidic conditions (pH $\leq$ 0.5). The mixed-phases of rutile and anatase were formed by the microwave-assisted hydrothermal metatitanic acid nanosheet colloidal solution within a certain pH range (0.5 < pH $\leq$ 1.0). The rutile phase disappeared when the pH increased to 1.5, indicating that the rutile phase was stable under strong acidic conditions and that its stability was greater than anatase. A pure anatase phase could be obtained in a wide pH range of 1.5 $\leq$ pH $\leq$ 13.5 at a certain temperature, and its crystallinity increased with increasing temperature and pH values (see Supplementary Materials, Figures S2–S8). The unreacted layered compound was observed at pH $\geq$6.5 and at low temperatures (see Supplementary Materials, Figures S2, S5, and S8) and was formed through the restacking of the exfoliated metatitanic acid nanosheets, implying that higher pH values and lower temperatures are not conductive to a transformation from nanosheets to anatase nanocrystals.

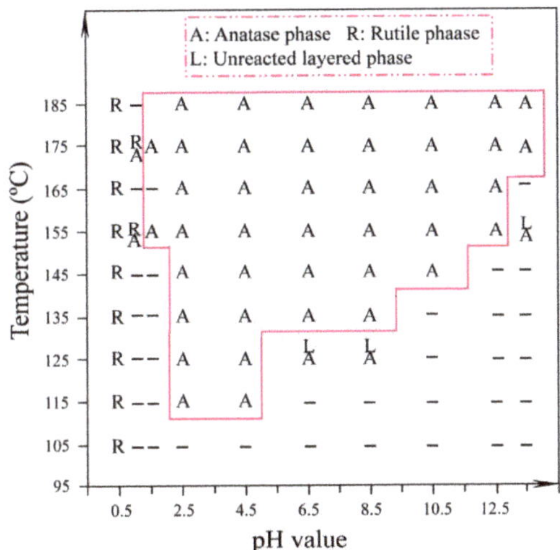

Figure 3. Changes in the crystal form of titanium dioxide with temperature and pH value.

Figure 4 shows the XRD patterns of the typical samples at different pH values varying from 0.5 to 13.5. Apparently, for the pH0.5-T175 sample, the diffraction peaks (2θ values) of 27.42°, 36.06°, 38.29°, 41.24°, 44.12°, 54.30°, 56.56°, 62.82°, 64.00°, and 69.04°, could be assigned to (110), (101), (200), (111), (210), (211), (220), (002), (310), and (301) crystalline surfaces of the rutile TiO_2 phase, respectively (JCPDS No. 21-1276, Figure 4a). With an increasing pH value in the suspension liquid, the pH1.0-T175 sample exhibited an anatase/rutile coexisting diffraction pattern, as shown in Figure 4b. The peaks at 2θ values of 27.40°, 36.04°, 41.28°, 54.34°, 56.70°, 62.88°, and 69.05° could be ascribed to (110), (101), (111), (211), (220), (002), and (301) crystalline surfaces of the rutile phase, respectively (JCPDS No. 21-1276), while other diffraction peaks at 25.46°, 37.82°, and 48.02° arose from the (101), (004), and (200) crystalline surfaces of the anatase TiO_2 phase, respectively (JCPDS No. 21-1272). The mass fraction of anatase and the rutile contents present in the sample could be accurately obtained from the following equations [43]:

$$W_A = \frac{1}{1 + 1.26[I_R(110)/I_A(101)]} \tag{1}$$

$$W_R = \frac{1}{1 + 0.8[I_A(101)/I_R(110)]} \tag{2}$$

where W_A and W_R represent the mass fraction of anatase and rutile in the mixed phase, respectively, and $I_A(101)$ and $I_R(110)$ represent the diffraction peak integral intensity of the anatase TiO_2 (101) crystal surface and the rutile TiO_2(110) crystal surface, respectively. As shown in Figure 4b, the diffraction peak integral intensity of the anatase TiO_2 (101) crystal surface and rutile TiO_2 (110) crystal surface was 60.7% and 100.0% in the mixed phase, respectively. Therefore, the mass fraction was 32.51% for anatase TiO_2 and 67.31% for rutile TiO_2. With the pH values further increasing to 1.5, the rutile phase decreased to almost null, which means that the H_2TiO_3 nanosheets were completely transformed to anatase nanocrystals and that the entire structure became anatase (Figure 4c–j), i.e, the higher pH inhibited the growth of rutile TiO_2 and promoted the growth of anatase TiO_2 during the microwave-assisted hydrothermal reaction process. The various diffraction peaks observed at ~25.66° (101), ~38.22° (004), ~48.32° (200), ~54.10° (105), ~54.50° (211), ~62.86° (204), and ~68.67° (116) were the characteristic peaks of anatase TiO_2 (JCPDS No. 21-1272). It could be found that the broadening diffraction peaks became sharper with increasing pH values, demonstrating that the grain size of the anatase matrix increased with increasing pH values. The increase in grain size was attributed to the Ostwald ripening phenomena during the particle growth process [43].

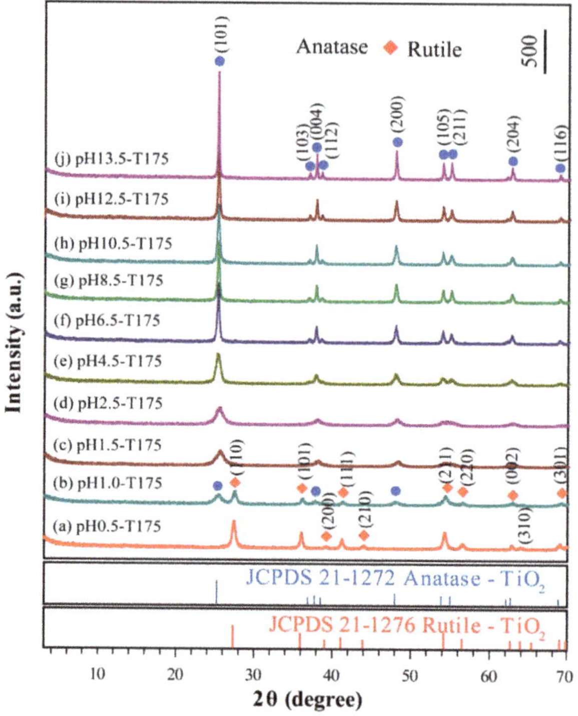

Figure 4. Evolution of the XRD patterns of the TiO_2 nanocrystal samples synthesized at various pH values with the indicated lattice phases of (●) anatase and the (◆) rutile phases.

3.3. Morphology and Exposed Crystal Facets of TiO_2 Nanocrystals

Figure 5a depicts a TEM image of the pH1.5-T175 nanocrystals derived from microwave-assisted hydrothermal treatment of the exfoliated H_2TiO_3 nanosheet colloidal solution at 175 °C with a reaction time of 2 h. As shown, a large number of rod-shaped nanocrystals with a size of about 10–30 nm and a small number of rhombic nanocrystals with a size of ~10 nm were observed. The corresponding high-resolution TEM (HRTEM) images (Figure 5b,c) showed unparallel (101) and (011) atomic planes of the rod-shaped nanocrystals with a lattice spacing of 3.44 and 3.47 Å (or 3.42 Å), respectively, and an interfacial angle of 82°, which matches well with the theoretical value for the angle between the {101} and {011} facets of anatase TiO_2 nanocrystals [14]. Therefore, the dominant exposed crystal plane of the pH1.5-T175 nanocrystals was perpendicular to the above {101} and {011} crystal facets, i.e., the crystal plane was vertical to the [111] crystal zone axis. Since the crystal plane perpendicular to the {101} and {011} crystal facets was uncertain, for convenience, we express it as an [111] facet. The exposed rod-shaped nanocrystal [111] crystal facet could be further confirmed through a fast-Fourier-transform (FFT) diffraction pattern (Figure 5c inset). It should be noted that the [111] facet was different from the {111} facet, because anatase belongs to a tetragonal system, not a cubic system. In a cubic crystal system, the crystal plane perpendicular to [111] the crystal zone axis is the {111} facet. According to the above analysis, the rod-shaped nanocrystals co-exposed in the crystal planes were [111] facets and {101} facets, i.e., {101}/[111] facets. Figure 5d shows the unparallel (101) and (002) planes of the rhombic nanocrystals with a lattice spacing of 3.38 and 4.62 Å, respectively. Moreover, the interfacial angle of ca. 68.3° between the aforementioned atomic planes also matches well with the theoretical value of anatase TiO_2 nanocrystals [15,44]. Therefore, the above rhombic nanocrystal was confirmed to be anatase TiO_2 with dominant {010} facets on the two basal surfaces and {101} facets on the four lateral surfaces. Figure 5e–h presents HRTEM images of pH2.5-T175. As shown, nanocrystals with short rod-shaped, square-prism-shaped, and some irregular morphology were observed. Along the [111] zone axis, the HRTEM image in Figure 5e,f,h displays (101) and (011) crystal facets of the rod-shaped (or square-prism-shaped) nanocrystal with an interfacial angle of 82°, which is in good agreement with the theoretic value. Therefore, the above rod-shaped and square-prism-shaped anatase TiO_2 nanocrystals co-exposed {101}/[111] facets. Furthermore, lattice fringes with a spacing of 3.50 Å could be assigned to the {101} facets of the irregularly shaped anatase nanoparticles. The FFT diffraction pattern of the yellow dashed lines region (Figure 5e,g,h inset) further indicated that the rod-shaped, square-prism-shaped, and irregular TiO_2 nanocrystals were single-crystalline.

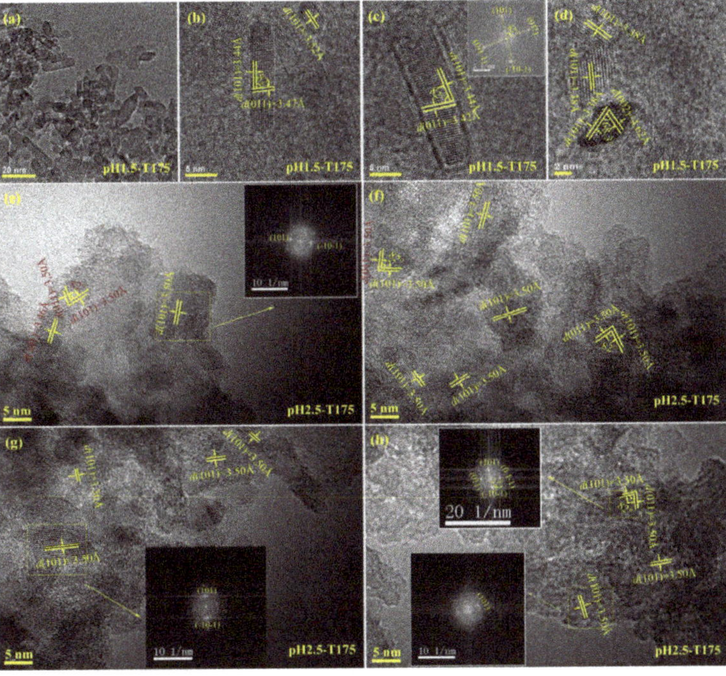

Figure 5. (**a**) TEM and (**b–d**) high-resolution (HR)TEM images of (**a–d**) pH1.5-T175 and (**e–h**) HRTEM images of pH2.5-T175 samples prepared by microwave-assisted hydrothermal treatment of the exfoliated H_2TiO_3 nanosheet colloidal solution at 175 °C with a reaction time of 2 h.The insets in (c,e,g,h) are fast-Fourier-transform (FFT) diffraction patterns.

As shown in Figure 6a,b, the synthesized pH4.5-T175 nanocrystals (anatase phase) mainly had two morphologies, a square prism and a rhombus. Moreover, a small number of anatase nanocrystals with irregular morphologies were also observed in the above sample. Figure 6c,d shows an HRTEM image of typical square-prism-shaped anatase TiO_2 nanocrystals. The lattice fringes of 3.52 and 3.52 Å corresponded to {101} and {011} lattice spacing, respectively, and the angle of 82° measured between the {101} and {011} facets matched closely with the theoretical value, corroborating that the exposed crystal facets were [111] facets and {101} facets (or {011} facets) on the two basal surfaces and the four lateral surfaces of the square-prism-shaped anatase TiO_2 nanocrystals, respectively. According to the above discussion, the irregular anatase TiO_2 nanocrystals in Figure 6e co-exposing the crystal facets were also {101}/[111] facets. The lattice fringes with spacing of 3.51 Å could be assigned to the {101} facet of the approximately rhombic anatase TiO_2 nanocrystal (Figure 6f), which was parallel to the lateral surface, indicating that the lateral surface had exposed {101} facets.

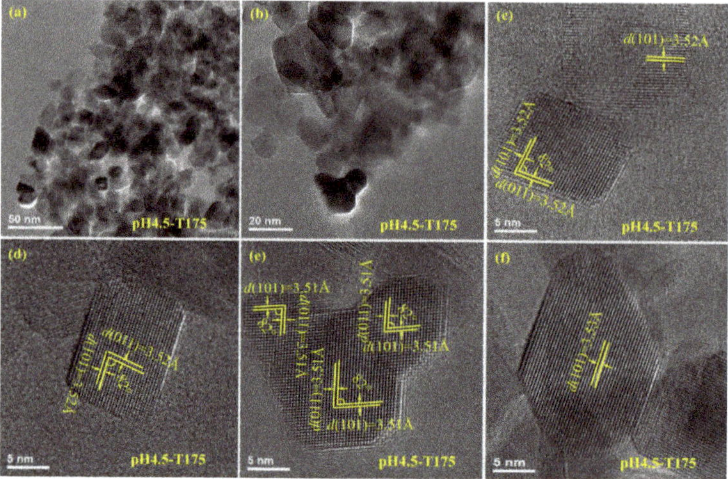

Figure 6. (**a**,**b**) TEM images and (**c**–**f**) corresponding HRTEM images of the pH4.5-T175 sample prepared by microwave-assisted hydrothermal treatment of the exfoliated H_2TiO_3 nanosheet colloidal solution at 175 °C with a reaction time of 2 h.

Figure 7a,c shows a TEM image of the product obtained at pH = 6.5 and t = 175 °C with a reaction time of 2 h. As shown, a large number of well-defined spindle-shaped nanocrystals 70–130 nm in length and 15–30 nm widewere observed. In addition, some square-prism-shaped nanocrystals 20–50 nm in length and 20–30 nm wide were also observed. Figure 7b,d includes corresponding HRTEM images taken from the marked area of the TEM images in Figure 7a,c, respectively. As shownin Figure 7b, the spindle-shaped nanocrystal displayed the (101), (10−1), and (002) atomic planes with lattice spacings of 3.53, 3.53 and 4.76 Å, and the angles α, β, γ between the (101) and (002), (10−1) and (002), and (101) and (10−1) facets were 68.3°, 68.3°, and 43.4°, respectively, the same as the theoretical values [45]. These facts prove that the spindle-shaped anatase nanocrystal exhibited four flat facets {010} facets on the four vertical surfaces, eight inclined {101} facets on the slant surfaces, and two parallel {001} facets on the top and bottom surfaces. Along the [010] zone axis, the HRTEM image in Figure 7d also displays (101) and (002) crystal facets with an interfacial angle of 68.3°, indicating a spindle-shaped anatase nanocrystal with exposed {010} facets on the four vertical surfaces.

The well-defined nanospindle structure (with a particle size of about 70–140 × 17–35 nm^2) of the pH8.5-T175 sample was further confirmed by the TEM image, as shown in Figure 8a. Two magnified TEM images of the uniform anatase nanospindles are presented in Figure 8c,e. Figure 8b,d,f shows HRTEM images marked from the yellow dotted frame area indicated in Figure 8a,c,e, respectively. There were three types of lattice with fringe spacing of 3.53, 4.74, and 3.53 Å (or 3.49, 4.78, and 3.49 Å) and interfacial angles of 68.3°, 68.3°, and 43.4°, which agreed well with the (101), (002), and (10−1) lattice planes of anatase TiO_2. The {001} facets were perpendicular to the long axis (c axis) direction and parallel to the transverse axis (b axis) direction of the spindle-shaped nanocrystal, respectively, indicating the oriented growth of spindle-shaped nanocrystals in the [001] crystallographic direction. Therefore, the exposed lattice facets of the nanospindles were mainly {010} facets (based on the HRTEM analysis).

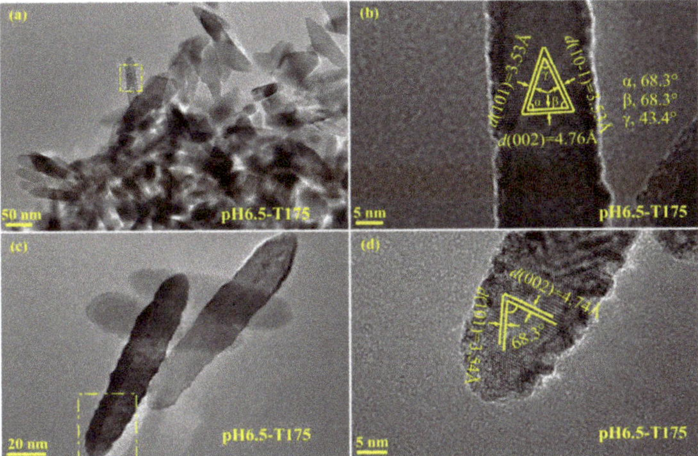

Figure 7. (**a**,**c**) TEM images and (**b**,**d**) corresponding HRTEM images of the pH6.5-T175 sample derived from microwave-assisted hydrothermal treatment of the exfoliated H_2TiO_3 nanosheet colloidal solution at 175 °C with a reaction time of 2 h.

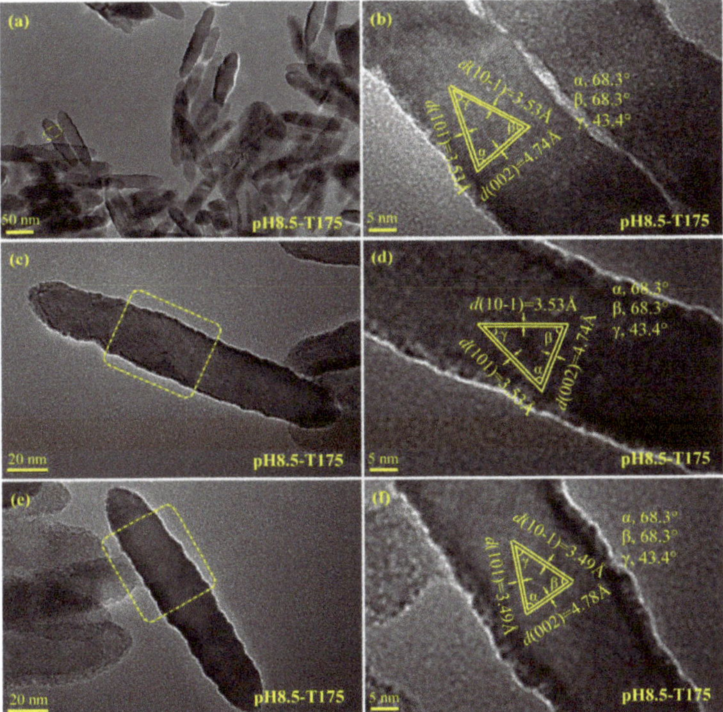

Figure 8. (**a**,**c**,**e**) TEM images and (**b**,**d**,**f**) corresponding HRTEM images of the pH8.5-T175 sample derived from microwave-assisted hydrothermal treatment of the exfoliated H_2TiO_3 nanosheet colloidal solution at 175 °C with a reaction time of 2 h.

Figure 9 displays typical FESEM images of the samples derived from microwave-assisted hydrothermal treatment of the exfoliated H_2TiO_3 nanosheet colloidal solution at various pH values (0.5–2.5) and temperatures (155–185 °C). The pH0.5-Tx (x = 155, 165, 175, or 185) sample exhibited quadrangular prism morphology with a length of 50–100 nm and a width (or thickness) of 15–30 nm, which corresponded with the rutile TiO_2 phase and the elongation direction of the nanocrystals along the [001] crystallographic direction (Figure 9a–d) [46]. The average size of the quadrangular prism nanocrystals did not change obviously with an increase in the temperature, indicating that an increase in temperature mainly increased the crystallinity of the nanocrystals during the microwave-assisted hydrothermal reaction, which was consistent with the results of the XRD analysis. As shown in Figure S1, the intensity of the diffraction peaks of the rutile phase at 2θ = 27.42° (110), 36.06 (101), 38.29° (200), 41.24° (111), 54.30° (211), and 56.56° (220) increased with an increase in the reaction temperature (from 105 to 185 °C), indicating that the crystallinity of rutile nanocrystals improved. For the pH1.0-T175 sample, some bigger rod-shaped nanocrystals (rutile phase) and lots of smaller nanocrystals (anatase phase) with various morphologies were observed (Figure 9e). With a further increase in the pH value, rod-shaped (or square-prism-shaped) nanocrystals with a size of about several tens of nanometers were observed, all of them belonging to the anatase phase (Figure 9f–i). The results showed that the pH value of the nanosheet colloidal solution had a great influence on the crystal structure and morphology of the produced TiO_2.

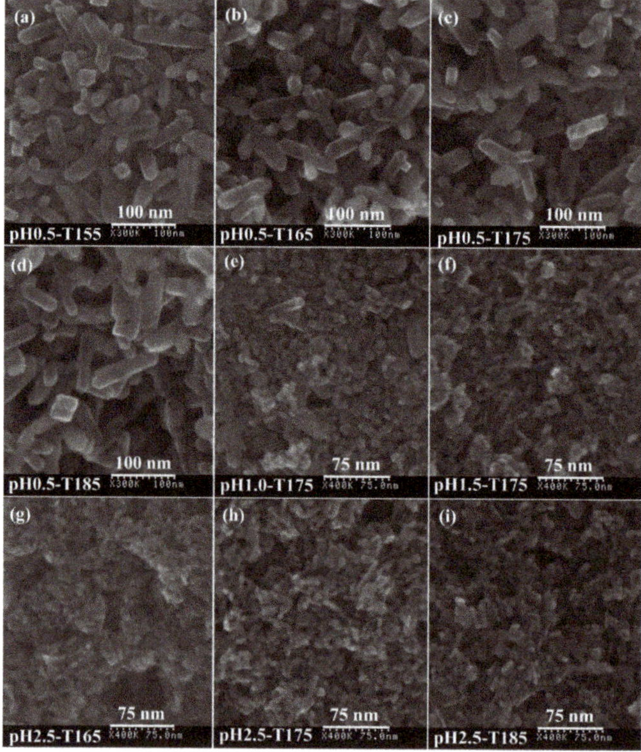

Figure 9. FESEM images of TiO_2 nanocrystals synthesized under different pH values and temperatures: (**a**) pH0.5-T155, (**b**) pH0.5-T165, (**c**) pH0.5-T175, (**d**) pH0.5-T185, (**e**) pH1.0-T175, (**f**) pH1.5-T175, (**g**) pH2.5-T165, (**h**) pH2.5-T175, and (**i**) pH2.5-T185.

As shown in Figure 10a–c, the pH4.5-Ty (y = 165, 175, or 185) sample mainly exhibited square-prism and rhombus morphologies with a particle size of ~15–30 nm. With the increase of temperature, the particle size did not increase significantly, which indicated that the increase of temperature only increased the crystallinity of the particles, which was consistent with the results of the XRD analysis. As shown in Figure S4, the intensity of the characteristic diffraction peaks of the anatase phase at 2θ = 25.66° (101), 38.22° (004), 48.32° (200), 54.10° (105), 54.50° (211), and 62.86° (204) increased with increases in the reaction temperature (from 115 to 185 °C), indicating that the crystallinity of anatase nanocrystals increased with an increase in the reaction temperature. Figure 10d–j presents the FE-SEM images of pHx-Ty (x = 6.5, 8.5, 10.5, or 12.5; y =125, 165, 175, or 185) products, which revealed that the obtained TiO_2 products consisted of uniform, well-defined, spindle-shaped structures with an axial length of 75–155 nm and a transverse width of 20–35 nm. Interestingly, a series of directional streaks could be observed on the surface of the spindle-shaped crystal, which were perpendicular to the long axis (c axis)direction and parallel to the transverse axis (b axis) direction of the spindle-shaped nanocrystals, indicating that the orientation growth was in the [001] crystallographic direction. The crystal size increased with an increase in the pH value (from 6.5 to 12.5), especially for the [001] orientation of the spindle-shaped nanocrystals. Arrow-shaped crystals 500–700 nm in length were formed at pH13.5, as shown in Figure 10l. Spindle-shaped, arrow-shaped nanocrystals and unreacted nanosheets were observed at lower temperatures, indicating that a lower temperature was not conducive to the fracturing of the nanosheet (Figure 10g,k).

Figure 10. FESEM images of the TiO_2 nanocrystals synthesized under different pH values and temperatures: (**a**) pH4.5-T165, (**b**) pH4.5-T175, (**c**) pH4.5-T185, (**d**) pH6.5-T165, (**e**) pH6.5-T175, (**f**) pH6.5-T185, (**g**) pH8.5-T125, (**h**) pH8.5-T175, (**i**) pH10.5-T185, (**j**) pH12.5-T185, (**k**) pH13.5-T155, and (**l**) pH13.5-T175.

3.4. Transformation Reaction from the Delaminated H_2TiO_3 Nanosheets to TO_2 Nanocrystals

The nanocrystal conversion process between the exfoliated H_2TiO_3 nanosheets and the TiO_2 nanocrystals can be described as the schematic diagram in Figure 11. On the basis of the discussion above, the transformation reaction mechanism from the exfoliated metatitanic nanosheets with distorted hexagonal cavities to TiO_2 nanocrystals can be described by the following chemical equations:

$$[TiO_3]^{2-} + 3H_2O \rightarrow [TiO_6]^{8-} + 6H^+ \tag{3}$$

$$[TiO_6]^{8-} + 8H^+ \rightarrow TiO_2 + 4H_2O \tag{4}$$

$$[TiO_6]^{8-} + 4H_2O \rightarrow TiO_2 + 8OH^- \tag{5}$$

First, the delaminated $[TiO_3]^{2-}$ nanosheets were split into many primary $[TiO_6]^{8-}$ monomers along the edge-shared oxygen atoms under microwave-assisted hydrothermal conditions. At this stage, the pH values of the $[TiO_3]^{2-}$ **nanosheet colloidal solution had an important influence on the reaction (3).** As shown in Figure 3, the acidic conditions were advantageous to the reaction (Equation (3)), which could be carried out at relatively lower temperatures, while the alkaline conditions were disadvantageous to the reaction (Equation (3)), which needed to be carried out at relatively higher temperatures. Furthermore, the rutile-type and anatase-type TiO_2 nuclei were formed by polymeric $[TiO_6]^{8-}$ monomers along the equatorial and apical edges at different pH values, respectively. Rutile nuclei could be formed under super acidic conditions (Equation (4)), and anatase nuclei could be formed under weak acidic, neutral, or alkaline conditions (Equations (4) and (5)). Finally, the different types of TiO_2 nanocrystals with various morphologies and exposed facets were formed through the directional growth of the TiO_2 nuclei.

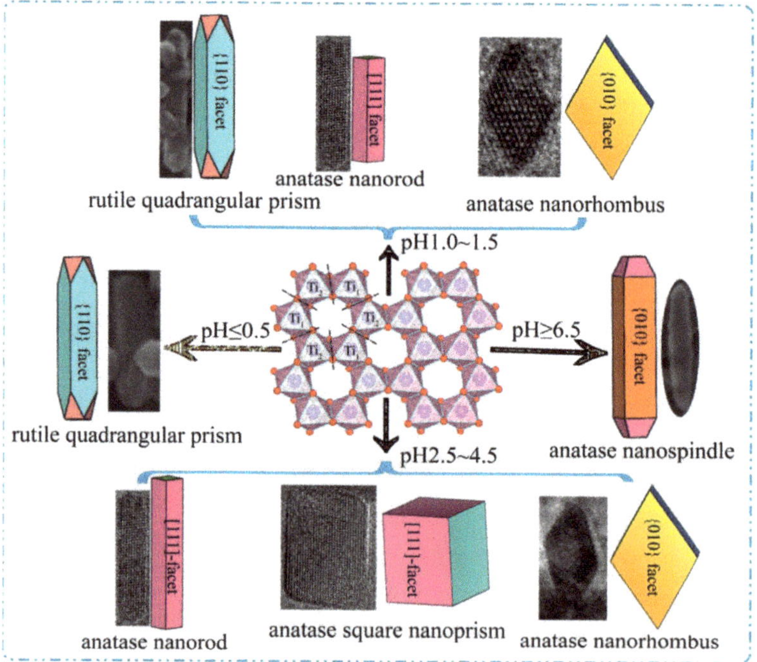

Figure 11. Schematic diagramof the formation mechanism of the TiO_2 nanocrystals with different morphologies from the exfoliated H_2TiO_3 nanosheets at different pH values.

The growth of the rutile-type and anatase-type TiO_2 nuclei along the different crystallographic directions caused the formation of a rutile quadrangular prism, an anatase nanorod, an anatase nanorhombus, an anatase square nanoprism, and an anatase nanospindle. As shown in Figure 11, a rutile quadrangular prism with exposed {110} facets on the four lateral surfaces was formed by the directional growth of the rutile-type TiO_2 nuclei along the [001] crystallographic direction under strong acidic conditions (pH = 0.5, 1.0). An anatase nanorod (or square nanoprism) with co-exposed {101}/[111] facets was formed by the directional growth of the anatase-type TiO_2 nuclei along the [101] and [011] crystallographic directions, and an anatase nanorhombus with dominant exposed {010} facets was formed by the directional growth of the nuclei of anatase-type TiO_2 along the [101] and [001] crystallographic directions at relatively low pH values (1.0 ≤ pH ≤ 4.5). At pH ≥ 6.5, an anatase nanospindle with dominant exposed {010} facets was formed by the directional growth of the nuclei of anatase-type TiO_2 along the [001] crystallographic direction.

3.5. Photocatalytic Activities of the As-Synthesized TiO_2 Nanocrystals

The photocatalytic activities of the TiO_2 nanocrystal samples synthesized in the present work were quantitatively evaluated via bleaching 7.5 mg/L MO solutions under UV light irradiation at room temperature. The photocatalytic degradation of MO can be considered as the following reactions [47,48]:

$$
\begin{gathered}
\mathrm{TiO_2} + h\nu \rightarrow \mathrm{TiO_2}(\mathrm{e^-} + \mathrm{h^+}) \\
\mathrm{h^+} + \mathrm{OH^-} \rightarrow \cdot\mathrm{OH} \\
\mathrm{e^-} + \mathrm{O_2} \rightarrow \mathrm{O_2}^- \\
\mathrm{O_2}^- + \mathrm{H_2O} \rightarrow \mathrm{HO_2}\cdot + \mathrm{OH^-} \\
\mathrm{HO_2}\cdot + \mathrm{H_2O} \rightarrow \mathrm{H_2O_2} + \cdot\mathrm{OH} \\
\mathrm{H_2O_2} \rightarrow 2\cdot\mathrm{OH}
\end{gathered}
$$

·OH + MO → peroxides or hydroxylated intermediates→ degraded or mineralized products.

The above reactions were selected because both the photoelectrons and photoholes eventually participate in the formation of the OH radical, which degrades MO. In order to have a better evaluation of the photocatalytic efficiency of the as-prepared $^-TiO_2$ nanocrystals, Degussa P25 (average particle size of 26.2 nm) was chosen as the photocatalytic reference. Figure 12a and Figure S9 show the UV-Vis absorption spectra of the centrifuged MO solutions at certain intervals during the photodegradation treatment with and without TiO_2 nanocrystals. As can be seen, the intensity of the peaks in the visible region gradually decreased with an extension of the irradiation time. Moreover, it can be observed that the strongest peaks at 464 nm shifted toward the shorter wavelength direction (i.e., a hypsochromic shift), which can be attributed to the demethylation of the MO [49]. Demethylation and cleavage of the MO chromophore ring structure occurred simultaneously during the initial period of photodegradation of MO, in which demethylation played a predominant role [49]. With the prolongation of UV light irradiation time, the demethylated MO intermediates could be further decomposed, which was indicated by the change in peak intensity at 435 nm (Figure 12a). However, the intensity of the peaks changed little after 90 min of UV irradiationin the absence of the TiO_2 sample (see Figure S9d), which indicated that the bleaching/degradation of the MO aqueous solution occurred on the surface of the TiO_2 photocatalyst [35].

Figure 12b shows the evolution of the photodegradation rate along with the irradiation time. It can be noticed that the pH2.5-T175 nanocrystal showed the highest decomposition efficiency among all the tested samples, reaching above 96.5% under UV irradiation for 90 min. The second fastest decomposition efficiency was obtained with the pH4.5-T175 nanocrystal, where 93.3% of MO was decomposed after 90 min of UV light irradiation. The order of decomposition efficiency observed for MO degradation in the other samples was P25 (81.1%) > pH6.5-T175 (44.7%) > blank sample (3.1%). In order to have a quantitative comparison of the decomposition efficiency of the above samples, the experimental data in Figure 12b were fitted according to the pseudo-first-order kinetic process, which can be expressed as follows [50]:

$$\ln(c_0/c_t) = kt$$

where k is the photocatalytic degradation rate constant ($\mathrm{min^{-1}}$), and t is the degradation time. The fitted curves are shown in Figure 12c, and the obtained k ($\mathrm{min^{-1}}$), as well as the correlative coefficient (R^2) of all the curves, is listed in Table 1. Obviously, the curves agreed well with the experimental data points. It can be seen that the pH2.5-T175 nanocrystal showed the highest k value of 0.0346 min^{-1}, almost 1.23, 1.85, 5.32, and 86.5 times as high as that of pH4.5-T175, P25, pH6.5-T175, and the blank sample (Table 1), respectively, which helped us to have a better understanding of the pH2.5-T175 nanocrystal having the highest performance among all of the tested samples.

It is well known that the photocatalytic activity of a photocatalyst can be strongly influenced by many factors, including phase structure, particle morphology, crystallite size, specific surface area, crystal composition, crystallinity, and crystal facets [35,51]. Generally speaking, the specific surface area increases with a decrease in the particle size. Figure S10 shows the particle size distributions of pH2.5-T175, pH4.5-T175, pH6.5-T175, and P25 measured from the enlarged photograph of FESEM images, and the corresponding parameters are listed in Table 1. Since degradation occurs on the surface of catalysts, the surface area is an important parameter for photocatalytic activity. The specific surface area was ranked in the order of pH2.5-T175 > pH4.5-T175 > P25 > pH6.5-T175, which was opposite to the increasing order of the average particle size (Table 1). The photocatalytic activity increased in the order of pH2.5-T175 > pH4.5-T175 > P25 > pH6.5-T175 > blank. The pH2.5-T175 sample exhibited the highest photocatalytic activity of the samples, while pH6.5-T175 showed the lowest photocatalytic activity of the samples. On the basis of the results, the order of the photocatalytic activity can be explained by considering the following factors: (1) A smaller particle size can provide a powerful redox capability in the photochemical reaction due to the quantum size effect, resulting in an increase in the rate of migration and a decrease in the rate of recombination of the photoelectrons and photogenerated holes, thereby improving the photocatalytic activity [52]. (2) The larger specific surface can provide more adsorption sites, resulting in an enhancement of MO adsorbed on the surface of TiO_2, thereby contributing to the improvement of the photocatalytic activity [53].

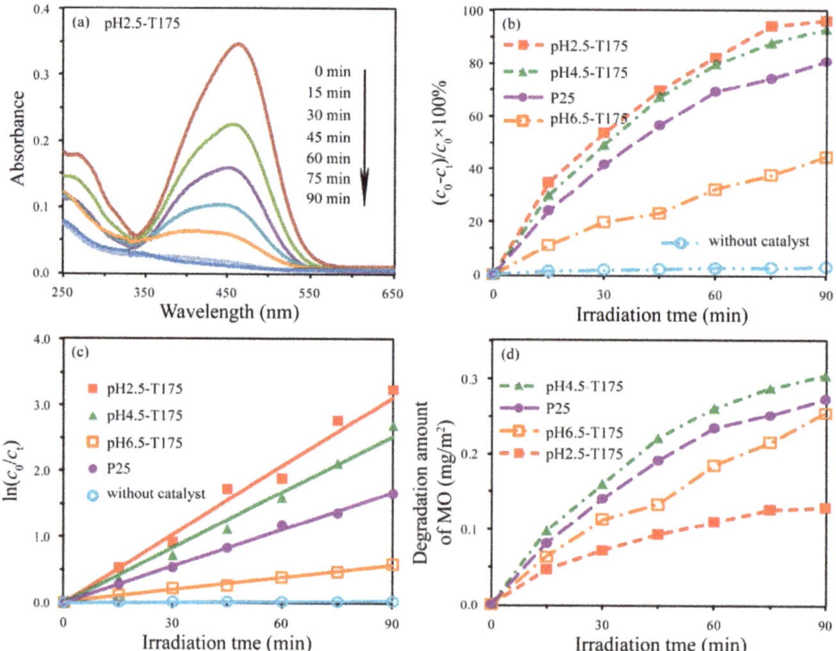

Figure 12. (**a**) UV-Vis spectral changes of the centrifuged methyl orange (MO) solutions at certain intervals during the photodegradation treatment with pH2.5-T180; (**b**) photocatalysis degradation profiles of MO under UV irradiation; (**c**) first-order kinetics fitting data for the photodegradation of the MO; (**d**) degradation amount of MO per unit area of TiO_2 nanocrystals.

In order to better understand the intrinsic photocatalytic activity of the different TiO_2 nanocrystals, we examined the degradation amount of MO per unit surface area of catalyst (mg (MO) per m^2 (TiO_2 surface area)), as shown in Figure 12d. The degradation amount of pH2.5-T175, pH6.5-T175, P25, and pH4.5-T175 was calculated to be 0.13, 0.25, 0.27, and 0.31 mg/m^2, respectively. On the basis of the above results and discussions, we know that the co-exposed facets of pH2.5-T175, pH4.5-T175, pH6.5-T175, and P25 were {101}/[111] facets, {101}/[111] facets, {010}/{001} facets, and {101}/[111] facets [54], respectively. Although the exposed crystal facets of the pH2.5-T175 and pH4.5-T175 nanocrystals were the same, their photocatalytic performances were quite different, indicating that the crystal facets were not the main factor affecting the photocatalytic performance. The degradation amount of MO per unit surface area of the pH2.5-T175 nanocrystals was the smallest among the samples, which can be attributed to them having the lowest crystallinity. Except for pH2.5-T175, theother samples all had high crystallinity, so their photocatalytic activity depended on the crystal plane. For the low-index facets of anatase, the average surface energy (γ) can be arranged in the following order: $\gamma_{\{111\}}$ (1.61 J/m^2) > $\gamma_{\{110\}}$ (1.09 J/m^2) > $\gamma_{\{001\}}$ (0.90 J/m^2) > $\gamma_{\{010\}}$ (0.53 J/m^2) > $\gamma_{\{101\}}$ (0.44 J/m^2) [55]. The P25 sample contained~87% anatase nanocrystals (partially co-exposed in the {101}/[111]-facets) and ~13% rutile nanocrystals, with an average size of ~26.2 nm [54]. The degradation amounts of MO per unit surface area of the pH4.5-T175 and P25 nanocrystals were higher than that of the pH6.5-T175 nanocrystal, which can be attributed to the existence of oxygen vacancies (active reaction sites) on the [111] facets of the TiO_2 nanocrystals [14]. The pH4.5-T175 nanocrystal exhibited higher degradation amounts than did the P25 nanocrystal due to the larger proportion of [111] facets exposed on the surface of TiO_2, which could create more active reaction sites in the process of photocatalytic reaction.

Table 1. specific surface area, average particle size (nm), *k*, and R^2 values of the TiO_2 nanocrystals.

Samples	Specific Surface Area (m^2/g)	Average Particle Size (nm)	Degradation Rate Constant k (min^{-1})	Correlative Coefficient (R^2)
pH2.5-T175	135.6	8.3	0.0346	0.9502
pH4.5-T175	54.8	13.3	0.0281	0.9846
pH6.5-T175	32.3	87.0	0.0065	0.9920
P25	52.5	23.9	0.0187	0.9971
blank			0.0004	0.9595

3.6. Photovoltic Performances of As-Synthesized TiO_2 Nanocrystals

The photocurrent–voltage characteristics of the DSSCs with the as-synthesized TiO_2 nanocrystals and commercial Degussa P25 nanoparticles in the photoelectrodes are shown in Figure 13, and the detailed photovoltaic parameters of the four DSSCs, including open-circuit (V_{OC}) and short-circuit photocurrent densities (J_{SC}), the fill factor (*FF*), and the overall conversion efficiency (η), are summarized in Table 2. For comparison, the thicknesses of the TiO_2 porous films were kept almost the same, at about 20 μm. The value of J_{SC} increased in the order of pH6.5-T175 < P25 < pH4.5-T175 < pH2.5-T175, which agreed with the decreasing order of the crystal size, while the value of V_{OC} increased in the order of P25 < pH2.5-T175 < pH4.5-T175 < pH6.5-T175, which was almost consistent with the increasing order of the crystal size, except for the P25 sample. The value of *FF* increased in the order of P25 < pH4.5-T175 < pH2.5-T175 < pH6.5-T175, which was almost the same as the V_{OC} increasing order, except for the pH2.5-T175 sample. The value of η increased in the same order as the J_{SC} value, i.e., the cell made of pH2.5-T175 nanocrystals showed the highest performance for the DSSCs, which represented an enhancement 20.7%, 25.1%, and 33.5% compared to the pH4.5-T175, P25, and pH6.5-T175 samples, respectively. It is well known that the particle size, specific surface area, film thickness, exposed crystal facets, crystal structure, and particle morphology of photoanodes are essential factors for enhancing photovoltaic properties [56,57]. The photoanode films of DSSCs that are prepared by smaller nanoparticles often possess a large internal surface area, which is beneficial to the adsorption of dye, resulting in an increase in the photocurrent and energy conversion efficiency [57]. The cells with the pH6.5-T175 electrode exhibited the lowest value of J_{SC} and η (Table 2) due to having the largest particle size and the lowest specific area (Table 1), which reduced the available internal surface area for the adsorption of the dye molecules, resulting in a decrease in light-harvesting efficiency [58]. The cells made of pH2.5-T175 possessed the highest J_{SC} and η values, which can be attributed to their minimum particle size and maximum surface area.

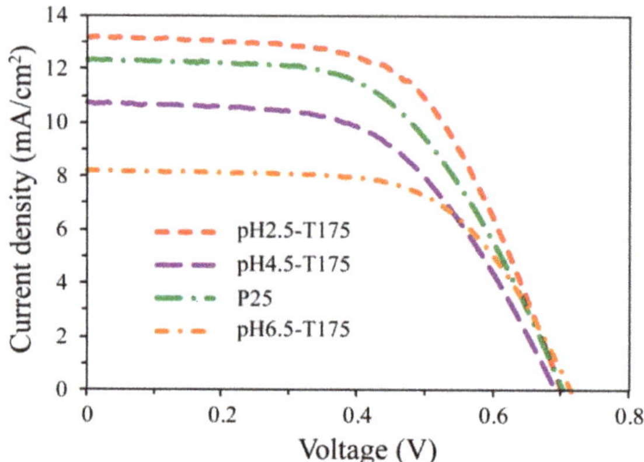

Figure 13. Current–voltage characteristic curves of dye-sensitized solar cells (DSSCs) fabricated using pH2.5-T175, pH4.5-T175, pH6.5-T175, and P25 samples.

Table 2. Cell performance parameters of DSSCs fabricated using different TiO_2 samples.

Photoelectrode	Film Thickness (μm)	J_{SC} (mA/cm^2)	V_{OC} (V)	*FF*	η (%)
pH2.5-T175	19.17	13.19	0.701	0.595	5.50
pH4.5-T175	20.03	12.32	0.704	0.556	4.83
pH6.5-T175	19.11	8.22	0.715	0.622	3.66
P25	20.40	10.73	0.693	0.554	4.12

In accordance with the previous discussion, the exposed crystal facets of the pH2.5-T175, pH4.5-T175, pH6.5-T175, and P25 nanocrystals were {101}/[111] facets, {101}/[111] facets, {101}/{010} facets, and {101}/[111] facets (only a small fraction), respectively. It has been reported that the adsorption equilibrium constant and the surface uptake density of N719 dye molecules on different crystal facets of the TiO_2 surface increase in the order of the following: without specific exposed facets < {101} facets < [111] facets < {010} facets on the surface of anatase TiO_2. In other words, the strong adsorption and surface uptake density of N719 dye molecules on the surface of anatase TiO_2 can increase the light-harvesting efficiency, resulting in an improvement in the photovoltaic performance [59,60]. However, the {101}/{010} facets co-exposed in the pH6.5-T175 photoanode exhibited the lowest J_{SC} value, which implies that compared to the specific area, the crystal facets were not the most important effecting factor in this research. The cells with a pH6.5-T175 photoanode showed the highest V_{OC} and *FF* values, which may be attributed to the fact that the high surface uptake density can decrease the charge recombination at the TiO_2/electrolyte interface [56], and the oriented anatase TiO_2 nanospindle structure can improve the charge transport properties in the porous TiO_2 film [57]. The photoanode made of pH2.5-T175 nanocrystals (co-exposed {101} and [111] facets) possessed a higher J_{SC} value than did the one made of pH4.5-T175 nanocrystals (co-exposed {101} and [111] facets) or P25 nanocrystals (partial co-exposed {101} and [111] facets) at a similar film thickness, which can be attributed to the higher specific surface area enhancing N719 dye adsorption, resulting in an increase in the J_{SC} value.

4. Conclusions

Microwave-assisted hydrothermal synthesis is an effective synthesis route to produce TiO_2 nanocrystals with different morphologies and high-energy surfaces at relatively low temperatures, which can shorten the reaction time, reduce the energy consumption, and enhance the purity and crystallinity of the produced products compared to the traditional hydrothermal method. Titanate nanosheets were exfoliated from layered porous metatitanic acid with a lepidocrocite-type structure, which was suitable as a building block for the assembly of different structured TiO_2 nanocrystals. The pH values of the nanosheets had a significant effect on the crystallinity, crystallite, phase structure, morphology, and exposure facets of the prepared TiO_2 nanocrystals. Rutile TiO_2 quadrangular prisms with dominant {110} facets, anatase TiO_2 nanorods and square nanoprisms with co-exposed {101}/[111] facets, anatase TiO_2 nanorhombuses with co-exposed {101}/{010} facets, and anatase TiO_2 nanospindles with dominant {010} facets were synthesized through a facile green approach with the use of exfoliated porous metatitanic acid nanosheets as the precursor at various pH values. The morphology and exposed crystal facets of the obtained TiO_2 nanocrystals could be controlled by adjusting the pH value of the nanosheet solution. The transformation reaction mechanism from the exfoliated metatitanic nanosheets with distorted hexagonal cavities to TiO_2 nanocrystals could include a dissolution reaction of the nanosheets, nucleation of the primary $[TiO_6]^{8-}$ monomers, and the growth of rutile-type and anatase-type TiO_2 nuclei. The {101}/[111]-faceted pH2.5-T175 nanocrystal showed the highest photocatalytic and photovoltaic performance compared to the other TiO_2 samples, which could be attributed mainly to its minimum particle size and maximum specific surface area.

Supplementary Materials: The following are available online at http://www.mdpi.com/1996-1944/12/21/3614/s1, Figures S1–S8: Evolution of the XRD patterns of the TiO_2 nanocrystal samples synthesized at pH0.5–13.5 and various temperatures; Figure S9: UV-Vis spectral changes of MO solutions as a function of UV irradiation time in the presence of (**a**) pH4.5-T175, (**b**) pH6.5-T175, and (**c**) P25 photocatalysts and in the (**d**) absence of photocatalysts; Figure S10: Particle size distributions of (**a**) pH2.5-T175, (**b**) pH4.5-T175, (**c**) pH6.5-T175, and (**d**) P25.

Author Contributions: Conceptualization, Y.-e.D. and X.N.; methodology, Y.-e.D. and X.N.; formal analysis, Y.-e.D. and J.A.; investigation, Y.-e.D. and W.L.; resources, Y.-e.D., Y.L., and Y.C.; data curation, Y.-e.D.; writing—original draft preparation, Y.-e.D.; writing—review and editing, Y.-e.D., X.Y., and Q.F.; supervision, P.W., X.Y., and Q.F.; project administration, Y.-e.D.; funding acquisition, Y.L., Y.C., and Y.-e.D.

Acknowledgments: This work was supported by the Grants-in-Aid for Doctor Research Funds, the higher education reform and innovation project of Shanxi (J2019186), the education reform and innovation project of Jinzhong University (J201903), the fund for the Shanxi "1331 Project" Key Innovation Team (PY201817), the fund for the Jinzhong University "1331 Project" Key Innovation Team (jzxycxtd2017004), the National Science Foundation of China (no. 51272030 and 51572031), and the Grants-in-Aid for Scientific Research (B) (grant number 26289240) from the Japan Society for the Promotion of Science and Kagawa University.

Conflicts of Interest: The authors declare no conflicts of interest.

References

1. Wu, L.; Yang, B.X.; Yang, X.H.; Chen, Z.G.; Li, Z.; Zhao, H.J.; Gong, X.G.; Yang, H.G. On the synergistic effect of hydrohalic acids in the shape-controlled synthesis of anatase TiO_2 single crystals. *Cryst. Eng. Comm.* **2013**, *15*, 3252–3255. [CrossRef]
2. Wen, C.Z.; Jiang, H.B.; Qiao, S.Z.; Yang, H.G.; Lu, G.Q. Synthesis of high-reactive facets dominated anatase TiO_2. *J. Mater. Chem.* **2011**, *21*, 7052–7061. [CrossRef]
3. Zhang, J.; Zhang, L.L.; Shi, Y.X.; Xu, G.L.; Zhang, E.P.; Wang, H.B.; Kong, Z.; Xia, J.H.; Ji, Z.G. Anatase TiO_2 nanosheets with coexposed {101} and {001}facets coupled with ultrathin SnS_2 nanosheets as a face-to-face n-p-n dual heterojunction photocatalyst for enhancing photocatalytic activity. *Appl. Surf. Sci.* **2017**, *420*, 839–848. [CrossRef]
4. Regue, M.; Sibby, S.; Ahmet, I.Y.; Friedrich, D.; Abdi, F.F.; Johnson, A.L.; Eslava, S. TiO_2 photoanodes with exposed {010} facets grown by aerosol-assisted chemical vapor deposition of a titanium oxo/alkoxy cluster. *J. Mater. Chem. A* **2019**, *7*, 19161–19172. [CrossRef]
5. Kashiwaya, S.; Olivier, C.; Majimel, J.; Klein, A.; Jaegermann, W.; Toupance, T. Nickel oxide selectively depositedon the {101} facet of anatase TiO_2 nanocrystal bipyramids for enhanced photocatalysis. *ACS Appl. Nano Mater.* **2019**, *2*, 4793–4803. [CrossRef]

6. He, J.; Du, Y.-E.; Bai, Y.; An, J.; Cai, X.M.; Chen, Y.Q.; Wang, P.F.; Yang, X.J.; Feng, Q. Facile formation of anatase/rutile TiO_2 nanocomposites with enhanced photocatalytic activity. *Molecules* **2019**, *24*, 2996. [CrossRef]
7. Ding, Y.; Zhang, T.T.; Liu, C.; Yang, Y.; Pan, J.H.; Yao, J.X.; Hu, L.H.; Dai, S.Y. Shape-controlled synthesis of single-crystalline anatase TiO_2 micro/nanoarchitectures for efficient dye-sensitized solar cells. *Sustain. Energy Fuels* **2017**, *1*, 520–528. [CrossRef]
8. Ong, W.J.; Tan, L.L.; Chai, S.P.; Yong, S.T.; Mohamed, A.R. Highly reactive {001} facets of TiO_2-based composites: synthesis, formation mechanism and characterization. *Nanoscale* **2014**, *6*, 1946–2008. [CrossRef]
9. Du, Y.-E.; Niu, X.J.; Bai, Y.; Qi, H.X.; Guo, Y.Q.; Chen, Y.Q.; Wang, P.F.; Yang, X.J.; Feng, Q. Synthesis of anatase TiO_2 nanocrystals with defined morphologies from exfoliated nanoribbons: photocatalytic performance and application in dye-sensitized solar cell. *ChemistrySelect* **2019**, *4*, 4443–4457. [CrossRef]
10. Yu, S.L.; Liu, B.C.; Wang, Q.; Gao, Y.X.; Shi, Y.; Feng, X.; An, X.T.; Liu, L.X.; Zhang, J. Ionic liquid assisted chemical strategy to TiO_2 hollow nanocube assemblies with surface-fluorination and nitridation and high energy crystal facet exposure for enhanced photocatalysis. *ACS Appl. Mater. Interfaces* **2014**, *6*, 10283–10295. [CrossRef]
11. Yu, J.G.; Wang, G.H.; Cheng, B.; Zhou, M.H. Effects of hydrothermal temperature and time on the photocatalytic activity and microstructures of bimodal mesoporous TiO_2 powders. *Appl. Catal. B* **2007**, *69*, 171–180. [CrossRef]
12. Wang, Y.; Zhang, H.M.; Han, Y.H.; Liu, P.R.; Yao, X.D.; Zhao, H.J. A selective etching phenomenon on {001} faceted anatase titanium dioxide single crystal surfaces by hydrofluoric acid. *Chem. Commun.* **2011**, *47*, 2829–2831. [CrossRef] [PubMed]
13. Han, X.G.; Kuang, Q.; Jin, M.S.; Xie, Z.X.; Zheng, L.S. Synthesis of titania nanosheets with a high percentage of exposed (001) facets and related photocatalytic properties. *J. Am. Chem. Soc.* **2009**, *131*, 3152–3153. [CrossRef] [PubMed]
14. Xu, H.; Reunchan, P.; Ouyang, S.X.; Tong, H.; Umezawa, N.; Kako, T.; Ye, J.H. Anatase TiO_2 single crystals exposed with high-reactive {111} facets toward efficient H_2 evolution. *Chem. Mater.* **2013**, *25*, 405–411. [CrossRef]
15. Yang, H.G.; Sun, C.H.; Qiao, S.Z.; Zou, J.; Liu, G.; Smith, S.C.; Cheng, H.M.; Lu, G.Q. Anatase TiO_2 single crystals with a large percentage of reactive facets. *Nature* **2008**, *453*, 638–641. [CrossRef]
16. Wen, P.H.; Itoh, H.; Tang, W.P.; Feng, Q. Single nanocrystals of anatase-type TiO_2 prepared from layered titanate nanosheets: Formation mechanism and characterization of surface properties. *Langmuir* **2007**, *23*, 11782–11790. [CrossRef]
17. Yang, H.G.; Liu, G.; Qiao, S.Z.; Sun, C.H.; Jin, Y.G.; Smith, S.C.; Zou, J.; Cheng, H.M.; Lu, G.Q. Solvothermal Synthesis and photoreactivity of anatase TiO_2 nanosheets with dominant {001} facets. *J. Am. Chem. Soc.* **2009**, *131*, 4078–4083. [CrossRef]
18. Liu, G.; Yu, J.C.; Lu, G.Q.; Cheng, H.M. Crystal facet engineering of semiconductor photocatalysts: motivations, advances and unique properties. *Chem. Commun.* **2011**, *47*, 6763–6783. [CrossRef]
19. Ye, L.Q.; Mao, J.; Liu, J.Y.; Jiang, Z.; Peng, T.Y.; Zan, L. Synthesis of anatase TiO_2 nanocrystals with {101}, {001} or {010} single facets of 90% level exposure and liquidphase photocatalytic reduction and oxidation activity orders. *J. Mater. Chem. A* **2013**, *1*, 10532–10537. [CrossRef]
20. Li, T.; Shen, Z.L.; Shu, Y.L.; Li, X.G.; Jiang, C.J.; Chen, W. Facet-dependent evolution of surface defects in anatase TiO_2 by thermal treatment: Implications for environmental applications of photocatalysis. *Environ. Sci. Nano* **2019**, *6*, 1740–1753. [CrossRef]
21. Feng, J.Y.; Yin, M.C.; Wang, Z.Q.; Yan, S.C.; Wan, L.J.; Li, Z.S.; Zou, Z.G. Facile synthesis of anatase TiO_2 mesocrystal sheets with dominant {001} facets based on topochemical conversion. *CrystEngComm* **2010**, *12*, 3425–3429. [CrossRef]
22. Illa, S.; Boppella, R.; Manorama, S.V.; Basak, P. Mesoporous assembly of cuboid anatase nanocrystals into hollow spheres: Realizing enhanced photoactivity of high energy {001} facets. *J. Phys. Chem. C* **2016**, *120*, 18028–18038. [CrossRef]
23. Khalil, M.; Anggraeni, E.S.; Ivandini, T.A.; Budianto, E. Exposing TiO_2 (001) crystal facet in nano Au-TiO_2 heterostructures for enhanced photodegradation of methylene blue. *Appl. Surf. Sci.* **2019**, *487*, 1376–1384. [CrossRef]

24. Ye, L.Q.; Liu, J.Y.; Tian, L.H.; Peng, T.Y.; Zan, L. The replacement of {101} by {010} facets inhibits the photocatalytic activity of anatase TiO_2. *Appl. Catal., B* **2013**, *134–135*, 7515–7519. [CrossRef]
25. Amoli, V.; Bhat, S.; Maurya, A.; Banerjee, B.; Bhaumik, A.; Sinha, A.K. Tailored synthesis of porous TiO_2nanocubes and nanoparallelepipeds with exposed {111} facets and mesoscopic void space: A superior candidate for efficient dye-sensitized solar cells. *ACS Appl. Mater. Interfaces* **2015**, *7*, 26022–26035. [CrossRef]
26. Niu, X.J.; Du, Y.E.; Liu, Y.F.; Qi, H.X.; An, J.; Yang, X.J.; Feng, Q. Hydrothermal synthesis and formation mechanism of the anatase nanocrystals with co-exposed highenergy {001}, {010} and [111]-facets for enhanced photocatalytic performance. *RSC Adv.* **2017**, *7*, 24616–24627. [CrossRef]
27. Liu, Y.F.; Du, Y.-E.; Bai, Y.; An, J.; Li, J.Q.; Yang, X.J.; Feng, Q. Facile synthesis of {101}, {010} and [111]-Faceted anatase-TiO_2 nanocrystals derived from porous metatitanic acid H_2TiO_3 for enhanced photocatalytic performance. *ChemistrySelect* **2018**, *3*, 2867–2876. [CrossRef]
28. Gerbec, J.A.; Magana, D.; Washington, A.; Strouse, G.F. Microwave-enhanced reaction rates for nanoparticle synthesis. *J. Am. Chem. Soc.* **2005**, *127*, 15791–15800. [CrossRef]
29. Yoon, S.; Lee, E.-S.; Manthiram, A. Microwave-solvothermal synthesis of various polymorphs of nanostructured TiO_2 in different alcohol media and their lithium ion storage properties. *Inorg. Chem.* **2012**, *51*, 3505–3512. [CrossRef]
30. Krishnapriya, R.; Praneetha, S.; Murugan, A.V. Microwave-solvothermal synthesis of different TiO_2 nanomorphologies with perked up efficiency escorted by incorporating Ni nanoparticle in electrolyte for dye sensitized solar cells. *Inorg. Chem. Front.* **2017**, *4*, 1665–1678. [CrossRef]
31. Kokel, A.; Schäfer, C.; Török, B. Application of microwave-assisted heterogeneous catalysis in sustainable synthesis design. *Green Chem.* **2017**, *19*, 3729–3751. [CrossRef]
32. Nunes, D.; Pimentel, A.; Santos, L.; Barquinha, P.; Fortunato, E.; Rodrigo Martins, R. Photocatalytic TiO_2 nanorod spheres and arrays compatible with flexible applications. *Catalysts* **2017**, *7*, 60. [CrossRef]
33. Shi, X.C.; Zhang, Z.B.; Zhou, D.F.; Zhang, L.F.; Chen, B.Z.; Yu, L.L. Synthesis of Li^+ adsorbent (H_2TiO_3) and its adsorption properties. *Trans. Nonferrous Met. Soc. China* **2013**, *23*, 253–259. [CrossRef]
34. Wang, C.-C.; Li, J.-R.; Lv, X.-L.; Zhang, Y.-Q.; Guo, G.S. Photocatalytic organic pollutants degradation in metal–organic frameworks. *Energy Environ. Sci.* **2014**, *7*, 2831–2867. [CrossRef]
35. Liao, Y.L.; Que, W.X.; Jia, Q.Y.; He, Y.C.; Zhang, J.; Zhong, P. Controllable synthesis of brookite/anatase/rutile TiO_2 nanocomposites and single-crystalline rutile nanorods array. *J. Mater. Chem.* **2012**, *22*, 7937–7944. [CrossRef]
36. Zheng, J.-Y.; Bao, S.-H.; Guo, Y.; Jin, P. Anatase TiO_2 films with dominant {001} facets fabricated by direct-current reactive magnetron sputtering at room temperature: oxygen defects and enhanced visible-light photocatalytic behaviors. *ACS Appl. Mater. Interfaces* **2014**, *6*, 5940–5946. [CrossRef]
37. Azuma, K.; Dover, C.; Grinter, D.C.; Grau-Grespo, R.; Almora-Barrios, N.; Thornton, G.; Oda, T.; Tanaka, S. Scanning tunneling microscopy and molecular dynamics study of the Li_2TiO_3 (001) surface. *J. Phys. Chem. C* **2013**, *117*, 5126–5131. [CrossRef]
38. Chitrakar, R.; Makita, Y.; Ooi, K.; Sonoda, A. Lithium recovery from salt brine H_2TiO_3. *Dalton Trans.* **2014**, *43*, 8933–8939. [CrossRef]
39. Yu, C.-L.; Yanagisawa, K.; Kamiya, S.; Kozawa, T.; Ueda, T. Monoclinic Li_2TiO_3nano-particles via hydrothermal reaction: processing and structure. *Ceram. Int.* **2014**, *40*, 1901–1908. [CrossRef]
40. Hou, W.H.; Chen, Y.S.; Guo, C.X.; Yan, Q.J. Synthesis of porous chromia-pillared tetratitanate. *J. Solid State Chem.* **1998**, *136*, 320–321. [CrossRef]
41. Liu, Z.-H.; Ooi, K.; Kanoh, H.; Tang, W.-P.; Tomida, T. Swelling and delamination behaviors of birnessite-type manganese oxide by intercalation of tetraalkylammonium ions. *Langmuir* **2000**, *16*, 4154–4164. [CrossRef]
42. Allen, M.R.; Thibert, A.; Sabio, E.M.; Browning, N.D.; Larsen, D.S.; Osterloh, F.E. Evolution of physical and photocatalytic properties in the layered titanates $A_2Ti_4O_9$ (A = K, H) and nanosheets derived by chemical exfoliation. *Chem. Mater.* **2010**, *22*, 1220–1228. [CrossRef]
43. Pal, S.; Laera, A.M.; Licciulli, A.; Catalano, M.; Taurino, A. Biphase TiO_2 microspheres with enhanced photocatalytic activity. *Ind. Eng. Chem. Res.* **2014**, *53*, 7931–7938. [CrossRef]
44. Liu, S.W.; Yu, J.G.; Jaroniec, M. Tunable photocatalytic selectivity of hollow TiO_2 microspheres composed of anatase polyhedral with exposed {001} facets. *J. Am. Chem. Soc.* **2010**, *132*, 11914–11916. [CrossRef]

45. Yang, W.G.; Wang, Y.L.; Shi, W.M. One-step synthesis of single-crystal anatase TiO_2 tetragonal faceted-nanorods for improved-performance dye-sensitized solar cells. *CrystEngComm* **2012**, *14*, 230–234. [CrossRef]
46. Zuo, F.; Dillon, R.J.; Wang, L.; Smith, P.; Zhao, X.; Bardeen, C.; Feng, P. Active facets on titanium(III)-doped TiO_2: A effective strategy to improve the visible-light photocatalytic. *Angew. Chem., Int. Ed.* **2012**, *51*, 6223–6226. [CrossRef]
47. Liu, B.; Khare, A.; Aydil, E.S. TiO_2-B/anatase core-shell heterojunction nanowires for photocatalysis. *ACS Appl. Mater. Inter.* **2011**, *3*, 4444–4450. [CrossRef]
48. Wu, T.X.; Liu, G.M.; Zhao, J.C. Photoassisted degradation of dye pollutants. v. self-photosensitized oxidative transformation of rhodamine B under visible light irradiation in aqueous TiO_2 dispersions. *J. Phys. Chem. B* **1998**, *102*, 5845–5851. [CrossRef]
49. Zhang, T.; Oyama, T.; Aoshima, A.; Hidaka, H.; Zhao, J.; Serpone, N. Photooxidative N-demethylation of methylene blue in aqueous TiO_2 dispersions under UV irradiation. *J. Photochem. Photobiol. A Chem.* **2001**, *140*, 163–172. [CrossRef]
50. Liu, X.; Li, Y.; Deng, D.; Chen, N.; Xing, X.; Wang, Y. A one-step nonaqueous sol–gel route to mixed-phase TiO_2 with enhanced photocatalytic degradation of rhodamine B under visible light. *CrystEngComm.* **2016**, *18*, 1964–1975. [CrossRef]
51. Zhang, H.M.; Liu, P.R.; Li, F.; Liu, H.W.; Wang, Y.; Zhang, S.Q.; Guo, M.X.; Cheng, H.M.; Zhao, H.J. Facile fabrication of anatase TiO_2 microspheres on solid substrates and surface crystal facet transformation from {001} and {100}. *Chem. Eur. J.* **2011**, *17*, 5949–5957. [CrossRef] [PubMed]
52. Fu, W.W.; Li, G.D.; Wang, Y.; Zeng, S.J.; Yan, Z.J.; Wang, J.W.; Xin, S.G.; Zhang, L.; Wu, S.W.; Zhang, Z.T. Facile formation of mesoporous structured mixed-phase (anatase/rutile) TiO_2 with enhanced visible light photocatalytic activity. *Chem. Commun.* **2018**, *54*, 58–61. [CrossRef] [PubMed]
53. Hu, D.W.; Zhang, W.X.; Tanaka, Y.; Kusunose, N.; Peng, Y.; Feeng, Q. Mesocrystalline nanocomposition of TiO_2 polymorphs: Topochemical mesocrystal conversion, characterization, and photocatalytic response. *Cryst. Growth Des.* **2015**, *15*, 1214–1225. [CrossRef]
54. Du, Y.-E.; Bai, Y.; Liu, Y.F.; Guo, Y.Q.; Cai, X.M.; Feng, Q. One-pot synthesis of [111]-/{010} facets coexisting anatase nanocrystals with enhanced dye-sensitized solar cell performance. *ChemistrySelect* **2016**, *1*, 6632–6640. [CrossRef]
55. Liu, M.; Piao, L.Y.; Zhao, L.; Ju, S.T.; Yan, Z.J.; He, T.; Zhou, C.L.; Wang, W.J. Anatase TiO_2 single crystals with exposed {001} and {110} facets: facile synthesis and enhanced photocatalysis. *Chem. Commun.* **2010**, *46*, 1664–1666. [CrossRef] [PubMed]
56. Wen, P.H.; Xue, M.; Ishikawa, Y.; Itoh, H.; Feng, Q. Relationships between cell parameters of dye-sensitized solar cells and dye-adsorption parameters. *ACS Appl. Mater. Interfaces* **2012**, *4*, 1928–1934. [CrossRef] [PubMed]
57. Wang, H.E.; Zheng, L.X.; Liu, C.P.; Liu, Y.K.; Luan, C.Y.; Cheng, H.; Li, Y.Y.; Martinu, L.; Zapien, J.A.; Bello, I. Rapid microwave synthesis of porous TiO_2 spheres and their applications in dye-sensitized solar cells. *J. Phys. Chem. C* **2011**, *115*, 10419–10425. [CrossRef]
58. Chen, J.Z.; Li, B.; Zheng, J.F.; Jia, S.P.; Zhao, J.H.; Jing, H.W.; Zhu, Z.P. Role of one-dimensional ribbonlike nanostructures in dye-sensitized TiO_2-based solar cells. *J. Phys. Chem. C* **2011**, *115*, 7104–7113. [CrossRef]
59. Chen, C.D.; Sewvandi, G.A.; Kusunose, T.; Tanaka, Y.; Nakanishi, S.; Feng, Q. Synthesis of {010}-faceted anatase TiO_2 nanoparticles from layered titanate for dye-sensitized solar cells. *CrystEngComm* **2014**, *16*, 8885–8895. [CrossRef]
60. Du, Y.-E.; Feng, Q.; Chen, C.D.; Tanaka, Y.; Yang, X.J. Photocatalytic and dye-sensitized solar cell performances of {010}-faceted and [111]-faceted anatase TiO_2 nanocrystals synthesized from tetratitanate nanoribbons. *ACS Appl. Mater. Interfaces* **2014**, *6*, 16007–16019. [CrossRef]

Article

Photocatalytic Activity and Mechanical Properties of Cements Modified with TiO_2/N

Magdalena Janus [1,2,*], Szymon Mądraszewski [2], Kamila Zając [1], Ewelina Kusiak-Nejman [3], Antoni W. Morawski [3] and Dietmar Stephan [2]

1 Faculty of Civil Engineering and Architecture, West Pomeranian University of Technology, Szczecin, al. Piastów 50, 70-311 Szczecin, Poland; kamila.zajac@zut.edu.pl
2 Building Materials and Construction Chemistry, Technische Universität Berlin, Gustav-Meyer-Allee 25, 13355 Berlin, Germany; szymon.madraszewski@tu-berlin.de (S.M.); stephan@tu-berlin.de (D.S.)
3 Faculty of Chemical Technology and Engineering, West Pomeranian University of Technology, Szczecin, ul. Pułaskiego 10, 70-310 Szczecin, Poland; ewelina.kusiak@zut.edu.pl (E.K.-N.); amor@zut.edu.pl (A.W.M.)
* Correspondence: mjanus@zut.edu.pl; Tel.: +48-91-449-4083

Received: 30 September 2019; Accepted: 12 November 2019; Published: 14 November 2019

Abstract: In this paper, studies of the mechanical properties and photocatalytic activity of new photoactive cement mortars are presented. The new building materials were obtained by the addition of 1, 3, and 5 wt % (based on the cement content) of nitrogen-modified titanium dioxide (TiO_2/N) to the cement matrix. Photocatalytic active cement mortars were characterized by measuring the flexural and the compressive strength, the hydration heat, the zeta potential of the fresh state, and the initial and final setting time. Their photocatalytic activity was tested during NOx decomposition. The studies showed that TiO_2/N gives the photoactivity of cement mortars during air purification with an additional positive effect on the mechanical properties of the hardened mortars. The addition of TiO_2/N into the cement shortened the initial and final setting time, which was distinctly observed using 5 wt % of the photocatalyst in the cement matrix.

Keywords: photoactive cement; TiO_2/N; NOx decomposition; mechanical properties

---

1. Introduction

In the last few decades, nanoparticles have been considered as an additive to the concrete and related cement products in order to improve the properties of building materials [1]. The first documented addition of the nanoparticles to a cement-based system occurred in 1964 when the nano-SiO_2 facilitated a faster and more complete hydration of cement [2]. However, the application of various nanoparticles, such as nano-TiO_2, nano-Al_2O_3, and nano-Fe_2O_3 in cement and concrete materials has developed intensively since circa 2004 [3–5]. Combining TiO_2 nanoparticles with cementitious binders appeared to be one of the most promising ways to obtain environmentally friendly products [6]. Namely, a TiO_2 photocatalyst, when activated by the suitable light, is capable of supporting the chemical reactions, which can degrade an atmospheric pollutant and give a self-cleaning property [7]. It is worth pointing out that the building surfaces are exposed to the highest levels of air pollution and at the same time to the solar radiation, which is necessary in photocatalytic processes.

In the urban areas, NOx (NO + NO_2) is one of the most common pollutants from the external sources (traffic, industry) [8]. NOx contributes to the formation of the photochemical smog, and it is associated with lung problems and asthma [9]. The potential of cementitious materials containing photocatalysts to decrease the NOx concentration was proven many times [10–12]. For example, Lee et al. [13] studied changes in NO and NO_2 concentration using TiO_2-containing cement-based materials during UV irradiation, suggesting that the materials are capable of oxidizing both gases efficiently. It was observed the similar amounts of NO and NO_2 gases were degraded at 3 h, regardless of the variations in the water/cement ratio. The mechanism of NOx degradation consists of a series of reactions that take place during the photocatalytic process. Typically, it can be described as a sequential oxidation process, as follows: $NO \rightarrow HNO_2 \rightarrow NO_2^- \rightarrow NO_3^-$ [8,11].

Many works in the photocatalytic branch are directed at modification of the base TiO_2 structure through doping of the photocatalyst with the non-metals or the metal ions [14,15]. Mainly, it can enhance its activation in the visible light [16], but other advantages are also observed. In the treatment of NOx, modifications of TiO_2 can improve the catalytic selectivity toward nitrate rather than the more toxic NO_2 [7].

Although the photocatalytic cements and concretes have been extensively studied [17], it is still controversial whether the added photocatalyst enhances building properties. On the one hand, the presence of TiO_2 nanoparticles can have a positive filler effect in cement mortars, increasing the mechanical strengths. One of the reported best performance enhancement results of the inclusion of TiO_2 nanoparticles in the cementitious materials included a 45% increase in the compressive strength [18] and an 87% increase in the flexural strength [19]. Yang et al. [20] indicated that the addition of 0.5 wt % TiO_2 to cement slag pastes allowed achieving approximately 10%, 15%, and 9% higher compressive strengths in comparison to the reference material at 3 d, 7 d, and 28 d, respectively. Meanwhile, the flexural strengths of the same photocatalytic materials were 25%, 25%, and 38% higher than the reference sample after the same range time of the curing. On the other hand, researchers also observed [21,22] a slight decrease in the mechanical strength of the photocatalytic cement mortars, which has been attributed to the decline of the sample's homogeneity and the formation of weak zones in the structure of hardened mortars.

Some authors [7] indicated that the effective use of photocatalysts in cement is highly connected with the assurance of the optimized dispersion of TiO_2 particles in the cement matrix. The agglomeration of TiO_2 particles can interfere not only with the mechanical strength, but also block access to an internal surface of TiO_2, limiting the photocatalytic efficiency. The degree of repelling between particles of cement mortars has a direct relationship with "zeta potential", which shows the electrokinetic behavior of particles and gives a valuable indication of the surface charge state, achieving values from −30 mV to +30 mV [23,24]. When the constituent particles of mortars have the same charge, they tend to repeal each other, and no agglomeration occurs. Up to now, the zeta potential measurements of cementitious materials have been performed with a low fraction of solid. Therefore, it is challenging to obtain information about the zeta potential values of real fresh cement mortars and even more so referring to photocatalytic cement mortars. Lowke and Gehlen [24] considered the zeta potential of Portland cement and mineral additions in cement suspensions with high solid fractions. They found a continuous increase in the zeta potential of cementitious suspensions with the increasing w/c ratio (water/cement). Moreover, the determining factor on the zeta potential appeared to be the molar Ca/SO_4 concentration ratio, which was more crucial than the effect of the type of addition.

The absolute values of zeta potential may vary not only with mortar composition or w/c ratio but also with the time of hydration [25]. The hydration mechanism of cement consists of the reactions of cement components (e.g., alite or tricalcium silicate, belite or dicalcium silicate) with water. The formed crystalline calcium hydroxide and calcium–silicate–hydrate (C-S-H) comprise over 60 wt % of the hydration products in the total mass [26]. As the reactions continue with time, the hydration products gradually bind together and with other components of concrete to form a solid mass. The hydration of cement is an exothermic chemical reaction. The generation of heat is highly determined by the chemical composition of the cement mixture. It was reported that nanoparticles of TiO_2 could significantly change the hydration of cement and influence the rates of heat evolution [27,28]. The cement hydration is directly related to the setting time of cement mortars. Mostly, the photocatalytic cements showed a shortened initial and final setting time for the samples with higher TiO_2 contents, which is attributed to the acceleration of the hydration rate [29].

The aim of this paper is to present the results of our study on the influence of a prepared TiO_2/N photocatalyst on the properties of the fresh and hardened cement mortars. TiO_2/N was chosen as the photocatalyst because there is the possibility of producing this material in the amount of 0.5 kg per day. Moreover, the technological project of installation for TiO_2/N production exists and may be used for photocatalyst production at a large scale. The photocatalyst has been added in different dosages (1, 3, and 5 wt % to cement mass) to cement mortars. The measurements of the initial and the final setting time, the flexural and the compressive strength, the hydration heat, and the zeta potential were conducted. The photocatalytic activity was monitored during the NOx degradation process.

2. Materials and Methods

2.1. Materials

Ordinary Portland cement CEMI 42.5 N from Holcim, Germany was used in this study. Standard sand, according to EN 196-1, was used for all mortars.

The preparation of the photocatalyst (TiO_2/N) was carried out using HEL Ltd. "Autolab" E746 installation. The commercial titanium dioxide supplied by Grupa Azoty Zakłady Chemiczne 'Police' S.A. (Poland) was used as a starting material. First, 600 g of TiO_2 and 350 mL of NH_4OH with a concentration of 2.5% were placed in an autoclave. The reactor was closed, and the mixture was blended using a magnetic stirrer and heated up to 100 °C for 4 h. Afterwards, the catalyst was dried in air for 4 h at 100 °C. Finally, the obtained photocatalyst TiO_2/N was ground with mortar to form a fine powder. The structural and the textural parameters on N-modified TiO_2 in Table 1 were placed. The results of TEM (transmission electron microscope), XRD (X-ray powder diffraction), FTIR/DRS (fourier transform infrared spectroscopy/diffuse reflectance), XPS (X-ray photoelectron spectroscopy), and Raman spectroscopy in our earlier publication were presented [30]. The presence of nitrogen in the modified titania sample was confirmed by FTIR analysis. The narrow bands at 1640 cm^{-1} and 1440 cm^{-1} are attributed to the hydroxyl (OH) and ammonium (NH_4^+) groups, respectively, while the band at 1536 cm^{-1} could be assigned to either NH_2 or NO_2 and NO groups. The sample was also studied by Raman spectroscopy. The Raman spectra of the sample exhibit four distinct peaks located at 145 cm^{-1}, 393 cm^{-1}, 514 cm^{-1}, and 646 cm^{-1}; those bands correspond to the anatase phase of TiO_2 [30].

Table 1. Structural and textural parameters of N-modified TiO_2.

Photocatalyst	Local Mean Crystallite Size According to TEM [nm]	Global Mean Crystallite Size According to XRD [nm]	Mean Particle Size According to DLS (Dynamic Light Scattering) [nm]	S_{BET} [m^2/g]
TiO_2/N	6.1	10.8	167.6	235

2.2. Specimen Preparation

The specimens 40·40·160 mm^3 and 80·40·10 mm^3 were produced according to EN 196-1 with a water to binder ratio (w/b) = 0.4 and cement to standard sand ratio of 1:3. Cement was replaced by catalyst in 1, 3, and 5 wt % by mass of cement. Samples without replacement were produced as a reference. For each type of mortar, 6 specimens were produced. Masses needed for the preparation of 3 specimens are presented in Table 2.

Table 2. Mass of materials used for the production of three 40·40·160 mm^3 mortar specimens.

Materials	Mass of Used Materials [g]		
	1%	3%	5%
CEM I 42.5 N	444.5	436.5	427.5
TiO_2/N	4.5	13.5	22.5
Standard sand	1350	1350	1350
Water	180	180	180

A standard mixer with a stainless steel bowl with a capacity of 5 dm^3 according to EN 196-1 was used. First, water was poured into a bowl, and cement was added. The mixer was started immediately at low speed (rotation 140 min^{-1}) and after 30 s, standard sand was steadily added for the next 30 s. Afterwards, the mixer was switched to the higher speed (285 min^{-1}) for an additional 30 s. To remove all the mortar adhering to the wall and the bottom part of the bowl, the mixer was stopped for 90 s. In the end, the mixing was continued at high speed for 60 s. The specimens were molded immediately after the preparation. The first layer of mortar was poured into the mold situated on the jolting table and then compacted. The second layer of mortar was poured on the first layer and compacted. The excess mortar was struck off with the straight metal edge. Casting molds containing fresh samples were wrapped with stretch film and stored at room conditions for 24 h. All specimens were demolded after 1 day and were cured in tap water for the next 27 days.

2.3. Compressive and Flexural Strength Measurements

After 28 d, specimens were tested for their flexural and the compressive strength. The flexural and the compressive strength measurements were carried out following EN 196-1. For each mortar type, six 40·40·160 mm^3 specimens were tested for the flexural strength. The prism halves (after the test of flexural strength) were tested for compressive strength, so for each mortar type, 12 specimens were tested. A standard testing machine (ToniNORM 2010.040, Toni/Technik, Berlin, Germany) was used both for the flexural and the compressive strength measurements.

2.4. Setting Time (Vicat Needle Test)

The setting of cement and its rate affects the open time of the mortar. In this study, the influence of the addition of the catalyst on the setting time of cement was tested. The Vicat Apparatus is a device that is used to determine the setting time of the cement paste. In this study, an automatic device ToniSET COMPACT version 05/00, which did 6 parallel tests, was used for determining the setting time. For each mortar type, 2 specimens were tested. Mortar preparation and the setting time measurements were run according to the EN 196-3 standard. During the measurement, the specimens were kept at 20 °C. The water to binder ratio of paste used for the setting time test was w/b = 0.3. The time when the needle stopped 6 mm from the base plate was recorded as the time for the initial setting. The final setting was defined as the time when the needle only made a 0.5 mm mark on the surface.

2.5. Hydration Heat Measurements

Calorimetry data were obtained from externally mixed pastes containing 40 g of cement and 16 g of water, in at least a twofold determination. Data points were recorded every 60 s at 20 °C (Isothermal heat flow calorimeter MC-CAL100, *C3 Analysentechnik*, C3 Prozess und Analystechnik, Haar, Germany).

2.6. Zeta Potential Measurements

A Zeta and Titration 310 instrument from Dispersion Technology was used for the zeta potential measurements without the dilution of samples, which to some extent avoided the differences in the hydration and surface properties between diluted and original samples. First, 16 g of water was added to 40 g of cement and mixing for 20 s. As the background, the centrifuged water from such prepared mortars was used. The average particle sizes of cement amounted to 9 μm.

2.7. NOx Decomposition

The photocatalytic activity of the prepared plates of cement mortar toward the degradation of air pollutions was also proved. In our previous works [31,32], the gaseous NO (1.989 vppm ± 0.040 ppm, Air Liquide) was used as model pollution in photocatalytic tests. NOx removal was evaluated using the experimental installation, whose scheme is presented in Figure 1.

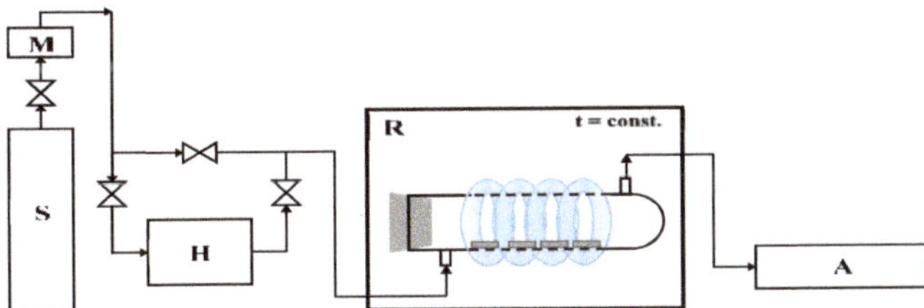

Figure 1. The scheme of installation to the photocatalytic removal of NOx (S—the source of pollution; M—mass flower; H—humidifier; R—photocatalytic reactor with irradiation source; A—NOx analyzer).

The studied plate of cement mortar (one at dimensions of 80 × 40 × 10 mm^3) was placed in the central part of a cylindrical reactor (Pyrex glass; Ø × H = 9 × 32 cm^2), and the reactor was tightly closed. The NO was diluted with humidified synthetic air in a ratio of 1:1. The oxygen and water molecules were necessary for the formation of oxidative species, which are essential in the photocatalytic reactions. The polluted air flowed through the reactor continuously with a rate of 500 cm^3/min. At the beginning of the process, the dark conditions were maintained until NO concentration reached equilibrium (about 1 ppm during about 35 min). Then, the UV lamps were turned on for 30 min. The irradiation sources surrounded the reactor and were characterized by the cumulative intensity of 100 W/m^2 UV and 4 W/m^2 VIS. The temperature of the whole system was stable at the level of 22 °C by using a thermostatic chamber. The NO and NO_2 concentrations were continuously measured in the outlet of the reactor using chemiluminescent NOx analyzer (T200, Teledyne). All measurements were repeated three times, and errors were 2%.

3. Results

3.1. Compressive and Flexural Strength

The compressive and the flexural strength of pure cement and cement with the addition of 1, 3 and 5 wt % TiO_2/N specimens were measured. The obtained results are presented in Figure 2a,b. As it can be seen in Figure 2a, the value of the compressive strength of unmodified cement amounted to 53 MPa (red line), while the addition of 1, 3, and 5 wt % of TiO_2/N increased the compressive strength of the specimens in all cases. The highest value of the compressive strength was observed for specimens with 1 wt % of TiO_2/N and amounted to 57.4 MPa. The lowest increase of the compressive strength was found for a specimen with the addition of 5 wt % of TiO_2/N. Similar behaviour occurred during the flexural strength measurements. As can be seen in Figure 2b, the value of the flexural strength of unmodified cement amounted to 6.92 MPa (red line). Analogous as in the case of the compressive strength, the addition of 1, 3 and 5 wt % of TiO_2/N increased the flexural strength of the specimens in all cases. The highest value of the flexural strength was observed for a specimen with 1 wt % of TiO_2/N and amounted to 7.60 MPa. The lowest increase of flexural strength was obtained using specimen with addition of 5 wt % of TiO_2/N.

The mechanical properties (the compressive and the flexural strength) of cement strongly depend on the amount of used titanium dioxide. Wang et al. [33] discovered that with the incorporation of TiO_2 nanoparticles, the strength firstly showed a fast increase compared with the ordinary mortar until the dosage of TiO_2 nanoparticles reached up to 2 wt %, and then the rate of this increase slowed down. The strength of the cement mortar is closely related to the amount of ettringite and C-S-H gels, and the existence of nanoparticles facilitates the cement hydration, thereby producing more hydration products. In addition to the filler property of nanoparticles to fill the pores in C-S-H gels, it is well known that nanoparticles have a large surface area to volume ratio, and hence, the additional surface area turns out to be an appropriate place for hydration products to precipitate. Nanoparticles enable the formation of a bond between itself and C-S-H gels. As a result, the strength can be accordingly improved. However, there is also an undesirable effect due to the large ratio of surface area to volume, since nanoparticles can glue together, many nanoparticle clusters, and the strength that can be generated is very weak, leading to a heterogeneous microstructure.

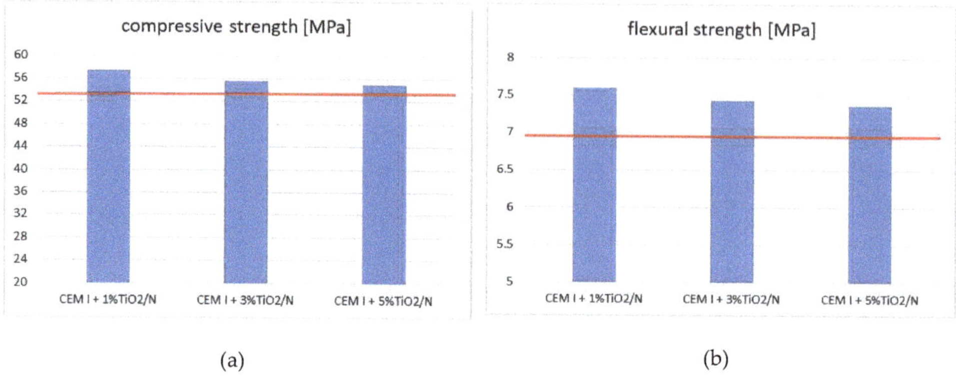

(a) (b)

Figure 2. (**a**) Compressive and (**b**) flexural strength of CEM I 42.5 N with the addition of 1, 3, and 5 wt % of photocatalyst TiO_2/N. In the red line, the compressive strength (53 MPa) and flexural strength (6.92 MPa) of pure CEM I 42.5 was presented.

Beyond 3 wt % nano-TiO_2, the cementing system seems to be saturated, and the poor dispersion of the nanoparticles generated by their high surface area may create weak zones in the system. In addition, it could also enhance the particle packing density of the blended cement by filling up the nanopores and reducing both the larger pores as well as the overall porosity of the mix. This decreased the total specific volume of the pores; the refinement of the pore structure when up to 3 wt % nano-TiO_2 is used as a partial replacement of cement was also reported by Praveenkumar et al. [34] and Nazari and Riahi [35,36].

3.2. Setting Time

The initial and the final setting time of tested specimens is presented in Table 3. As can be seen, with the increasing addition of TiO_2/N to cement, the initial setting time decreased. In the case of the specimens modified by the addition of 5 wt % of TiO_2/N, the initial setting time was 40 min faster than that for unmodified cement. The same behavior was observed for the final setting time. With an increasing addition of TiO_2/N to cement, the final setting time decreased. Specimens of cement with an addition of 5 wt % of TiO_2/N showed a final setting time that was about 57 min faster in comparison to the unmodified cement. A similar observation was made by Hernández-Rodríguez et al. [37]; they added commercial TiO_2 P25 to CEM I 52.5 R and the results showed that the photocatalysts act as a setting accelerator.

Table 3. The values of initial and final setting time of CEM I and CEM I with the addition of 1, 3, and 5 wt % of TiO_2/N photocatalysts.

Samples	The Initial Setting Time [min]	The Final Setting Time [min]
CEM I 42.5N	218	305
CEM + **1** wt %TiO_2/N	217	310
CEM + **3** wt %TiO_2/N	207	275
CEM + **5** wt %TiO_2/N	178	248

3.3. Hydration Heat

In Figure 3, the isothermal calorimetry results of unmodified cement and cement modified by the addition of 1, 3, and 5 wt % of TiO_2/N photocatalysts were presented.

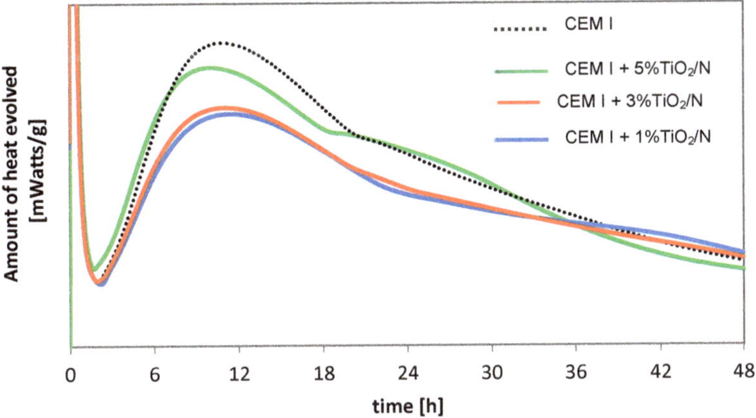

Figure 3. Isothermal calorimetry results for cement modified by the addition of 1, 3, and 5 wt % of TiO_2/N to deionized water at a water to binder ratio (w/b) = 0.4.

According to the literature, there are five stages of heat for a typical Portland cement [38,39]. The addition of modified titanium dioxide into the cement influences hydration heat; the paste with the addition of TiO_2/N showed less heat generated up to 20 h compared to the unmodified cement.

3.4. Zeta Potential Measurements

The average value of zeta potential amounted to −5.01 mV for unmodified cement mortar and −4.90 mV, −4.69 mV and −5.94 mV for cement mortars modified by the addition of 1, 3, and 5 wt % of TiO_2/N, respectively.

It is worth pointing out that TiO_2 photocatalysts are characterized by a negative charge in high pH medium. In our previous work [40], it was proven that the point of zero charge of TiO_2/N is about 5.8. Namely, the TiO_2/N surface appeared to be positively charged at pH < 5.8, whereas it was negatively charged at pH > 5.8. The application of TiO_2/N with highly alkaline cement resulted in the presence of a negatively charged form of TiO_2/N particles.

Zingg et al. [41] concluded that the phases C_3S and C-S-H are positively charged, whereas the ettringite is negatively charged. During the initial stage of cement hydration, the aluminate reacts with water and sulfate, forming a gel-like material (ettringite) surrounding the cement grains. The negative values of zeta potential at the beginning of the hydration process confirmed it.

3.5. NOx Decomposition

In Figure 4, the photocatalytic activity of unmodified and modified cements is presented. The activity of obtained materials during NO removal was tested. The mechanism of photocatalytic NO removal is as follows [42]. Initially, active oxidizing groups are generated at the TiO_2 surface (reactions 1–3):

$$O_2 + e^- \rightarrow O_2^- \tag{1}$$

$$OH^- + h^+ \rightarrow \cdot OH \tag{2}$$

$$H^+ + O_2^- \rightarrow HO_2\cdot \tag{3}$$

The action of these moieties on NO molecules leads to their oxidation to the form of NO_2, followed by the formation of nitric(III) and (V) acids (reactions 4–6):

$$NO + HO_2^- \rightarrow NO_2 + \cdot OH \tag{4}$$

$$NO_2 + \cdot OH \rightarrow HNO_3 \quad (5)$$

$$NO + \cdot OH \rightarrow HNO_2. \quad (6)$$

In Figure 4a, the decreasing of NOx concentration [ppm] is presented. During the first 40 min, the equilibrium of NO was obtained. After 40 min, the UV light was switch on, and it is possible to observe that the NOx concertation decreased. The irradiation takes 30 min, and after this time, the light was switched off. Figure 4b presents the NOx degradation in percent after 30 min of UV light irradiation. The reference sample, pure CEM I, showed the removal of NOx on the level of about 6.3%. The same observation concerning the blank sample was presented by Xu et al. [43]. They found that using reference cement composites without any TiO_2, the NOx concentration slowly decreased by 6% during 15 min of irradiation. It is worth pointing out that in our studies, the photolysis of tested gas amounted to 1.3% under the same conditions and the same irradiation source. The application of nitrogen-modified TiO_2 in cement mortars involved the degradation of NOx on the photocatalytic path, which can be observed as the unambiguous decrease of NOx concentration directly after turning on the irradiation. The increase of TiO_2/N loading in cement matrix caused the increase of the NOx degradation rate from 14.2% for CEM I + 1 wt %TiO_2/N to 22.9% for CEM I + 5 wt %TiO_2/N. Apart from the influence of the photocatalyst dose in the cement matrix, the accessible surface area of the photocatalyst is essential for the photocatalytic effectiveness [10]. Therefore, we did not observe a proportional increase of NOx degradation rate with the higher TiO_2/N loading. However, it appeared that the nitrogen-modified photocatalyst might be used as an additive to cement materials to increase its air purification properties. Moreover, in Table 4, the NO removal and NO_2 formation during the photocatalytic oxidation of NO are presented.

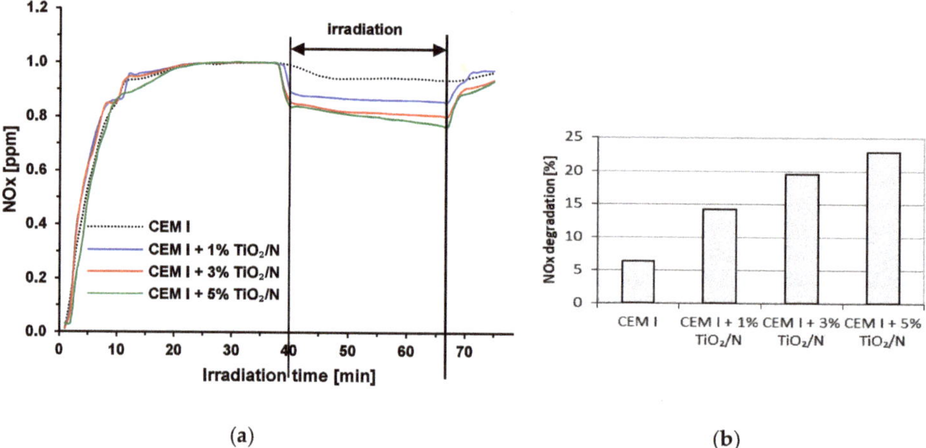

(a) (b)

Figure 4. (**a**) Graph of NOx [ppm] decomposition and (**b**) NOx degradation [%] on CEM I samples, and cements modified by the addition of 1, 3, and 5 wt % of TiO_2/N under UV light irradiation.

Table 4. The NO removal and NO_2 creation during NO photooxidation with cement modified by TiO_2/N.

Sample	NO Removal [ppm]	NO_2 Formation [ppm]	NOx Removal [ppm]
Photolysis	0.023	0.013	0.010
CEM I	0.057	0.009	0.048
CEM I + 1% TiO_2/N	0.141	0.030	0.111
CEM I + 3% TiO_2/N	0.179	0.025	0.154
CEM I + 5% TiO_2/N	0.211	0.032	0.179

In Table 5, the initial photodegradation rates are presented. It was calculated 5 min after switching on the UV light. This value was calculated as µg of NO removal, NO_2 creation, and NOx total removal on the surface of modified cement plates [cm^2] during the time of UV light irradiation [h]. As it can be seen, the highest vales of NO removal, NO_2 creation, and NOx total removal were when the cement was modified by the addition of 5 wt % of TiO_2/N.

Table 5. The initial photodegradation rate for modified cement during NO removal, NO creation, and NOx total removal.

Sample	NO Removal [µg/cm^2/h]	NO_2 Formation [µg/cm^2/h]	NOx Total Removal [µg/cm^2/h]
Photolysis	0.289	0.038	0.251
CEM I	0.315	0.091	0.224
CEM I + 1% TiO_2/N	2.530	0.622	1.908
CEM I + 3% TiO_2/N	3.145	0.496	2.649
CEM I + 5% TiO_2/N	3.403	0.651	2.752

The similar results of NOx photocatalytic degradation on cement materials were observed by other authors as well. It was reported [13] that 5% TiO_2 replacement by the mass of cement in cement pastes allowed decreasing the NO concentration from 1 ppm to about 0.7 ppm. The results were calculated after 3 h of exposure to UV irradiation, because it was the necessary time to achieve the relative stasis in NO concentration. Jimenez-Relinque et al. [21] applied 2% of commercial TiO_2 with different types of cement in normalized mortars. NO gas diluted in the air was used as model pollutant with an initial concentration of 1 ppm ± 50 ppb. After 1 h of UV irradiation, they obtained NO photocatalytic degradation on the level of 15–30% and NOx removal in the range of 18–25%, depending on the applied cement type.

In Figure 5, the lifetime of tested modified cement plates was presented. As it can be seen, there was no decrease in the photocatalytic activity of the modified cement plates. NO removal and total NOx removal are on the same level. There are only small differences between NO and NOx concentration, and this behavior suggests that with time (increasing the number of cycles), more NO_2 is produced.

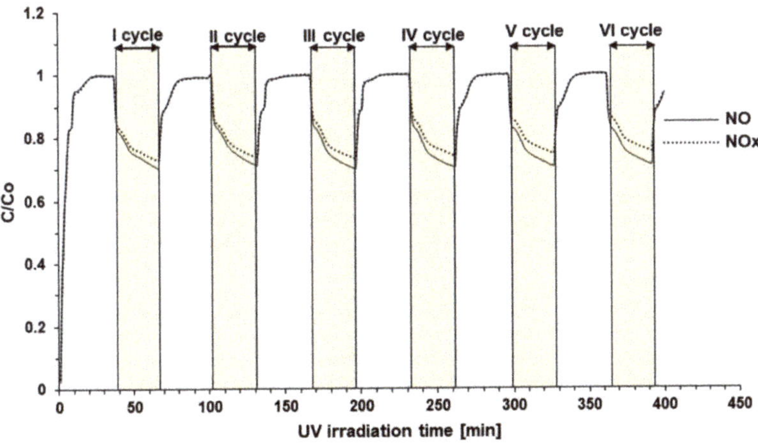

Figure 5. The lifetime of CEM I + 5 wt %TiO_2/N under six cycles of irradiation.

4. Conclusions

The nitrogen-modified titanium dioxide (TiO_2/N) may be used as an additive to cement mortars to produce the cement with photocatalytic properties. All photocatalytic samples degraded regarding the NOx concentration during irradiation time, achieving a higher NOx removal rate with a higher TiO_2/N dosage in cement materials. The addition of TiO_2/N up to 5 wt % into the cement mortar did not decrease the mechanical properties but even slightly increased the compressive and the flexural strength.

Nanoparticles of TiO_2/N appeared to have an influence on the cement hydration. Acceleration of the initial and the final setting time indicated that the photocatalytic particles might act as seeds for the precipitation of C-S-H. The addition of 5 wt % of TiO_2/N into the cement mortar shortened the setting time by about 57 min. Moreover, the presence of TiO_2/N in the cement matrix caused less heat to be generated during the hydration process.

The negative charge of high solid cement mortar, which was determined based on the zeta potential, was amplified using a higher amount of TiO_2/N photocatalyst, from −4.3 mV to −5.5 mV at the beginning of hydration. High TiO_2/N loading in the cement matrix resulted in more negative zeta potential, because the very fine TiO_2 is negatively charged at a high pH.

Author Contributions: Conceptualization, M.J. and S.M.; methodology, M.J. and S.M.; investigation, M.J., S.M. and K.Z.; writing—original draft preparation, M.J., S.M. and K.Z.; photocatalyst preparation, A.W.M. writing—review and editing, M.J., S.M., K.Z. and D.S.; supervision, M.J and D.S.; funding acquisition, M.J. and E.K.-N.

Funding: This research was funded by the Polish National Agency for Academic Exchange within the Bekker program.

Conflicts of Interest: The authors declare no conflict of interest.

References

1. Reches, Y. Nanoparticles as concrete additives: Review and perspectives. *Constr. Build. Mater.* **2018**, *175*, 483–495. [CrossRef]
2. Stein, H.; Stevels, J. Influence of silica on the hydration of 3 CaO, SiO_2. *J. Appl. Chem.* **1964**, *14*, 338–346. [CrossRef]
3. Li, H.; Xiao, H.-G.; Ou, J.-P. A study on mechanical and pressure-sensitive properties of cement mortar with nanophase materials. *Cem. Concr. Res.* **2004**, *34*, 435–438. [CrossRef]
4. Li, H.; Xiao, H.-G.; Yuan, J.; Ou, J. Microstructure of cement mortar with nano-particles. *Compos. B Eng.* **2004**, *35*, 185–189. [CrossRef]
5. Land, G.; Stephan, D. Controlling cement hydration with nanoparticles. *Cem. Concr. Compos.* **2015**, *57*, 64–67. [CrossRef]
6. Karapati, S.; Giannakopoulou, T.; Todorova, N.; Boukos, N.; Antiohos, S.; Papageorgiou, D.; Chaniotakis, E.; Dimotikali, D.; Trapalis, C. TiO_2 functionalization for efficient NOx removal in photoactive cement. *Appl. Surf. Sci.* **2014**, *319*, 29–36. [CrossRef]
7. Macphee, D.E.; Folli, A. Photocatalytic concretes—The interface between photocatalysis and cement chemistry. *Cem. Concr. Res.* **2016**, *85*, 48–54. [CrossRef]
8. Lucas, S.S.; Ferreira, V.M.; de Aguiar, J.B. Incorporation of titanium dioxide nanoparticles in mortars—Influence of microstructure in the hardened state properties and photocatalytic activity. *Cem. Concr. Res.* **2013**, *43*, 112–120. [CrossRef]
9. Cárdenas, C.; Tobón, J.I.; García, C.; Vila, J. Functionalized building materials: Photocatalytic abatement of NOx by cement pastes blended with TiO_2 nanoparticles. *Constr. Build. Mater.* **2012**, *36*, 820–825. [CrossRef]
10. Seo, D.; Yun, T.S. NOx removal rate of photocatalytic cementitious materials with TiO_2 in wet condition. *Build. Environ.* **2017**, *112*, 233–240. [CrossRef]
11. Yang, L.; Hakki, A.; Wang, F.; Macphee, D.E. Photocatalyst efficiencies in concrete technology: The effect of photocatalyst placement. *Appl. Catal. B Environ.* **2018**, *222*, 200–208. [CrossRef]
12. Guo, M.-Z.; Chen, J.; Xia, M.; Wang, T.; Poon, C.S. Pathways of conversion of nitrogen oxides by nano TiO_2 incorporated in cement-based materials. *Build. Environ.* **2018**, *144*, 412–418. [CrossRef]
13. Lee, B.Y.; Jayapalan, A.R.; Bergin, M.H.; Kurtis, K.E. Photocatalytic cement exposed to nitrogen oxides: Effect of oxidation and binding. *Cem. Concr. Res.* **2014**, *60*, 30–36. [CrossRef]
14. Chen, J.; Qiu, F.; Xu, W.; Cao, S.; Zhu, H. Recent progress in enhancing photocatalytic efficiency of TiO_2-based materials. *Appl. Catal. A Gen.* **2015**, *495*, 131–140. [CrossRef]

15. Agbe, H.; Nyankson, E.; Raza, N.; Dodoo-Arhin, D.; Chauhan, A.; Osei, G.; Kumar, V.; Kim, K.-H. Recent advances in photoinduced catalysis for water splitting and environmental applications. *J. Ind. Eng. Chem.* **2019**, *72*, 31–49. [CrossRef]
16. Rimoldi, L.; Pargoletti, E.; Meroni, D.; Falletta, E.; Cerrato, G.; Turco, F.; Cappelletti, G. Concurrent role of metal (Sn, Zn) and N species in enhancing the photocatalytic activity of TiO_2 under solar light. *Catal. Today* **2018**, *313*, 40–46. [CrossRef]
17. Li, Z.; Ding, S.; Yu, X.; Han, B.; Ou, J. Multifunctional cementitious composites modified with nano titanium dioxide: A review. *Compos. Part A Appl. Sci. Manuf.* **2018**, *111*, 115–137. [CrossRef]
18. Rahim, A.; Nair, S.R. Influence of nano-materials in high strength concrete. *J. Chem. Pharm. Sci.* **2016**, *974*, 15–21.
19. Han, B.; Li, Z.; Zhang, L.; Zeng, S.; Yu, X.; Han, B.; Ou, J. Reactive powder concrete reinforced with nano SiO_2-coated TiO_2. *Constr. Build. Mater.* **2017**, *148*, 104–112. [CrossRef]
20. Yang, L.Y.; Jia, Z.J.; Zhang, Y.M.; Dai, J.G. Effects of nano-TiO_2 on strength, shrinkage and microstructure of alkali activated slag pastes. *Cem. Concr. Comp.* **2015**, *57*, 1–7. [CrossRef]
21. Jimenez-Relinque, E.; Rodriguez-Garcia, J.R.; Castillo, A.; Castellote, M. Characteristics and efficiency of photocatalytic cementitious materials: Type of binder, roughness and microstructure. *Cem. Concr. Res.* **2015**, *71*, 124–131. [CrossRef]
22. Zhao, A.; Yang, J.; Yang, E.-H. Self-cleaning engineered cementitious composites. *Cem. Concr. Comp.* **2015**, *64*, 74–83. [CrossRef]
23. Yousefi, A.; Allahverdi, A.; Hejazi, P. Effective dispersion of nano-TiO_2 powder for enhancement of photocatalytic properties in cement mixes. *Constr. Build. Mater.* **2013**, *41*, 224–230. [CrossRef]
24. Lowke, D.; Gehlen, C. The zeta potential of cement and additions in cementitious suspensions with high solid fraction. *Cem. Concr. Res.* **2017**, *95*, 195–204. [CrossRef]
25. Plank, J.; Hirsch, C. Impact of zeta potential of early cement hydration phases on superplasticizer adsorption. *Cem. Concr. Res.* **2007**, *37*, 537–542. [CrossRef]
26. Jennings, H.M. Refinements to colloid model of C-S-H in cement: CM-II. *Cem. Concr. Res.* **2008**, *38*, 275–289. [CrossRef]
27. Paul, S.C.; van Rooyen, A.S.; van Zijl, G.P.; Petrik, L.F. Properties of cement-based composites using nanoparticles: A comprehensive review. *Constr. Build. Mater.* **2018**, *189*, 1019–1034. [CrossRef]
28. Chen, J.; Kou, S.C.; Poon, C.S. Hydration and properties of nano-TiO_2 blended cement composites. *Cem. Concr. Comp.* **2012**, *34*, 642–649. [CrossRef]
29. Zhang, R.; Cheng, X.; Hou, P.; Ye, Z. Influences of nano-TiO_2 on the properties of cement-based materials: Hydration and drying shrinkage. *Constr. Build. Mater.* **2015**, *81*, 35–41. [CrossRef]
30. Bubacz, K.; Choina, J.; Dolat, D.; Borowiak-Paleń, E.; Moszyński, D.; Morawski, A.W. Studies on nitrogen modified TiO_2 photocatalyst prepared in different conditions. *Mater. Res. Bull.* **2010**, *45*, 1085–1091. [CrossRef]
31. Zając, K.; Janus, M.; Kuźmiński, K.; Morawski, A.W. Preparation of gypsum building materials with photocatalytic properties. A strong emphasis on waste gypsum from flue gas desulfurization. *Przemysł Chem.* **2016**, *95*, 2222–2226.
32. Janus, M.; Zając, K.; Ehm, C.; Stephan, D. Fast Method for Testing the Photocatalytic Performance of Modified Gypsum. *Catalysts* **2019**, *9*, 693–701. [CrossRef]
33. Wang, L.; Zhang, H.; Gao, Y. Effect of TiO_2 Nanoparticles on Physical and Mechanical Properties of Cement at Low Temperatures. *Adv. Mater. Sci. Eng.* **2018**. [CrossRef]
34. Praveenkumar, T.R.; Vijayalakshim, M.M.; Meddah, M.S. Strengths and durability performances of blended cement concrete with TiO_2 nanoparticles and rice husk ash. *Constr. Build. Mater.* **2019**, *217*, 343–351. [CrossRef]
35. Nazari, A.; Riahi, S. The effect of TiO_2 nanoparticles on water permeability and thermal and mechanical properties of high strength self-compacting concrete. *Mater. Sci. Eng. A* **2010**, *528*, 756–763. [CrossRef]
36. Nazari, A.; Riahi, S. The effects of TiO_2 nanoparticles on physical, thermal and mechanical properties of concrete using ground granulated blast furnace slag as binder. *Mater. Sci. Eng. A* **2011**, *528*, 2085–2092. [CrossRef]

37. Hernández-Rodríguez, M.J.; Santana Rodríguez, R.; Darias, R.; González Díaz, O.; Pérez Luzardo, J.M.; Doña Rodríguez, J.M.; Pulido Melián, E. Effect of TiO_2 Addition on Mortars: Characterisation and Photoactivity. *Appl. Sci.* **2019**, *9*, 2598–2611. [CrossRef]
38. Taylor, P.C.; Kosmatka, G.F.; Voigt, G.F. Integrated Materials and Constructions Practices for Concrete Pavement: A State-of-the-Practice Manual. 2006. Available online: https://intrans.iastate.edu/app/uploads/2018/03/imcp_manual_october2007.pdf (accessed on 15 June 2019).
39. Bullard, J.W.; Jennings, H.M.; Livingston, R.A.; Nonat, A.; Scherer, G.W.; Schweitzer, J.S.; Scrivener, K.L.; Thomas, J.J. Mechanisms of cement hydration. *Cem. Concr. Res.* **2011**, *41*, 1208–1223. [CrossRef]
40. Mozia, S.; Bubacz, K.; Janus, M.; Morawski, A.W. Decomposition of 3-chlorophenol on nitrogen modified TiO2 photocatalysts. *J. Hazard. Mater.* **2012**, *203*, 128–136. [CrossRef]
41. Zingg, A.; Winnefeld, F.; Holzer, L.; Pakusch, J.; Becker, S.; Gauckler, L.J. Adsorption of polyelectrolytes and its influence on the rheology, zeta potential, and microstructure of various cement and hydrate phases. *Colloid Interface Sci.* **2008**, *323*, 301–312. [CrossRef]
42. Ballari, M.M.; Yu, Q.L.; Brouwers, H.J.H. Experimental study of the NO and NO_2 degradation by photocatalytically active concrete. *Catal. Today* **2011**, *161*, 175–180. [CrossRef]
43. Xu, M.; Bao, Y.; Wu, T.; Xia, T.; Clark, H.L.; Shi, H.; Li, V.C. Influence of TiO_2 incorporation methods on NOx abatement in Engineered Cementitious Composites. *Const. Build. Mater.* **2019**, *221*, 375–383. [CrossRef]

Article

Silver-Copper Oxide Heteronanostructures for the Plasmonic-Enhanced Photocatalytic Oxidation of N-Hexane in the Visible-NIR Range

Hugo Suarez [1], Adrian Ramirez [1,2], Carlos J. Bueno-Alejo [1,3] and Jose L. Hueso [1,3,4,*]

1 Institute of Nanoscience of Aragon (INA) and Department of Chemical and Environmental Engineering, C/Poeta Mariano Esquillor, s/n; Campus Rio Ebro, Edificio I+D, 50018 Zaragoza, Spain; hasuarezo2@gmail.com (H.S.); adrian.galilea@kaust.edu.sa (A.R.); carlosj_bueno@yahoo.es (C.J.B.-A.)

2 KAUST Catalysis Center (KCC), King Abdullah University of Science and Technology (KAUST), 23955 Thuwal, Saudi Arabia

3 Networking Research Center on Bioengineering, Biomaterials and Nanomedicine (CIBER-BBN), 28029 Madrid, Spain

4 Instituto de Ciencia de Materiales de Aragon (ICMA), Consejo Superior de Investigaciones Cientificas (CSIC-University of Zaragoza), 50018 Zaragoza, Spain

* Correspondence: jlhueso@unizar.es

Received: 14 October 2019; Accepted: 20 November 2019; Published: 22 November 2019

Abstract: Volatile organic compounds (VOCs) are recognized as hazardous contributors to air pollution, precursors of multiple secondary byproducts, troposphere aerosols, and recognized contributors to respiratory and cancer-related issues in highly populated areas. Moreover, VOCs present in indoor environments represent a challenging issue that need to be addressed due to its increasing presence in nowadays society. Catalytic oxidation by noble metals represents the most effective but costly solution. The use of photocatalytic oxidation has become one of the most explored alternatives given the green and sustainable advantages of using solar light or low-consumption light emitting devices. Herein, we have tried to address the shortcomings of the most studied photocatalytic systems based on titania (TiO_2) with limited response in the UV-range or alternatively the high recombination rates detected in other transition metal-based oxide systems. We have developed a silver-copper oxide heteronanostructure able to combine the plasmonic-enhanced properties of Ag nanostructures with the visible-light driven photoresponse of CuO nanoarchitectures. The entangled Ag-CuO heteronanostructure exhibits a broad absorption towards the visible-near infrared (NIR) range and achieves total photo-oxidation of n-hexane under irradiation with different light-emitting diodes (LEDs) specific wavelengths at temperatures below 180 °C and outperforming its thermal catalytic response or its silver-free CuO illuminated counterpart.

Keywords: plasmonic photocatalysis; silver-copper oxide; VOCs remediation; full-spectrum photoresponse

---

1. Introduction

Global warming, massive deforestation for urbanization and increasing contamination caused by mankind practices are contributing to the alarming rise of pollutant emission levels worldwide. Among these contaminants, the exposure to volatile organic compounds (hereafter VOCs) is recognized as a serious hazard to human health contributing to skin, respiratory, and cancer diseases [1–3]. Even if the exposure dose is very low, it has become an issue of increasing interest since many of the VOCs emitting sources are not only stemming from big factories or production plants. Indoor sources such as tobacco smoke, solvents, paints, furniture, computer, personal use products, etc. are continuously contributing to the VOCs emissions in indoor habitats [1–5]. There are currently different exploring technologies devoted to VOCs remediation including the use of plasma discharges [6–9], microwaves combining absorption–desorption–combustion steps [10–12], photodegradation [2,5,13–20], and adsorption/catalytic oxidation [2,21–29]. Total oxidation of VOCs promoted by conventional catalysts represents one of the most appealing alternatives. Noble metals are able to completely oxidize VOCs into CO_2 and H_2O at mild reaction temperatures [22,23,25,30–39]. Transition metal oxides and complex metal oxides (i.e., rare earth element-based perovskites) are also excellent VOCs oxidation candidates that operate at relatively mild temperatures without incurring in the burdening costs of noble metals [3,8,16,37–45]. Alternatively, the use of inexpensive arrays of photocatalysts based on titania (TiO_2) has become one of the most important research fields towards the sustainable remediation of VOCs [5,20,46–51]. The advantages of using solar light or low consume artificial lights to promote VOCs oxidation at room temperature is being actively pursued. Current limitations are found either in the weak response of the most active semiconductor photocatalysts (i.e., TiO_2, ZnO) beyond the UV range (that only represents 4%–5% of the full solar spectrum) or in the rapid electron-hole recombination rates detected in transition metal oxide semiconducting photocatalysts with expanded absorption capacities towards the visible-near infrared (NIR) ranges (i.e., MO_x, M = Cu, Fe, Mn, Co) [2,43,52–58].

To overcome these drawbacks, current research interests in VOCs remediation are focused on the development of hybrid nanomaterials combining metal oxides, metal transition oxides and/or noble metals with photocatalytic response expanded towards the visible-NIR ranges [16,21,59–69]. Metallic nanoparticles can play a determining role in expanding the absorption range of regular metal oxides such as titania (TiO_2) and/or reducing the electron-hole recombination rates of metal transition oxides (i.e., MO_x (M = Fe, Mn, Co, Cu) [2,3,19,58,59,70–90]. Furthermore, metallic nanoparticles have become particularly relevant due to their plasmonic properties [2,76,81,86,91–96]. The localized surface plasmon resonance (LSPR) is a unique characteristic of these materials (normally Au, Ag, Pt, or other noble metals), which can extend the absorption of light towards the visible light spectrum [69,87–90]. Thus, LSPR greatly supports the utilization of the solar spectrum [69,81,92,97–104]. In addition, plasmonic nanoparticles may play active roles as sensitizers (via antenna effects) or accommodate charges from semiconductors upon forming effective Schottky metal-semiconductor junctions as in well-established noble metal-TiO_2 hybrid systems [69,92,97–100].

In the present work, we aimed at exploring the synthesis of a hybrid heterostructured catalyst combining the plasmonic properties of Ag and the p-type semiconductor capabilities of CuO [3,101–104]. Both materials are abundant, affordable, and exhibit a strong potential for full-range photocatalytic applications. Previous studies based on plasmonic silver nanostructures [76,92], Cu_xO_y systems [105–107] or in the combination of silver-copper alloys [108–112], silver-copper oxide heterostructures [109,113–115] or even silver-copper oxide decorating TiO_2 [70,108] have already proved their potential not only in visible-NIR expanded photocatalysis [70,114,116], but also in solar harvesting, electrocatalysis, bacteria disinfection, field emission enhancement or the formation of novel superconducting structures [55,108,109,111,114–120]. Herein, we have demonstrated that these Ag-CuO heterostructures with an intertwined configuration maximize the plasmon-semiconductor interaction. As a result, a very active hybrid heterostructure with full-spectrum LED-driven photoresponse towards the total oxidation of n-hexane has been developed. Its photocatalytic response becomes especially photoactive upon irradiation with LED wavelengths of 460 nm. The heterostructure fully photooxidizes n-hexane at temperatures below 180 °C and outperforms its silver-free CuO counterpart thereby outlining the relevance of the silver nanostructures entangled with the CuO nanotubes in the Ag-CuO hybrid. To the best of our knowledge, this study presents the first example of full-spectrum photocatalytic assisted VOCs oxidation in a diluted gas phase with this kind of Ag-CuO configuration.

2. Materials and Methods

2.1. Synthesis of the Photocatalysts

The Ag-CuO heteronanostructures were synthesized following a protocol reported elsewhere [121]. $Cu(NO_3)_2{\cdot}3H_2O$ (3.2 mmol, Aldrich, Saint Louis, MO, USA, 99.9%) and $AgNO_3$ (3.1 mmol, Aldrich, Saint Louis, MO, USA, 99.9%) were dissolved in 3 mL of deionized water. The resulting silver-copper solution was added to an aqueous solution of NaOH 3 M (4 mL, Aldrich) with vigorous stirring for 6 h under an inert Ar atmosphere. The resulting black adduct that was vacuum filtered, washed with water, and dried at 100 °C for 2 h. The solid was calcined at 350 °C for 6 h. In order to obtain silver-free CuO nanostructures, an analogous synthesis protocol was followed but skipping the addition of the silver salt precursor. The synthesis of the photocatalysts has been performed at the platform of Production of Biomaterials and Nanoparticles of the NANBIOSIS ICTS, Spain, more specifically by the Nanoparticle Synthesis Unit of the CIBER in BioEngineering, Biomaterials & Nanomedicine (CIBER-BBN).

2.2. Characterization Techniques

Transmission electron microscopy (TEM) analysis was carried out using a T20–FEI microscope (Hillsboro, OR, USA). Aberration corrected scanning transmission electron microscopy (STEM) images were acquired using a high angle annular dark field detector in a FEI XFEG TITAN microscope (Hillsboro, OR, USA) at 300 kV equipped with a CETCOR Cs-probe corrector. High-resolution transmission electron microscopy (HR-TEM) images were acquired with the aid of a FEI TITAN^3 electron microscope operated at 200 kV. Elemental analysis was carried out with Energy Dispersive Spectroscopy (EDS) (EDAX, Mahwah, NJ, USA) detector using single point and scanning profiles. The samples were drop-casted onto Ni mesh grids. The N_2 adsorption/desorption analyses were performed with the aid of a Micromeritics ASAP2020 analyzer (Norcross, GA, USA). 80–100 mg of the catalyst was degassed at 90 °C for 12 h. The surface area was determined using the Brunauer–Emmett–Teller (BET) method rendering a value of 4.7 $m^2{\cdot}g^{-1}$ for the Ag-CuO tubular heterostructures. Scanning electron microscopy (SEM) analysis was carried out with FEI-Inspect S50 equipment (Hillsboro, OR, USA). X-ray diffraction patterns were obtained in a PANalytical Empyrean equipment (Malvern, UK) in Bragg Brentano configuration using Cu-K radiation and equipped with a PIXcel1D detector. The absorption spectra were acquired with a JASCO V-670 UV-VIS-NIR spectrophotometer (Tokyo, Japan) and the aid of an integrated sphere accessory.

2.3. Photocatalytic Reaction Setup

The experimental setup designed to carry out the photocatalytic degradation of n-hexane has been previously described elsewhere [20,122]. Briefly, the reaction was conducted in a home-made system comprising a quartz cell (50 × 10 × 5 mm^3 (height × width × length). The cell was illuminated by two high power LEDs (LedEngin) cooled with the aid of custom-designed fans. Different LEDs wavelengths were individually tested. Each LED operated with different power: 405 nm (3.9 W), 460 nm (2.2 W), and 940 nm (3.2 W). Selected light irradiances ranged from 7 to 11,000 mW/cm^2 (based on LED specifications and experimental setup) and LED powers were programmed with the aid of an external power supply unit (ISO-TECH, IPS-405, 0–40 V). The temperature of the catalytic bed containing 300 mg of Ag-CuO during irradiation was monitored with the aid of a K-type thermocouple. The photocatalytic experiments were performed with a total flow of 50 $mL{\cdot}min^{-1}$ of gas containing 200 ppm of n-hexane (space velocity = 10,000 h^{-1}). Conventional heating experiments were carried out with a home-built system consisting of an aluminium holder designed to heat the same reactor area as in the quartz reactor used for the photocatalytic test [20]. The inlet final concentration of n-hexane was achieved upon mixing with the proper flow rates, n-hexane in N_2, O_2, and synthetic air (all purchased from PRAXAIR España S.L.U., Madrid, Spain), in order to get the different total flow rates assessed. After an equilibration period of 30 min that served us to evaluate the adsorption of n-hexane in the dark, the LED lights were turned on for different time intervals and the gas effluent outlet analyzed by gas chromatography (Agilent 3000 Micro GC, Santa Clara, CA, USA). An OV-1 and a PPQ column in line with a thermal conductivity detector (TCD) were employed to separate and detect the different gas compounds. The steady state final concentration achieved was always ≤10 ppm of n-hexane when maximum LED power was used. This steady state was always achieved within minutes regardless of the experimental settings. Under the conditions used, the n-hexane detection limit was 3 ppm and CO_2 was the only oxidation product detected. Maximum error in the mass balance closures for carbon and oxygen in this work was ±2%.

3. Results

3.1. Characterization of the Silver-Copper Oxide Plasmonic Photocatalyst

The morphological evaluation of the Ag-CuO heterostructures by SEM revealed the presence of tubular-shaped structures (Figure 1a) [121,123,124]. A more detailed analysis by HAADF-STEM in combination with EDX analysis confirmed the presence of both Ag and Cu species as segregated elements. Figure 1b–d reveal the corresponding analysis of the outer surfaces of the tubular structures. Small Ag nanoparticles are supported onto the Cu-based surface (Figure 1b,d). An extended EDX line profile analysis across two individual nanotubes further confirmed the alternating presence of either silver or copper elements (see Figure 1e–f).

It is also worth mentioning that Ag was identified with different morphologies, including small segregated nanoparticles, rod-shaped anisotropic structures (Figure 1g), and non-uniform aggregates dispersed along the tubular-shaped CuO matrixes (Figure 1i). HR-TEM analysis of the Cu-based regions confirmed the presence of a well-defined orientation corresponding to a CuO crystalline phase (Figure 1h). The FFT inset in Figure 1g corresponded to the orientation of the CuO fraction in the [0-1-1] direction. The (200), (11-1), and (-11-1) planes were identified and matched with a C2/c monoclinic system. EDX mapping of the intensities of Ag-L and Cu-L signals further assessed the entangled distribution of Cu (Figure 1j) and Ag phases (Figure 1k).

XRD analysis also corroborated the presence of both silver and copper oxide crystalline phases assigned to a cubic (Fm3m) and a monoclinic (C2/c) system, respectively (Figure 2a). The optical characterization of the Ag-CuO nanohybrids by UV-Vis-NIR spectroscopy revealed a broad absorption spectrum expanding towards the visible and near-infrared (NIR) range (Figure 2c). The silver-free CuO nanotubes synthesized as control (Figure 2b) exhibited similar optical absorption properties in the visible range, although it did not expand beyond the visible range and decayed in the NIR region (Figure 2c). The energy band gap for both the CuO and Ag-CuO nanostructures was determined from the optical absorption near the band edge using the classical Tauc approach and assuming an indirect band gap semiconductor system where $\alpha \cdot E_{photon} = K\,(E_{photon} - E_g)^{1/2}$ (being E_{photon} and E_g the discrete photon energy and the band gap energy, respectively). The estimated band gap energy was calculated at 1.33 eV for the CuO nanostructures and 0.71 eV for the Ag-CuO hybrids (Figure 2d). These values were lower than the band gap reported for bulk CuO structures (typically 1.4 eV) and confirmed the potential optical response of these structures in the visible-NIR range [53].

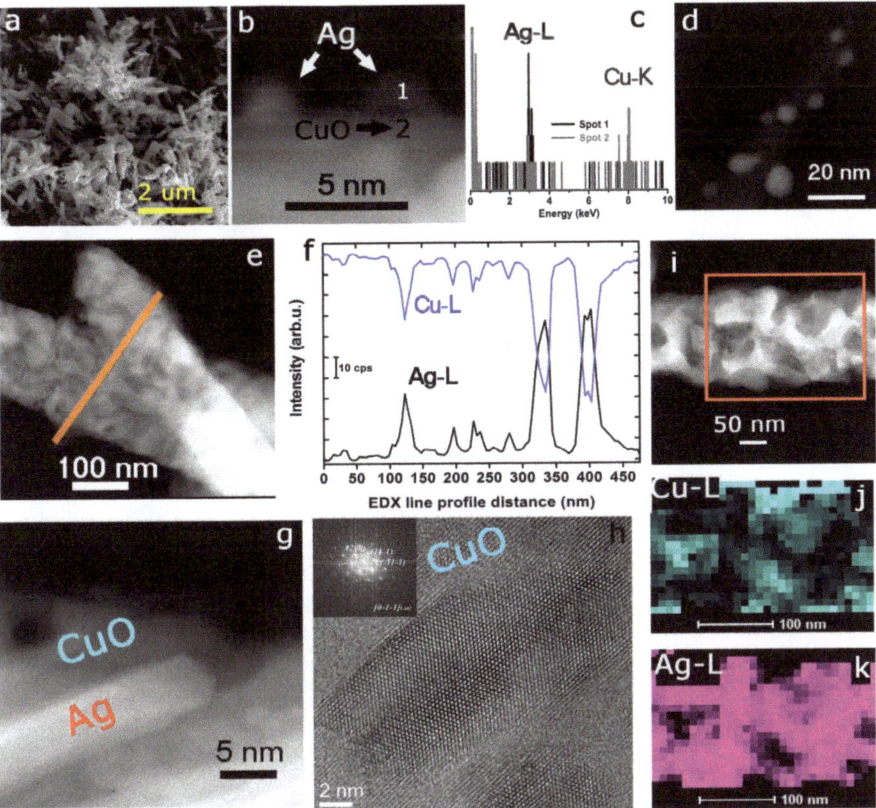

Figure 1. Morpho-chemical characterization of the silver-copper oxide photocatalyst: (**a**) SEM representative image accounting for the tubular shape of the Ag-CuO hybrids; (**b**) High Angle Annular Dark-Field (HAADF)-STEM image of small Ag nanoparticles (NPs) in the outer area of the nanotubes dispersed in a Cu-based matrix, the numbers refer to specific areas for EDX spectra acquisition; (**c**) EDX analysis of selected spots in (**b**) accounting for the specific present of Ag or Cu; (**d**) HAADF-STEM image of a CuO nanotube with Ag NPs decorating in the external region; (**e**) STEM image of individual nanotubes and EDX line profile analysis performed and plotted in (**f**); (**f**) evolution of Ag-L and Cu-L intensities across the EDX line profile analysis depicted in (**e**); (**g**) HAADF-STEM image accounting for the presence of anisotropic Ag shapes embedded within the Cu-based matrix; (**h**) HR-TEM image corresponding to the Cu-enriched region accounting for the presence of a monoclinic CuO phase (inset: Fast Fourier Transform (FFT) image with indexed CuO planes in the [0-1-1] direction); (**i**) HAADF-STEM image of a fraction of Ag-CuO nanotube containing bigger aggregates (the square accounts for the selected area for EDX mapping analysis); (**j**) EDX map accounting for the Cu-L edge (wt %) intensity in the selected area of (**i**); (**k**) EDX map accounting for the distribution of the Ag-L edge intensity (wt %) in the selected area of (**i**).

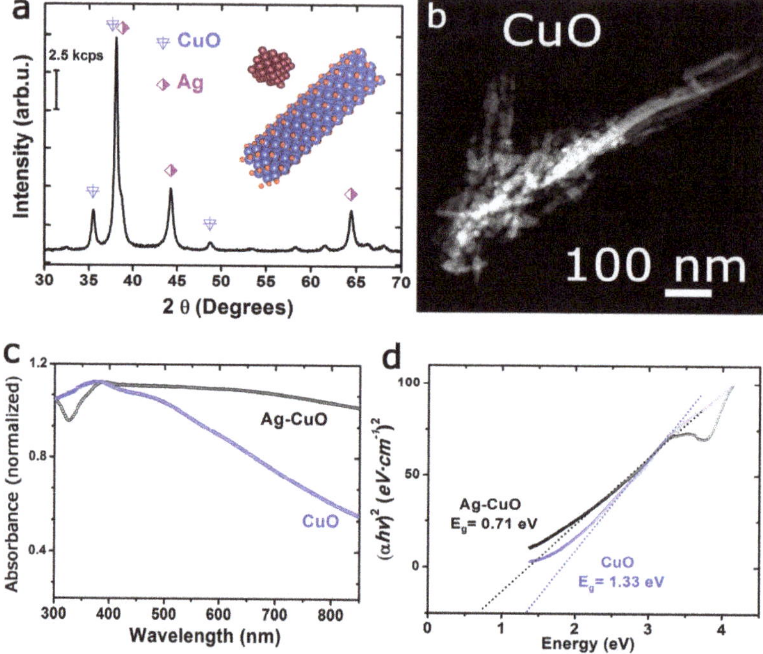

Figure 2. Additional characterization of the photocatalytic materials: (**a**) X-ray diffractogram of the Ag-CuO hybrid material accounting for the presence of both cubic and monoclinic crystallographic phases for silver and copper oxide, respectively; (**b**) HAADF-STEM representative image of the silver-free CuO nanostructures; (**c**) UV-Vis-Near Infrared absorption spectra of the Ag-CuO and CuO nanomaterials; (**d**) Tauc plots for the determination of the band gap energies for Ag-CuO and CuO structures assuming an indirect transition.

3.2. Photocatalytic Performance of the Ag-CuO Heterostructures for N-Hexane Total Oxidation

Figure 3 shows the photocatalytic response of the Ag-CuO hybrid towards the oxidation of n-hexane under illumination with a high irradiance LED emitting at 405 nm (see inset in Figure 3b). Total oxidation was achieved at temperatures below 180 °C. Remarkably, light-off oxidation curves started at temperatures below 50 °C and T_{50} (Temperature of reaction required to reach 50% of conversion) remained below 100 °C. These results contrast with the photocatalytic behavior identified for the silver-free CuO counterpart under similar LED irradiation conditions. In this latter case, temperatures above 200 °C were necessary to reach match T_{50} and a complete n-hexane oxidation was not achieved (maximum 90% conversion, see Figure 3a). Finally, it is worth mentioning that the thermal catalytic experiment (in the absence of LED irradiation) with the Ag-CuO catalyst yielded higher n-hexane conversion levels than the CuO catalyst but were less effective than the photocatalytic experimental conditions (Figure 3a).

Additional photocatalytic experiments were carried out with the Ag-CuO heterostructure under illumination with different LED wavelengths at 460 and 940 nm, respectively. The use of LEDs emitting in the visible and NIR ranges rendered equivalent n-hexane photo-oxidation levels and analogous overlapping light-off curves to the one displayed in Figure 3a (data not shown). The major differences were observed in terms of the LED power density required in each experiment to achieve those conversion levels. Figure 3b summarizes the LED power irradiance (expressed in W/cm^2) required at 405, 460, and 940 nm, respectively. Upon comparison of the three LEDs, it became clear that the photocatalytic efficiency was higher at 460 nm. The irradiation under the LED emitting at 405 nm required almost double irradiance to reach full photo-conversion of n-hexane. We observed a stable photo-response after multiple cumulative reaction runs performed under different LED wavelengths and no evidences of deactivation.

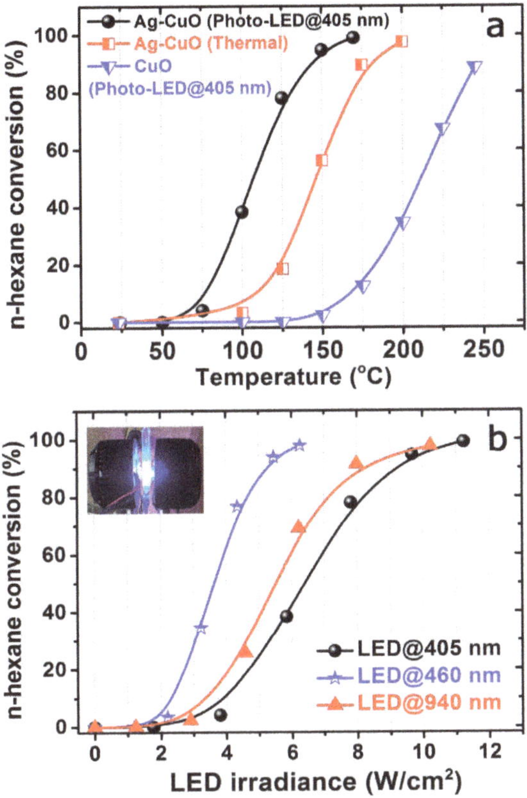

Figure 3. LED-driven photocatalytic oxidation of n-hexane: (**a**) n-hexane conversion curves obtained after photocatalytic activation of Ag-CuO (spherical symbols), CuO (triangle symbols) with a LED emitting at 405 nm, and alternatively after thermal heating of Ag-CuO with a conventional heating setup (square symbols); (**b**) n-hexane light-off curves under different irradiation wavelengths as a function of the irradiance (in W/cm^2) specifically required for each LED; (inset: Digital image of the 405 nm LEDs simultaneously irradiating the quartz cuvette reactor).

4. Discussion

The positive photocatalytic response towards n-hexane oxidation of the present Ag-CuO hybrid structures can be justified in terms of the synergetic combination of plasmonic silver and the visible-light response of the p-type semiconductor CuO. In contrast to previous Ag-Cu systems [55,108,109,111,114,117,119], our synthesis methodology enables the generation of a well entangled hybrid system where Ag domains of different sizes and morphologies are perfectly encapsulated within CuO nanotubes (Figures 1 and 2). The most plausible mechanism for the formation of this hybrid is the thermal decomposition of an unstable silver-copper mixed oxide $Ag_xCu_yO_z$ that evolves into the corresponding silver and copper oxides counterparts [125–127]. The thermal treatment at high temperature during the preparation also favored the subsequent thermal decomposition of Ag_2O into metallic Ag and oxygen at temperatures in the range of 195–205 °C and enabling partial mobility of silver throughout the CuO tubular scaffold (Figure 1) [76,128].

To justify the full-range response under different LED wavelengths (Figure 3b), a combination of different photo-excitation and charge-transfer mechanisms can be taking place [68,69,81,99,129,130]. The different photocatalytic response observed after comparison between Ag-CuO and CuO irradiated with the 405 nm LED (Figure 3a) clearly demonstrates the positive influence of the silver entangled nanostructures. Given the heterogeneous disposition of silver entities, different photo-activation pathways can be simultaneously occurring in our catalysts. First of all, a fraction of smaller silver domains (Figure 1b–d) with the proper energy levels can be acting as sinks or trap centers for the electrons photogenerated by the CuO semiconductor fraction (Figure 4a). As a result, the expected high electron-hole recombination rates of CuO can be inhibited and/or partially delayed. Therefore, the unpaired holes remain available in the valence band of CuO to readily oxidize n-hexane molecules (Figure 4a). Likewise, electrons in the Ag surface can participate in the formation of reactive superoxide anions or relax via thermal energy dissipation [95,131]. The superoxide anions may subsequently react after their photo-induced dissociation (Figure 4a) and contribute to the oxidation of the n-hexane molecules [5,68,69,76,81,105,130,132].

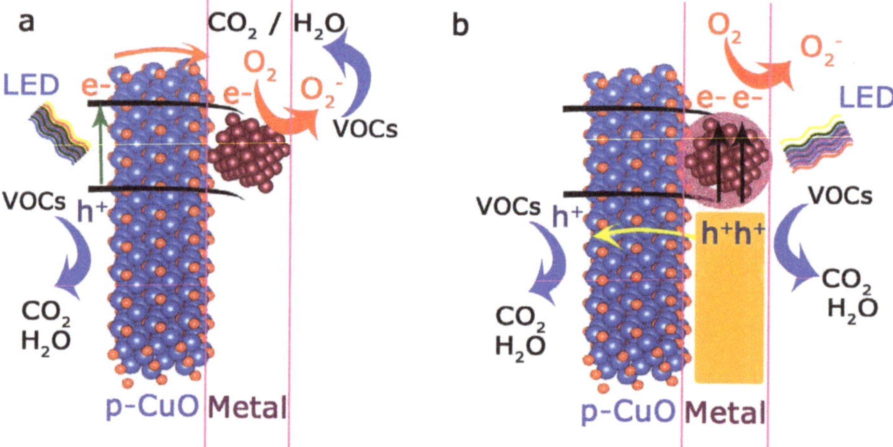

Figure 4. Schematic diagram illustrating the most plausible charge-transfer and photocatalytic mechanisms in the metal/p-type semiconductor Ag-CuO hybrids: (**a**) If the LED excitation wavelength is larger than the energy band gap of CuO, electrons from the valence band can be excited to the conduction of CuO and subsequently transferred and trapped by Ag energy levels; (**b**) Ag plasmon-induced charge transfer by hot holes injection into the p-type CuO energy levels.

Another fraction of silver structures with different sizes and anisotropic shapes (i.e., rod-like) that remain embedded within the CuO tubular matrix (Figure 1e,g,i) can provide additional plasmon-driven photo-excitation pathways. Metallic silver nanostructures are considered as excellent plasmonic materials [68,95,133]. The valence electron clouds present in their metal surfaces can oscillate and resonate at different frequencies in the UV-Vis-NIR ranges generating localized surface plasmons (LSPR). The surface plasmons can interact with the CuO nanotubes via radiative damping mechanisms that imply the reemission and/or trapping of light from the metal to the surrounding semiconductor matrix [95,133]. This approximation would be more likely to occur with the larger Ag domains isolated within the CuO nanotubes (Figure 1e–i) where scattering phenomena would be more plausible (Figure 2c) [95,131]. The photocatalytic response under 940 nm wavelengths can be also tentative attributed in part to the presence of rod-shaped Ag structures entangled within the CuO matrix (Figures 1g and 3b). These anisotropic silver nanostructures exhibit plasmon absorbance at longer wavelengths than spheres or cubes (Figure 1g) [68]. The presence of silver expands the absorption and light-trapping capabilities in the visible-NIR range, thereby expanding the potential exploitation of the solar energy (more than 80% of the solar spectrum range) [68,92].

Surface plasmons can alternatively decay via non-radiative pathways involving the generation of electron-hole pairs by interband and/or intraband excitations [68,81,92,95]. In this scenario, Ag and CuO entangled interfaces can form a metal/p-type semiconductor Schottky barrier for holes after matching their Fermi levels [3,55,119]. The excitation with sufficiently high energy LEDs (i.e., 405 and 460 nm, Figures 3 and 4b) enables the plasmon-induced injection of hot holes from the silver bands into the valence band of the CuO p-type semiconductor (Figure 4b). These high energetic holes are able to pass the Schottky barrier and rapidly react with the n-hexane. The reduced dimensions of the heteronanostructures minimize the probability of undergoing another relaxation/recombination process [68]. While the injection of hot electrons (using n-type semiconductors such as TiO_2) is the most accepted plasmon-driven charge transfer mechanism, there exist recent interesting studies claiming the importance of hot holes in other plasmonic-based systems [73,86,92,97,134–137] such as Au-NiO_x, Au-pGaN [73], Au nanorods coated with a CoO nanoshell [138], Au nanostructures [101,139,140], or Ag-BiOCl hybrids [86,137].

We tentatively propose a combination of the different photocatalytic mechanisms given the diversity of Ag domains. Indeed, the better photoresponse of the Ag-CuO hybrid in comparison with the CuO nanostructures confirms the important role of silver as plasmonic structure to harvest light in the whole visible to NIR range. Likewise, the close contact between both metal and semiconductor phases has enabled a suitable interfacial contact to promote electron and holes mobilities and minimize undesired recombination and relaxation pathways. In summary, we can conclude that our Ag-CuO represents a very attractive metal/p-type semiconductor candidate with full-spectrum response that can be envisioned as an affordable alternative for green and sustainable photo-assisted chemistry, with special attention to energy and remediation processes.

Author Contributions: Conceptualization, C.J.B.-A. and J.L.H.; Data curation, C.J.B.-A.; Formal analysis, A.R., C.J.B.-A., and J.L.H.; Investigation, H.S., A.R., C.J.B.-A., and J.L.H.; Methodology, H.S. and J.L.H.; Supervision, J.L.H.; Validation, A.R. and C.J.B.-A.; Writing—original draft, J.L.H.; Writing—review and editing, A.R., C.J.B.-A., and J.L.H.

Funding: This research was funded by ARCADIA (grant CTQ2016-77147) and CADENCE (grant 742684) and the APC was waived by the journal.

Acknowledgments: The TEM measurements were conducted at the Laboratorio de Microscopias Avanzadas, Instituto de Nanociencia de Aragon, Universidad de Zaragoza, Spain. Fundacion Carolina is acknowledged for funding of a scholarship for H.S. The synthesis of materials has been performed by the Platform of Production of Biomaterials and Nanoparticles of the NANOBIOSIS ICTS, more specifically by the Nanoparticle Synthesis Unit of the CIBER in BioEngineering, Biomaterials & Nanomedicine (CIBER-BBN).

Conflicts of Interest: The authors declare no conflict of interest.

References

1. Kamal, M.S.; Razzak, S.A.; Hossain, M.M. Catalytic oxidation of volatile organic compounds (VOCs)—A review. *Atmos. Environ.* **2016**, *140*, 117–134. [CrossRef]
2. He, C.; Cheng, J.; Zhang, X.; Douthwaite, M.; Pattisson, S.; Hao, Z.P. Recent Advances in the Catalytic Oxidation of Volatile Organic Compounds: A Review Based on Pollutant Sorts and Sources. *Chem. Rev.* **2019**, *119*, 4471–4568. [CrossRef]
3. Huang, H.B.; Xu, Y.; Feng, Q.Y.; Leung, D.Y.C. Low temperature catalytic oxidation of volatile organic compounds: A review. *Catal. Sci. Technol.* **2015**, *5*, 2649–2669. [CrossRef]
4. Smielowska, M.; Marc, M.; Zabiegala, B. Indoor air quality in public utility environments—A review. *Environ. Sci. Pollut. Res.* **2017**, *24*, 11166–11176. [CrossRef] [PubMed]
5. Mamaghani, A.H.; Haghighat, F.; Lee, C.S. Photocatalytic oxidation technology for indoor environment air purification: The state-of-the-art. *Appl. Catal. B Environ.* **2017**, *203*, 247–269. [CrossRef]
6. Kim, H.H.; Teramoto, Y.; Negishi, N.; Ogata, A. A multidisciplinary approach to understand the interactions of nonthermal plasma and catalyst: A review. *Catal. Today* **2015**, *256*, 13–22. [CrossRef]

7. Zhang, Z.X.; Jiang, Z.; Shangguan, W.F. Low-temperature catalysis for VOCs removal in technology and application: A state-of-the-art review. *Catal. Today* **2016**, *264*, 270–278. [CrossRef]
8. Hueso, J.L.; Cotrino, J.; Caballero, A.; Espinos, J.P.; Gonzalez-Elipe, A.R. Plasma catalysis with perovskite-type catalysts for the removal of NO and CH4 from combustion exhausts. *J. Catal.* **2007**, *247*, 288–297. [CrossRef]
9. Mao, L.G.; Chen, Z.Z.; Wu, X.Y.; Tang, X.J.; Yao, S.L.; Zhang, X.M.; Jiang, B.Q.; Han, J.Y.; Wu, Z.L.; Lu, H.; et al. Plasma-catalyst hybrid reactor with CeO_2/gamma-Al_2O_3 for benzene decomposition with synergetic effect and nano particle by-product reduction. *J. Hazard. Mater.* **2018**, *347*, 150–159. [CrossRef]
10. Nigar, H.; Sturm, G.S.J.; Garcia-Banos, B.; Penaranda-Foix, F.L.; Catala-Civera, J.M.; Mallada, R.; Stankiewicz, A.; Santamaria, J. Numerical analysis of microwave heating cavity: Combining electromagnetic energy, heat transfer and fluid dynamics for a NaY zeolite fixed-bed. *Appl. Therm. Eng.* **2019**, *155*, 226–238. [CrossRef]
11. Nigar, H.; Julian, I.; Mallada, R.; Santamaria, J. Microwave-Assisted Catalytic Combustion for the Efficient Continuous Cleaning of VOC-Containing Air Streams. *Environ. Sci. Technol.* **2018**, *52*, 5892–5901. [CrossRef] [PubMed]
12. Nigar, H.; Navascues, N.; De la Iglesia, O.; Mallada, R.; Santamaria, J. Removal of VOCs at trace concentration levels from humid air by Microwave Swing Adsorption, kinetics and proper sorbent selection. *Sep. Purif. Technol.* **2015**, *151*, 193–200. [CrossRef]
13. Yang, X.G.; Wang, D.W. Photocatalysis: From Fundamental Principles to Materials and Applications. *ACS Appl. Energ. Mater.* **2018**, *1*, 6657–6693. [CrossRef]
14. Kumari, G.; Zhang, X.Q.; Devasia, D.; Heo, J.; Jain, P.K. Watching Visible Light-Driven CO_2 Reduction on a Plasmonic Nanoparticle Catalyst. *ACS Nano* **2018**, *12*, 8330–8340. [CrossRef] [PubMed]
15. Brigden, C.T.; Poulston, S.; Twigg, M.V.; Walker, A.P.; Wilkins, A.J.J. Photo-oxidation of short-chain hydrocarbons over titania. *Appl. Catal. B Environ.* **2001**, *32*, 63–71. [CrossRef]
16. Chen, J.Y.; He, Z.G.; Li, G.Y.; An, T.C.; Shi, H.X.; Li, Y.Z. Visible-light-enhanced photothermocatalytic activity of ABO(3)-type perovskites for the decontamination of gaseous styrene. *Appl. Catal. B Environ.* **2017**, *209*, 146–154. [CrossRef]
17. Deng, X.Y.; Yue, Y.H.; Gao, Z. Gas-phase photo-oxidation of organic compounds over nanosized TiO2 photocatalysts by various preparations. *Appl. Catal. B Environ.* **2002**, *39*, 135–147. [CrossRef]
18. Boulamanti, A.K.; Philippopoulos, C.J. Photocatalytic degradation of C-5-C-7 alkanes in the gas-phase. *Atmos. Environ.* **2009**, *43*, 3168–3174. [CrossRef]
19. Boyjoo, Y.; Sun, H.Q.; Liu, J.; Pareek, V.K.; Wang, S.B. A review on photocatalysis for air treatment: From catalyst development to reactor design. *Chem. Eng. J.* **2017**, *310*, 537–559. [CrossRef]
20. Bueno-Alejo, C.J.; Hueso, J.L.; Mallada, R.; Julian, I.; Santamaria, J. High-radiance LED-driven fluidized bed photoreactor for the complete oxidation of n-hexane in air. *Chem. Eng. J.* **2019**, *358*, 1363–1370. [CrossRef]
21. Liotta, L.F.; Ousmane, M.; Di Carlo, G.; Pantaleo, G.; Deganello, G.; Marci, G.; Retailleau, L.; Giroir-Fendler, A. Total oxidation of propene at low temperature over Co_3O_4-CeO_2 mixed oxides: Role of surface oxygen vacancies and bulk oxygen mobility in the catalytic activity. *Appl. Catal. A Gen.* **2008**, *347*, 81–88. [CrossRef]
22. Ousmane, M.; Liotta, L.F.; Di Carlo, G.; Pantaleo, G.; Venezia, A.M.; Deganello, G.; Retailleau, L.; Boreave, A.; Giroir-Fendler, A. Supported Au catalysts for low-temperature abatement of propene and toluene, as model VOCs: Support effect. *Appl. Catal. B Environ.* **2011**, *101*, 629–637. [CrossRef]
23. Scire, S.; Liotta, L.F. Supported gold catalysts for the total oxidation of volatile organic compounds. *Appl. Catal. B Environ.* **2012**, *125*, 222–246. [CrossRef]
24. Liotta, L.F.; Wu, H.J.; Pantaleo, G.; Venezia, A.M. Co_3O_4 nanocrystals and Co_3O_4-MOx binary oxides for CO, CH_4 and VOC oxidation at low temperatures: A review. *Catal. Sci. Technol.* **2013**, *3*, 3085–3102. [CrossRef]
25. Ousmane, M.; Liotta, L.F.; Pantaleo, G.; Venezia, A.M.; Di Carlo, G.; Aouine, M.; Retailleau, L.; Giroir-Fendler, A. Supported Au catalysts for propene total oxidation: Study of support morphology and gold particle size effects. *Catal. Today* **2011**, *176*, 7–13. [CrossRef]
26. Grabchenko, M.V.; Mikheeva, N.N.; Mamontov, G.V.; Salaev, M.A.; Liotta, L.F.; Vodyankina, O.V. Ag/CeO_2 Composites for Catalytic Abatement of CO, Soot and VOCs. *Catalysts* **2018**, *8*, 285. [CrossRef]
27. Sihaib, Z.; Puleo, F.; Pantaleo, G.; La Parola, V.; Valverde, J.L.; Gil, S.; Liotta, L.F.; Giroir-Fendler, A. The Effect of Citric Acid Concentration on the Properties of $LaMnO_3$ as a Catalyst for Hydrocarbon Oxidation. *Catalysts* **2019**, *9*, 226. [CrossRef]

28. Liu, L.C.; Corma, A. Metal Catalysts for Heterogeneous Catalysis: From Single Atoms to Nanoclusters and Nanoparticles. *Chem. Rev.* **2018**, *118*, 4981–5079. [CrossRef]
29. Yang, C.T.; Miao, G.; Pi, Y.H.; Xia, Q.B.; Wu, J.L.; Li, Z.; Xiao, J. Abatement of various types of VOCs by adsorption/catalytic oxidation: A review. *Chem. Eng. J.* **2019**, *370*, 1128–1153. [CrossRef]
30. Uson, L.; Hueso, J.L.; Sebastian, V.; Arenal, R.; Florea, I.; Irusta, S.; Arruebo, M.; Santamaria, J. In-situ preparation of ultra-small Pt nanoparticles within rod-shaped mesoporous silica particles: 3-D tomography and catalytic oxidation of n-hexane. *Catal. Commun.* **2017**, *100*, 93–97. [CrossRef]
31. Uson, L.; Colmenares, M.G.; Hueso, J.L.; Sebastian, V.; Balas, F.; Arruebo, M.; Santamaria, J. VOCs abatement using thick eggshell Pt/SBA-15 pellets with hierarchical porosity. *Catal. Today* **2014**, *227*, 179–186. [CrossRef]
32. Hueso, J.L.; Sebastian, V.; Mayoral, A.; Uson, L.; Arruebo, M.; Santamaria, J. Beyond gold: rediscovering tetrakis-(hydroxymethyl)-phosphonium chloride (THPC) as an effective agent for the synthesis of ultra-small noble metal nanoparticles and Pt-containing nanoalloys. *RSC Adv.* **2013**, *3*, 10427–10433. [CrossRef]
33. Kumar, G.; Nikolla, E.; Linic, S.; Medlin, J.W.; Janik, M.J. Multicomponent Catalysts: Limitations and Prospects. *ACS Catal.* **2018**, *8*, 3202–3208. [CrossRef]
34. Cellier, C.; Lambert, S.; Gaigneaux, E.M.; Poleunis, C.; Ruaux, V.; Eloy, P.; Lahousse, C.; Bertrand, P.; Pirard, J.P.; Grange, P. Investigation of the preparation and activity of gold catalysts in the total oxidation of n-hexane. *Appl. Catal. B Environ.* **2007**, *70*, 406–416. [CrossRef]
35. Liotta, L.F. Catalytic oxidation of volatile organic compounds on supported noble metals. *Appl. Catal. B-Environ.* **2010**, *100*, 403–412. [CrossRef]
36. Guo, J.H.; Lin, C.X.; Jiang, C.J.; Zhang, P.Y. Review on noble metal-based catalysts for formaldehyde oxidation at room temperature. *Appl. Surf. Sci.* **2019**, *475*, 237–255. [CrossRef]
37. Liotta, L.F.; Ousmane, M.; Di Carlo, G.; Pantaleo, G.; Deganello, G.; Boreave, A.; Giroir-Fendler, A. Catalytic Removal of Toluene over Co_3O_4-CeO_2 Mixed Oxide Catalysts: Comparison with Pt/Al_2O_3. *Catal. Lett.* **2009**, *127*, 270–276. [CrossRef]
38. Pereniguez, R.; Hueso, J.L.; Gaillard, F.; Holgado, J.P.; Caballero, A. Study of Oxygen Reactivity in La1-x Sr (x) CoO_3-delta Perovskites for Total Oxidation of Toluene. *Catal. Lett.* **2012**, *142*, 408–416. [CrossRef]
39. Pereniguez, R.; Hueso, J.L.; Holgado, J.P.; Gaillard, F.; Caballero, A. Reactivity of LaNi1-y Co (y) O_3-delta Perovskite Systems in the Deep Oxidation of Toluene. *Catal. Lett.* **2009**, *131*, 164–169. [CrossRef]
40. Szabo, V.; Bassir, M.; Gallot, J.E.; Van Neste, A.; Kaliaguine, S. Perovskite-type oxides synthesised by reactive grinding–Part III. Kinetics of n-hexane oxidation over LaCo(1-x)FexO3. *Appl. Catal. B Environ.* **2003**, *42*, 265–277. [CrossRef]
41. Rhodes, C.J. Perovskites - some snapshots of recent developments. *Sci. Prog.* **2018**, *101*, 384–396. [CrossRef]
42. Njagi, E.C.; Genuino, H.C.; King'ondu, C.K.; Dharmarathna, S.; Suib, S.L. Catalytic oxidation of ethylene at low temperatures using porous copper manganese oxides. *Appl. Catal. A Gen.* **2012**, *421*, 154–160. [CrossRef]
43. Genuino, H.C.; Dharmarathna, S.; Njagi, E.C.; Mei, M.C.; Suib, S.L. Gas-Phase Total Oxidation of Benzene, Toluene, Ethylbenzene, and Xylenes Using Shape-Selective Manganese Oxide and Copper Manganese Oxide Catalysts. *J. Phys. Chem. C* **2012**, *116*, 12066–12078. [CrossRef]
44. Cordi, E.M.; O'Neill, P.J.; Falconer, J.L. Transient oxidation of volatile organic compounds on a CuO/Al_2O_3 catalyst. *Appl. Catal. B-Environ.* **1997**, *14*, 23–36. [CrossRef]
45. Li, T.Y.; Chiang, S.J.; Liaw, B.J.; Chen, Y.Z. Catalytic oxidation of benzene over CuO/Ce1-xMnxO_2 catalysts. *Appl. Catal. B Environ.* **2011**, *103*, 143–148. [CrossRef]
46. Liu, B.S.; Wu, H.; Parkin, I.P. Gaseous Photocatalytic Oxidation of Formic Acid over TiO_2: A Comparison between the Charge Carrier Transfer and Light-Assisted Mars-van Krevelen Pathways. *J. Phys. Chem. C* **2019**, *123*, 22261–22272. [CrossRef]
47. Shah, K.W.; Li, W.X. A Review on Catalytic Nanomaterials for Volatile Organic Compounds VOC Removal and Their Applications for Healthy Buildings. *Nanomaterials* **2019**, *9*, 910. [CrossRef]
48. Kontos, A.G.; Katsanaki, A.; Maggos, T.; Likodimos, V.; Ghicov, A.; Kim, D.; Kunze, J.; Vasilakos, C.; Schmuki, P.; Falaras, P. Photocatalytic degradation of gas pollutants on self-assembled titania nanotubes. *Chem. Phys. Lett.* **2010**, *490*, 58–62. [CrossRef]
49. Van Gerven, T.; Mul, G.; Moulijn, J.; Stankiewicz, A. A review of intensification of photocatalytic processes. *Chem. Eng. Process.* **2007**, *46*, 781–789. [CrossRef]

50. Da Costa, B.M.; Araujo, A.L.P.; Silva, G.V.; Boaventura, R.A.R.; Dias, M.M.; Lopes, J.C.B.; Vilar, V.J.P. Intensification of heterogeneous TiO_2 photocatalysis using an innovative micro-meso-structured-photoreactor for n-decane oxidation at gas phase. *Chem. Eng. J.* **2017**, *310*, 331–341. [CrossRef]
51. Moulis, F.; Krysa, J. Photocatalytic degradation of several VOCs (n-hexane, n-butyl acetate and toluene) on TiO2 layer in a closed-loop reactor. *Catal. Today* **2013**, *209*, 153–158. [CrossRef]
52. Chen, J.; Li, Y.Z.; Fang, S.M.; Yang, Y.; Zhao, X.J. UV-Vis-infrared light-driven thermocatalytic abatement of benzene on Fe doped OMS-2 nanorods enhanced by a novel photoactivation. *Chem. Eng. J.* **2018**, *332*, 205–215. [CrossRef]
53. Wang, L.J.; Zhou, Q.; Zhang, G.L.; Liang, Y.J.; Wang, B.S.; Zhang, W.W.; Lei, B.; Wang, W.Z. A facile room temperature solution-phase route to synthesize CuO nanowires with enhanced photocatalytic performance. *Mater. Lett.* **2012**, *74*, 217–219. [CrossRef]
54. Wu, S.M.; Li, F.; Zhang, L.J.; Li, Z. Enhanced field emission properties of CuO nanoribbons decorated with Ag nanoparticles. *Mater. Lett.* **2016**, *171*, 220–223. [CrossRef]
55. Yang, J.B.; Li, Z.; Zhao, W.; Zhao, C.X.; Wang, Y.; Liu, X.Q. Controllable synthesis of Ag-CuO composite nanosheets with enhanced photocatalytic property. *Mater. Lett.* **2014**, *120*, 16–19. [CrossRef]
56. Yang, Y.; Li, Y.Z.; Zhang, Q.; Zeng, M.; Wu, S.W.; Lan, L.; Zhao, X.J. Novel photoactivation and solar-light-driven thermocatalysis on epsilon-MnO_2 nanosheets lead to highly efficient catalytic abatement of ethyl acetate without acetaldehyde as unfavorable by-product. *J. Mater. Chem. A* **2018**, *6*, 14195–14206. [CrossRef]
57. Chang, Y.C.; Guo, J.Y. Double-sided plasmonic silver nanoparticles decorated copper oxide/zinc oxide heterostructured nanomaces with improving photocatalytic performance. *J. Photochem. Photobiol. A Chem.* **2019**, *378*, 184–191. [CrossRef]
58. Zhang, X.D.; Yang, Y.; Li, H.X.; Zou, X.J.; Wang, Y.X. Non-TiO_2 Photocatalysts Used for Degradation of Gaseous VOCs. *Prog. Chem.* **2016**, *28*, 1550–1559.
59. Liu, X.Q.; Iocozzia, J.; Wang, Y.; Cui, X.; Chen, Y.H.; Zhao, S.Q.; Li, Z.; Lin, Z.Q. Noble metal-metal oxide nanohybrids with tailored nanostructures for efficient solar energy conversion, photocatalysis and environmental remediation. *Energy Environ. Sci.* **2017**, *10*, 402–434. [CrossRef]
60. Almquist, C.B.; Biswas, P. The photo-oxidation of cyclohexane on titanium dioxide: An investigation of competitive adsorption and its effects on product formation and selectivity. *Appl. Catal. A Gen.* **2001**, *214*, 259–271. [CrossRef]
61. Tsoncheva, T.; Issa, G.; Blasco, T.; Dimitrov, M.; Popova, M.; Hernandez, S.; Kovacheva, D.; Atanasova, G.; Nieto, J.M.L. Catalytic VOCs elimination over copper and cerium oxide modified mesoporous SBA-15 silica. *Appl. Catal. A Gen.* **2013**, *453*, 1–12. [CrossRef]
62. Carrillo, A.M.; Carriazo, J.G. Cu and Co oxides supported on halloysite for the total oxidation of toluene. *Appl. Catal. B Environ.* **2015**, *164*, 443–452. [CrossRef]
63. Cui, E.T.; Hou, G.H.; Chen, X.H.; Zhang, F.; Deng, Y.X.; Yu, G.Y.; Li, B.B.; Wu, Y.Q. In-situ hydrothermal fabrication of Sr_2FeTaO_6/$NaTaO_3$ heterojunction photocatalyst aimed at the effective promotion of electron-hole separation and visible-light absorption. *Appl. Catal. B Environ.* **2019**, *241*, 52–65. [CrossRef]
64. Li, J.J.; Yu, E.Q.; Cai, S.C.; Chen, X.; Chen, J.; Jia, H.P.; Xu, Y.J. Noble metal free, CeO_2/$LaMnO_3$ hybrid achieving efficient photo-thermal catalytic decomposition of volatile organic compounds under IR light. *Appl. Catal. B Environ.* **2019**, *240*, 141–152. [CrossRef]
65. Liu, Y.; Zhang, Z.Y.; Fang, Y.R.; Liu, B.K.; Huang, J.D.; Miao, F.J.; Bao, Y.A.; Dong, B. IR-Driven strong plasmonic-coupling on Ag nanorices/$W_{18}O_{49}$ nanowires heterostructures for photo/thermal synergistic enhancement of H-2 evolution from ammonia borane. *Appl. Catal. B Environ.* **2019**, *252*, 164–173. [CrossRef]
66. Lee, Y.E.; Chung, W.C.; Chang, M.B. Photocatalytic oxidation of toluene and isopropanol by $LaFeO_3$/black-TiO_2. *Environ. Sci. Pollut. Res.* **2019**, *26*, 20908–20919. [CrossRef]
67. Ray, C.; Pal, T. Recent advances of metal-metal oxide nanocomposites and their tailored nanostructures in numerous catalytic applications. *J. Mater. Chem. A* **2017**, *5*, 9465–9487. [CrossRef]
68. Valenti, M.; Jonsson, M.P.; Biskos, G.; Schmidt-Ott, A.; Smith, W.A. Plasmonic nanoparticle-semiconductor composites for efficient solar water splitting. *J. Mater. Chem. A* **2016**, *4*, 17891–17912. [CrossRef]
69. Wu, N.Q. Plasmonic metal-semiconductor photocatalysts and photoelectrochemical cells: A review. *Nanoscale* **2018**, *10*, 2679–2696. [CrossRef]

70. Mendez-Medrano, M.G.; Kowalska, E.; Lehoux, A.; Herissan, A.; Ohtani, B.; Bahena, D.; Briois, V.; Colbeau-Justin, C.; Rodriguez-Lopez, J.L.; Remita, H. Surface Modification of TiO_2 with Ag Nanoparticles and CuO Nanoclusters for Application in Photocatalysis. *J. Phys. Chem. C* **2016**, *120*, 5143–5154. [CrossRef]
71. Boriskina, S.V.; Ghasemi, H.; Chen, G. Plasmonic materials for energy: From physics to applications. *Mater. Today* **2013**, *16*, 375–386. [CrossRef]
72. Zhang, X.D.; Yang, Y.; Song, L.; Wang, Y.X.; He, C.; Wang, Z.; Cui, L.F. High and stable catalytic activity of Ag/Fe_2O_3 catalysts derived from MOFs for CO oxidation. *Mol. Catal.* **2018**, *447*, 80–89. [CrossRef]
73. DuChene, J.S.; Tagliabue, G.; Welch, A.J.; Cheng, W.H.; Atwater, H.A. Hot Hole Collection and Photoelectrochemical CO_2 Reduction with Plasmonic Au/p-GaN Photocathodes. *Nano Lett.* **2018**, *18*, 2545–2550. [CrossRef] [PubMed]
74. Kim, Y.; Torres, D.D.; Jain, P.K. Activation Energies of Plasmonic Catalysts. *Nano Lett.* **2016**, *16*, 3399–3407. [CrossRef]
75. Khiavi, N.D.; Katal, R.; Eshkalak, S.K.; Masudy-Panah, S.; Ramakrishna, S.; Hu, J.Y. Visible Light Driven Heterojunction Photocatalyst of CuO-Cu_2O Thin Films for Photocatalytic Degradation of Organic Pollutants. *Nanomaterials* **2019**, *9*, 1011. [CrossRef]
76. Christopher, P.; Xin, H.L.; Linic, S. Visible-light-enhanced catalytic oxidation reactions on plasmonic silver nanostructures. *Nat. Chem.* **2011**, *3*, 467–472. [CrossRef]
77. Leong, K.H.; Abd Aziz, A.; Sim, L.C.; Saravanan, P.; Jang, M.; Bahnemann, D. Mechanistic insights into plasmonic photocatalysts in utilizing visible light. *Beilstein J. Nanotechnol.* **2018**, *9*, 628–648. [CrossRef]
78. Ma, L.; Chen, S.; Shao, Y.; Chen, Y.L.; Liu, M.X.; Li, H.X.; Mao, Y.L.; Ding, S.J. Recent Progress in Constructing Plasmonic Metal/Semiconductor Hetero-Nanostructures for Improved Photocatalysis. *Catalysts* **2018**, *8*, 634. [CrossRef]
79. Truppi, A.; Petronella, F.; Placido, T.; Striccoli, M.; Agostiano, A.; Curri, M.L.; Comparelli, R. Visible-Light-Active TiO_2-Based Hybrid Nanocatalysts for Environmental Applications. *Catalysts* **2017**, *7*, 100. [CrossRef]
80. Zhang, Y.C.; He, S.; Guo, W.X.; Hu, Y.; Huang, J.W.; Mulcahy, J.R.; Wei, W.D. Surface-Plasmon-Driven Hot Electron Photochemistry. *Chem. Rev.* **2018**, *118*, 2927–2954. [CrossRef]
81. Tatsuma, T.; Nishi, H.; Ishida, T. Plasmon-induced charge separation: chemistry and wide applications. *Chem. Sci.* **2017**, *8*, 3325–3337. [CrossRef] [PubMed]
82. Erwin, W.R.; Zarick, H.F.; Talbert, E.M.; Bardhan, R. Light trapping in mesoporous solar cells with plasmonic nanostructures. *Energy Environ. Sci.* **2016**, *9*, 1577–1601. [CrossRef]
83. Araujo, T.P.; Quiroz, J.; Barbosa, E.C.M.; Camargo, P.H.C. Understanding plasmonic catalysis with controlled nanomaterials based on catalytic and plasmonic metals. *Curr. Opin. Colloid Interface Sci.* **2019**, *39*, 110–122. [CrossRef]
84. Mao, M.Y.; Li, Y.Z.; Lv, H.Q.; Hou, J.T.; Zeng, M.; Ren, L.; Huang, H.; Zhao, X.J. Efficient UV-vis-IR light-driven thermocatalytic purification of benzene on a Pt/CeO_2 nanocomposite significantly promoted by hot electron-induced photoactivation. *Environ. Sci. Nano* **2017**, *4*, 373–384. [CrossRef]
85. Fu, S.F.; Zheng, Y.; Zhou, X.B.; Ni, Z.M.; Xia, S.J. Visible light promoted degradation of gaseous volatile organic compounds catalyzed by Au supported layered double hydroxides: Influencing factors, kinetics and mechanism. *J. Hazard. Mater.* **2019**, *363*, 41–54. [CrossRef]
86. Ma, X.C.; Dai, Y.; Yu, L.; Huang, B.B. Energy transfer in plasmonic photocatalytic composites. *Light Sci. Appl.* **2016**, *5*, e16017. [CrossRef]
87. Gomez, L.; Hueso, J.L.; Ortega-Liebana, M.C.; Santamaria, J.; Cronin, S.B. Evaluation of gold-decorated halloysite nanotubes as plasmonic photocatalysts. *Catal. Commun.* **2014**, *56*, 115–118. [CrossRef]
88. Graus, J.; Bueno-Alejo, C.J.; Hueso, J.L. In-Situ Deposition of Plasmonic Gold Nanotriangles and Nanoprisms onto Layered Hydroxides for Full-Range Photocatalytic Response towards the Selective Reduction of p-Nitrophenol. *Catalysts* **2018**, *8*, 354. [CrossRef]
89. Uson, L.; Sebastian, V.; Mayoral, A.; Hueso, J.L.; Eguizabal, A.; Arruebo, M.; Santamaria, J. Spontaneous formation of Au-Pt alloyed nanoparticles using pure nano-counterparts as starters: A ligand and size dependent process. *Nanoscale* **2015**, *7*, 10152–10161. [CrossRef]
90. Zieba, M.; Hueso, J.L.; Arruebo, M.; Martinez, G.; Santamaria, J. Gold-coated halloysite nanotubes as tunable plasmonic platforms. *New J. Chem.* **2014**, *38*, 2037–2042. [CrossRef]

91. Halas, N.J.; Lal, S.; Chang, W.S.; Link, S.; Nordlander, P. Plasmons in Strongly Coupled Metallic Nanostructures. *Chem. Rev.* **2011**, *111*, 3913–3961. [CrossRef] [PubMed]
92. Aslam, U.; Rao, V.G.; Chavez, S.; Linic, S. Catalytic conversion of solar to chemical energy on plasmonic metal nanostructures. *Nat. Catal.* **2018**, *1*, 656–665. [CrossRef]
93. Kim, Y.; Smith, J.G.; Jain, P.K. Harvesting multiple electron-hole pairs generated through plasmonic excitation of Au nanoparticles. *Nat. Chem.* **2018**, *10*, 763–769. [CrossRef] [PubMed]
94. Bernardi, M.; Mustafa, J.; Neaton, J.B.; Louie, S.G. Theory and computation of hot carriers generated by surface plasmon polaritons in noble metals. *Nat. Commun.* **2015**, *6*, 7044. [CrossRef] [PubMed]
95. Brongersma, M.L.; Halas, N.J.; Nordlander, P. Plasmon-induced hot carrier science and technology. *Nat. Nanotechnol.* **2015**, *10*, 25–34. [CrossRef]
96. Kriegel, I.; Scotognella, F.; Manna, L. Plasmonic doped semiconductor nanocrystals: Properties, fabrication, applications and perspectives. *Phys. Rep. Rev. Sec. Phys. Lett.* **2017**, *674*, 1–52. [CrossRef]
97. Rao, V.G.; Aslam, U.; Linic, S. Chemical Requirement for Extracting Energetic Charge Carriers from Plasmonic Metal Nanoparticles to Perform Electron-Transfer Reactions. *J. Am. Chem. Soc.* **2019**, *141*, 643–647. [CrossRef]
98. Aslam, U.; Chavez, S.; Linic, S. Controlling energy flow in multimetallic nanostructures for plasmonic catalysis. *Nat. Nanotechnol.* **2017**, *12*, 1000–1005. [CrossRef]
99. Boerigter, C.; Aslam, U.; Linic, S. Mechanism of Charge Transfer from Plasmonic Nanostructures to Chemically Attached Materials. *ACS Nano* **2016**, *10*, 6108–6115. [CrossRef]
100. Linic, S.; Aslam, U.; Boerigter, C.; Morabito, M. Photochemical transformations on plasmonic metal nanoparticles. *Nat. Mater.* **2015**, *14*, 567–576. [CrossRef]
101. Gargiulo, J.; Berte, R.; Li, Y.; Maier, S.A.; Cortes, E. From Optical to Chemical Hot Spots in Plasmonics. *Acc. Chem. Res.* **2019**, *52*, 2525–2535. [CrossRef]
102. Linic, S.; Christopher, P.; Xin, H.L.; Marimuthu, A. Catalytic and Photocatalytic Transformations on Metal Nanoparticles with Targeted Geometric and Plasmonic Properties. *Acc. Chem. Res.* **2013**, *46*, 1890–1899. [CrossRef]
103. Brus, L. Noble Metal Nanocrystals: Plasmon Electron Transfer Photochemistry and Single-Molecule Raman Spectroscopy. *Acc. Chem. Res.* **2008**, *41*, 1742–1749. [CrossRef]
104. El-Sayed, M.A. Some interesting properties of metals confined in time and nanometer space of different shapes. *Acc. Chem. Res.* **2001**, *34*, 257–264. [CrossRef]
105. Marimuthu, A.; Zhang, J.W.; Linic, S. Tuning Selectivity in Propylene Epoxidation by Plasmon Mediated Photo-Switching of Cu Oxidation State. *Science* **2013**, *339*, 1590–1593. [CrossRef]
106. Wan, L.L.; Zhou, Q.X.; Wang, X.; Wood, T.E.; Wang, L.; Duchesne, P.N.; Guo, J.L.; Yan, X.L.; Xia, M.K.; Lie, Y.F.; et al. Cu_2O nanocubes with mixed oxidation-state facets for (photo)catalytic hydrogenation of carbon dioxide. *Nat. Catal.* **2019**, *2*, 889–898. [CrossRef]
107. Zhang, Q.B.; Zhang, K.L.; Xu, D.G.; Yang, G.C.; Huang, H.; Nie, F.D.; Liu, C.M.; Yang, S.H. CuO nanostructures: Synthesis, characterization, growth mechanisms, fundamental properties, and applications. *Prog. Mater. Sci.* **2014**, *60*, 208–337. [CrossRef]
108. Kumar, M.K.; Bhavani, K.; Naresh, G.; Srinivas, B.; Venugopal, A. Plasmonic resonance nature of Ag-Cu/TiO_2 photocatalyst under solar and artificial light: Synthesis, characterization and evaluation of H_2O splitting activity. *Appl. Catal. B Environ.* **2016**, *199*, 282–291.
109. Nguyen, N.L.; de Gironcoli, S.; Piccinin, S. Ag-Cu catalysts for ethylene epoxidation: Selectivity and activity descriptors. *J. Chem. Phys.* **2013**, *138*, 184707. [CrossRef]
110. Rapallo, A.; Rossi, G.; Ferrando, R.; Fortunelli, A.; Curley, B.C.; Lloyd, L.D.; Tarbuck, G.M.; Johnston, R.L. Global optimization of bimetallic cluster structures. I. Size-mismatched Ag-Cu, Ag-Ni, and Au-Cu systems. *J. Chem. Phys.* **2005**, *122*, 194308. [CrossRef]
111. Verma, A.; Gupta, R.K.; Shukla, M.; Malviya, M.; Sinha, I. Ag-Cu Bimetallic Nanoparticles as Efficient Oxygen Reduction Reaction Electrocatalysts in Alkaline Media. *J. Nanosci. Nanotechnol.* **2020**, *20*, 1765–1772. [CrossRef] [PubMed]
112. Piccinin, S.; Zafeiratos, S.; Stampfl, C.; Hansen, T.W.; Havecker, M.; Teschner, D.; Bukhtiyarov, V.I.; Girgsdies, F.; Knop-Gericke, A.; Schlogl, R.; et al. Alloy Catalyst in a Reactive Environment: The Example of Ag-Cu Particles for Ethylene Epoxidation. *Phys. Rev. Lett.* **2010**, *104*, 035503. [CrossRef] [PubMed]
113. Tchaplyguine, M.; Zhang, C.F.; Andersson, T.; Bjorneholm, O. Ag-Cu oxide nanoparticles with high oxidation states: towards new high T-c materials. *Dalton Trans.* **2018**, *47*, 16660–16667. [CrossRef] [PubMed]

114. Zhang, Y.Y.; Wang, L.L.; Kong, X.Y.; Jiang, H.Y.; Zhang, F.; Shi, J.S. Novel Ag-Cu bimetallic alloy decorated near-infrared responsive three-dimensional rod-like architectures for efficient photocatalytic water purification. *J. Colloid Interface Sci.* **2018**, *522*, 29–39. [CrossRef] [PubMed]
115. Walsh, D.; Arcelli, L.; Ikoma, T.; Tanaka, J.; Mann, S. Dextran templating for the synthesis of metallic and metal oxide sponges. *Nat. Mater.* **2003**, *2*, 386. [CrossRef] [PubMed]
116. Liang, Y.; Chen, Z.; Yao, W.; Wang, P.Y.; Yu, S.J.; Wang, X.K. Decorating of Ag and CuO on Cu Nanoparticles for Enhanced High Catalytic Activity to the Degradation of Organic Pollutants. *Langmuir* **2017**, *33*, 7606–7614. [CrossRef]
117. Elemike, E.E.; Onwudiwe, D.C.; Ogeleka, D.F.; Mbonu, J.I. Phyto-assisted Preparation of Ag and Ag-CuO Nanoparticles Using Aqueous Extracts of Mimosa pigra and their Catalytic Activities in the Degradation of Some Common Pollutants. *J. Inorg. Organomet. Polym. Mater.* **2019**, *29*, 1798–1806. [CrossRef]
118. Ji, W.K.; Shen, T.; Kong, J.J.; Rui, Z.B.; Tong, Y.X. Synergistic Performance between Visible-Light Photocatalysis and Thermocatalysis for VOCs Oxidation over Robust Ag/F-Codoped $SrTiO_3$. *Ind. Eng. Chem. Res.* **2018**, *57*, 12766–12773. [CrossRef]
119. Wan, X.; Yang, J.; Huang, X.Y.; Tie, S.L.; Lan, S. A high-performance room temperature thermocatalyst Cu_2O/Ag-0@Ag-NPs for dye degradation under dark condition. *J. Alloys Compd.* **2019**, *785*, 398–409. [CrossRef]
120. Kung, M.L.; Tai, M.H.; Lin, P.Y.; Wu, D.C.; Wu, W.J.; Yeh, B.W.; Hung, H.S.; Kuo, C.H.; Chen, Y.W.; Hsieh, S.L.; et al. Silver decorated copper oxide (Ag@CuO) nanocomposite enhances ROS-mediated bacterial architecture collapse. *Colloid Surf. B Biointerfaces* **2017**, *155*, 399–407. [CrossRef]
121. Ramirez, A.; Hueso, J.L.; Suarez, H.; Mallada, R.; Ibarra, A.; Irusta, S.; Santamaria, J. A Nanoarchitecture Based on Silver and Copper Oxide with an Exceptional Response in the Chlorine-Promoted Epoxidation of Ethylene. *Angew. Chem. Int. Edit.* **2016**, *55*, 11158–11161. [CrossRef] [PubMed]
122. Bottega-Pergher, B.; Graus, J.; Bueno-Alejo, C.J.; Hueso, J.L. Triangular and Prism-Shaped Gold-Zinc Oxide Plasmonic Nanostructures: In situ Reduction, Assembly, and Full-Range Photocatalytic Performance. *Eur. J. Inorg. Chem.* **2019**, *2019*, 3228–3234. [CrossRef]
123. Ramirez, A.; Hueso, J.L.; Mallada, R.; Santamaria, J. Ethylene epoxidation in microwave heated structured reactors. *Catal. Today* **2016**, *273*, 99–105. [CrossRef]
124. Ramirez, A.; Hueso, J.L.; Mallada, R.; Santamaria, J. In situ temperature measurements in microwave-heated gas-solid catalytic systems. Detection of hot spots and solid-fluid temperature gradients in the ethylene epoxidation reaction. *Chem. Eng. J.* **2017**, *316*, 50–60. [CrossRef]
125. Gomez-Romero, P.; Tejada-Rosales, E.M.; Palacin, M.R. $Ag_2Cu_2O_3$: The first silver copper oxide. *Angew. Chem. Int. Edit.* **1999**, *38*, 524–525. [CrossRef]
126. Tejada-Rosales, E.M.; Rodriguez-Carvajal, J.; Casan-Pastor, N.; Alemany, P.; Ruiz, E.; El-Fallah, M.S.; Alvarez, S.; Gomez-Romero, P. Room-temperature synthesis and crystal, magnetic, and electronic structure of the first silver copper oxide. *Inorg. Chem.* **2002**, *41*, 6604–6613. [CrossRef]
127. Carreras, A.; Conejeros, S.; Camon, A.; Garcia, A.; Casan-Pastor, N.; Alemany, P.; Canadell, E. Charge Delocalization, Oxidation States, and Silver Mobility in the Mixed Silver-Copper Oxide $AgCuO_2$. *Inorg. Chem.* **2019**, *58*, 7026–7035. [CrossRef]
128. Navaladian, S.; Viswanathan, B.; Viswanath, R.P.; Varadarajan, T.K. Thermal decomposition as route for silver nanoparticles. *Nanoscale Res. Lett.* **2007**, *2*, 44–48. [CrossRef]
129. Boerigter, C.; Campana, R.; Morabito, M.; Linic, S. Evidence and implications of direct charge excitation as the dominant mechanism in plasmon-mediated photocatalysis. *Nat. Commun.* **2016**, *7*, 10545. [CrossRef]
130. Zhang, T.; Wang, S.J.; Zhang, X.Y.; Su, D.; Yang, Y.; Wu, J.Y.; Xu, Y.Y.; Zhao, N. Progress in the Utilization Efficiency Improvement of Hot Carriers in Plasmon-Mediated Heterostructure Photocatalysis. *Appl. Sci.* **2019**, *9*, 2093. [CrossRef]
131. Zhou, L.A.; Swearer, D.F.; Zhang, C.; Robatjazi, H.; Zhao, H.Q.; Henderson, L.; Dong, L.L.; Christopher, P.; Carter, E.A.; Nordlander, P.; et al. Quantifying hot carrier and thermal contributions in plasmonic photocatalysis. *Science* **2018**, *362*, 69–72. [CrossRef] [PubMed]
132. Huang, Y.F.; Zhang, M.; Zhao, L.B.; Feng, J.M.; Wu, D.Y.; Ren, B.; Tian, Z.Q. Activation of Oxygen on Gold and Silver Nanoparticles Assisted by Surface Plasmon Resonances. *Angew. Chem. Int. Edit.* **2014**, *53*, 2353–2357. [CrossRef] [PubMed]

133. Kim, M.; Lee, J.H.; Nam, J.M. Plasmonic Photothermal Nanoparticles for Biomedical Applications. *Adv. Sci.* **2019**, *6*, 1900471. [CrossRef] [PubMed]
134. Brown, A.M.; Sundararaman, R.; Narang, P.; Goddard, W.A.; Atwater, H.A. Nonradiative Plasmon Decay and Hot Carrier Dynamics: Effects of Phonons, Surfaces, and Geometry. *ACS Nano* **2016**, *10*, 957–966. [CrossRef] [PubMed]
135. Atwater, H.A.; Polman, A. Plasmonics for improved photovoltaic devices. *Nat. Mater.* **2010**, *9*, 205–213. [CrossRef] [PubMed]
136. Jermyn, A.S.; Tagliabue, G.; Atwater, H.A.; Goddard, W.A.; Narang, P.; Sundararaman, R. Transport of hot carriers in plasmonic nanostructures. *Phys. Rev. Mater.* **2019**, *3*, 075201. [CrossRef]
137. Bai, S.; Li, X.Y.; Kong, Q.; Long, R.; Wang, C.M.; Jiang, J.; Xiong, Y.J. Toward Enhanced Photocatalytic Oxygen Evolution: Synergetic Utilization of Plasmonic Effect and Schottky Junction via Interfacing Facet Selection. *Adv. Mater.* **2015**, *27*, 3444–3452. [CrossRef]
138. Ghosh, P.; Kar, A.; Khandelwal, S.; Vyas, D.; Mir, A.; Chakraborty, A.L.; Hegde, R.S.; Sharma, S.; Dutta, A.; Khatua, S. Plasmonic CoO-Decorated Au Nanorods for Photoelectrocatalytic Water Oxidation. *ACS Appl. Nano Mater.* **2019**, *2*, 5795–5803. [CrossRef]
139. Al-Zubeidi, A.; Hoener, B.S.; Collins, S.S.E.; Wang, W.X.; Kirchner, S.R.; Jebeli, S.A.H.; Joplin, A.; Chang, W.S.; Link, S.; Landes, C.F. Hot Holes Assist Plasmonic Nanoelectrode Dissolution. *Nano Lett.* **2019**, *19*, 1301–1306. [CrossRef]
140. Pensa, E.; Gargiulo, J.; Lauri, A.; Schlucker, S.; Cortes, E.; Maier, S.A. Spectral Screening of the Energy of Hot Holes over a Particle Plasmon Resonance. *Nano Lett.* **2019**, *19*, 1867–1874. [CrossRef]

Article

Carbon/Graphene-Modified Titania with Enhanced Photocatalytic Activity under UV and Vis Irradiation

Kunlei Wang [1], Maya Endo-Kimura [1], Raphaëlle Belchi [2,3], Dong Zhang [4], Aurelie Habert [2], Johann Bouclé [3], Bunsho Ohtani [1,4], Ewa Kowalska [1,4,*] and Nathalie Herlin-Boime [2,*]

1 Institute for Catalysis (ICAT), Hokkaido University, N21 W10, Sapporo 001-0021, Japan; kunlei@cat.hokudai.ac.jp (K.W.); m_endo@cat.hokudai.ac.jp (M.E.-K.); ohtani@cat.hokudai.ac.jp (B.O.)

2 IRAMIS—NIMBE UMR 3685, Université Paris Saclay, CEA Saclay, 91191 Gif/Yvette CEDEX, France; raphaelle.belchi@cea.fr (R.B.); aurelie.habert@cea.fr (A.H.)

3 Univ. Limoges, CNRS, XLIM, UMR 7252, F-87000 Limoges, France; johann.boucle@unilim.fr

4 Graduate School of Environmental Science, Hokkaido University, Sapporo 060-0810, Japan; zhang.d@cat.hokudai.ac.jp

* Correspondence: kowalska@cat.hokudai.ac.jp (E.K.); nathalie.herlin@cea.fr (N.H.-B.)

Received: 4 November 2019; Accepted: 9 December 2019; Published: 11 December 2019

Abstract: Laser synthesis was used for one-step synthesis of titania/graphene composites (G-TiO_2(C)) from a suspension of 0.04 wt% commercial reduced graphene oxide (rGO) dispersed in liquid titanium tetraisopropoxide (TTIP). Reference titania sample (TiO_2(C)) was prepared by the same method without graphene addition. Both samples and commercial titania P25 were characterized by various methods and tested under UV/vis irradiation for oxidative decomposition of acetic acid and dehydrogenation of methanol (with and without Pt co-catalyst addition), and under vis irradiation for phenol degradation and inactivation of *Escherichia coli*. It was found that both samples (TiO_2(C) and G-TiO_2(C)) contained carbon resulting from TTIP and C_2H_4 (used as a synthesis sensitizer), which activated titania towards vis activity. The photocatalytic activity under UV/vis irradiation was like that by P25. The highest activity of TiO_2(C) sample for acetic acid oxidation was probably caused by its surface enrichment with hydroxyl groups. G-TiO_2(C) was the most active for methanol dehydrogenation in the absence of platinum (ca. five times higher activity than that by TiO_2(C) and P25), suggesting that graphene works as a co-catalyst for hydrogen evolution. High activity under both UV and vis irradiation for decomposition of organic compounds, hydrogen evolution and inactivation of bacteria suggests that laser synthesis allows preparation of cheap (carbon-modified) and efficient photocatalysts for broad environmental applications.

Keywords: carbon-doped titania; carbon-modified titania; graphene/titania; vis-active photocatalyst; antibacterial properties; laser pyrolysis

1. Introduction

Titania (titanium(IV) oxide; titanium dioxide) is probably the most intensively studied semiconductor for both environmental and energy application, i.e., water/air purification, wastewater treatment, self-cleaning surfaces, sterilization, photocurrent generation, water splitting and fuel production [1–6]. Although titania has many advantages, including low-cost, high activity and stability, two main shortcomings limit titania broad and common applications. First, quantum yields of photocatalytic reactions driven by titania are much lower than 100% since photogenerated charge carriers recombine fast, as typical for all semiconductors. Moreover, wide bandgap (ca. 3 eV, depending on polymorphic forms) of titania allowing good redox properties results in its activity only under UV irradiation, and thus main part of solar radiation cannot be used for photocatalytic process.

Therefore, various methods have been proposed for titania modifications, such as doping (substitutional or/and interstitial), surface modification or formation of composites with other materials (e.g., heterojunctions). The most famous and probably most efficient for activity enhancement under UV irradiation is surface modification of titania with noble metals, since noble metals work as an electron sink (larger work function than electron affinity of titania), and thus hindering charge carriers' recombination, as probably firstly reported by Kraeutler and Bard more than forty years ago [7]. Recently, another property of noble metals, i.e., surface plasmon resonance (SPR) at vis range has been studied to activate titania towards vis irradiation [8,9]. However, noble metals are quite expensive, and thus cheaper materials are more recommended. For example, titania modification with carbon seems to be more attractive for broad and commercial application.

Various sources of carbon have been proposed for titania modification, such as polyvinyl alcohol [10], n-hexane [11], alcohols (methyl, ethyl, isopropyl, n-butyl, 2-butyl and tert-butyl alcohols) [12], benzene [13], ethylene glycol and pentaerythritol [14] and glucose [15]. Interesting approach was proposed for carbon-doped titania nanostructures (micro- and nanospheres and nanotubes), synthesized via a single source vapor deposition in an inert atmosphere (Ar), where titanium butoxide (common organic precursor of titania) was used also as a carbon source [16]. Similar findings were reported for titania prepared from titanium isopropoxide by the sol–gel method and calcined at different temperatures (350, 450, 550, 650 and 750 °C) [17]. It was found that calcination at 350 °C resulted in the strongest vis absorption, correlating with the highest surface content of carbon and the highest photocatalytic activity under vis irradiation. However, surface modification (with C–C species) was suggested as the main reason of vis response, rather than C-doping (no bandgap narrowing).

Recently, 2D carbon structures, i.e., graphene (G), graphene oxide (GO) and reduced graphene oxide (rGO), have been proposed for titania modification, due to large specific surface area (efficient reagents adsorption), high conductivity (inhibited charge carriers' recombination by highly mobile electrons), flexible structure and high stability [18]. For example, (i) TiO_2-GO composite, prepared by thermal hydrolysis of suspension containing GO and titania peroxo-complex, was efficient for photocatalytic degradation of butene in the gas phase [19], (ii) TiO_2-rGO, synthesized by ionothermal method, was able to generate hydrogen [20], (iii) hydrothermally prepared TiO_2-rGO and TiO_2-G decomposed bisphenol A under both UV and vis irradiation [21] and 4-chlorophenol under solar radiation [22], respectively, (iv) ZnO-G, prepared by the Hummers/Offeman/hydrothermal method, caused efficient degradation of cyanide in water under UV, vis, solar and even NIR irradiation (plasmonic absorption of NIR through coupling of graphene) [23] and (v) α-Fe_2O_3-ZnO/rGO, synthesized by the Hummer method, redox replacement and electrochemical process, was able to capture and reduce CO_2 to CH_3OH under vis irradiation [24]. Accordingly, G, GO and rGO-modified titania samples are widely used for photocatalytic applications. However, the role of 2D carbon has not been completely clarified yet, as summarized in recent review paper by Giovannetti et al. [18]. It was proposed that under UV irradiation, electrons from conduction band (CB) of titania migrate to graphene, due to its more positive Fermi level, which hinders charge carriers' recombination. Whereas, under vis irradiation, the opposite direction of electrons' transfer has been proposed, i.e., from photoexcited state of graphene to CB of titania.

Our recent study on graphene-modified titania has shown that this material might be efficiently used for perovskite solar cells, where graphene presence enhanced power conversion efficiency significantly, as a result of photoluminescence quenching, due to enhanced electron migration [25]. Therefore, to further examine the potential and application possibility of graphene-modified titania, prepared by a novel and simple method using laser pyrolysis, more detailed characterization and photocatalytic activity for degradation of organic compounds, hydrogen evolution and microorganism inactivation under both UV and vis irradiation have been investigated in this study.

2. Materials and Methods

2.1. Synthesis

Titania (TiO_2(C)) and graphene-modified titania (G-TiO_2(C)) were prepared by laser pyrolysis. This synthesis technique has already been described in several publications, e.g., Pignon et al. for the synthesis of TiO_2 with controlled anatase to rutile ratio and Belchi et al. for the one-step synthesis of graphene/TiO_2 nanocomposites [25,26]. A schematic drawing of the experimental set-up is shown in Figure 1, and the most salient features of the technique are described hereafter. Briefly, liquid titanium(IV) tetraisopropoxide (TTIP) purchased from Sigma-Aldrich (≥97% purity) was used for the synthesis of titania (labeled as TiO_2(C)). A suspension of 0.04 wt% industrial graphene (reduced graphene oxide G-200 from SIMBATT Company (Shanghai, China), oxygen content of less than 8 at%, a few layers (<10 layers), leading to a high specific area (>600 m^2/g)) dispersed in TTIP was used for the synthesis of graphene-TiO_2(C) nanocomposite material labeled as G-TiO_2(C). Droplets of the liquid precursor or suspension were obtained with a pyrosol device (RBI, Meylan, France). The droplets were carried out with a carrier gas (Ar) to the reactor where they intersected with the beam of a high-power CO_2 laser (maximum emission at 10.6 µm). TTIP could not absorb well the laser radiation, therefore C_2H_4 was added as a sensitizer to the Ar carrier gas. After laser absorption, the reactive medium was rapidly thermalized by collisional transfer. Decomposition of precursors and growth of nanoparticles occurred with an appearance of a flame.

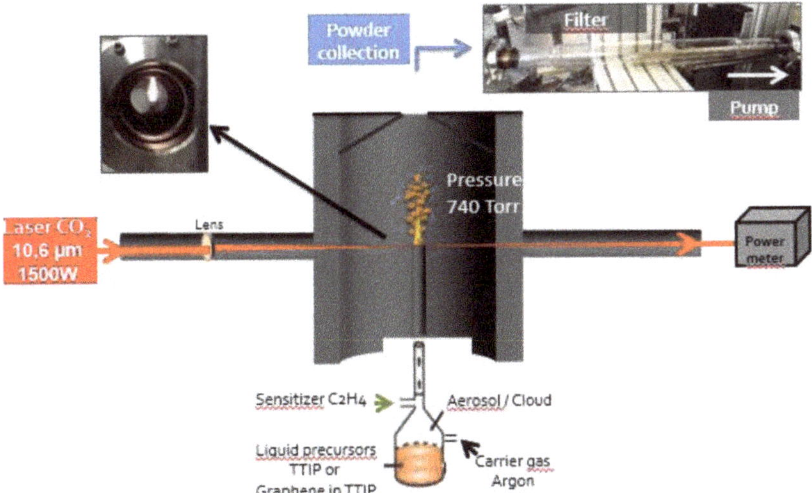

Figure 1. Schematic drawing of the laser pyrolysis experimental set-up for preparation of titania samples (TiO_2(C) from TTIP and G-TiO_2(C) from TTIP and graphene).

The laser internal power was set at 530 W corresponding to a 400 W laser power, measured under Ar atmosphere after the reaction zone. The laser power measured at the same place during the reaction was 230 W. The pressure was regulated at atmospheric pressure, i.e., 10^5 Pa. The ethylene (sensitizer) and argon (carrier) gas flow rates were 355 and 2000 sccm, respectively. The production rate was 1.3 g/h for the TiO_2(C) and 0.36 g/h for the G-TiO_2(C) sample. This difference is caused by the higher viscosity of the suspension in the presence of graphene, leading to reduced generation of droplets by the pyrosol device. Post-reaction annealing (6-h under air at 450 °C) was performed to remove amorphous carbon present in the powders, due to the carbon content in the precursors.

2.2. Characterization of Samples

The morphology of the powders was evaluated by a Carl Zeiss ULTRA55 scanning electron microscope (Carl Zeiss, Oberkochen, Germany) and by a JEOL 2010 high-resolution transmission electron microscope (Jeol, Tokyo, Japan) operated at 200 kV. For SEM analysis, the powder was directly observed on the carbon tape. For HRTEM measurements, the powder was dispersed in ethanol and nanoparticles were separated with intensive ultrasound radiation using a Hielscher Ultrasound Technology VialTweeter UIS250V (Teltow, Germany). Then, the dispersion was dropped on a grid made of a Lacey Carbon Film (300 mesh Copper, S166-3H, purchased from Oxford instruments SAS, Abingdon, Great Britain). The Raman spectra were acquired on an XploRA PLUS Horiba apparatus (Kyoto, Japan) using a laser emission at 532 nm for excitation. The composition of samples was estimated by an energy-dispersive X-ray spectroscopy (EDS; HD-2000, HITACHI, Tokyo, Japan).

Photoabsorption properties for samples before and after annealing were analyzed by diffuse reflectance spectroscopy (DRS; JASCO V-670 equipped with a PIN-757 integrating sphere, JASCO, LTD., Pfungstadt, Germany). Barium sulfate was used as reference for DRS analysis. Crystalline properties were analyzed by X-ray powder diffraction (XRD; Rigaku intelligent XRD SmartLab with a Cu target, Rigaku, LTD., Tokyo, Japan). Crystallite sizes of anatase, rutile and graphene were estimated using the Scherrer equation. Chemical composition of the surface (content and chemical state of elements, i.e., titanium, oxygen and carbon) was determined by X-ray photoelectron spectroscopy (XPS; JEOL JPC-9010MC with MgKα X-ray, JEOL, LTD., Tokyo, Japan).

Energy-resolved distribution of electron traps (ERDT) pattern and conduction band bottom (CBB) position were analyzed by reversed double-beam photoacoustic spectroscopy (RDB-PAS) and photoacoustic spectroscopy (PAS), respectively, and detailed procedure was described elsewhere [27]. In brief, for RDB-PAS measurement, as-prepared sample was filled in stainless-steel sample holder in a home-made PAS cell, equipped with an electret condenser microphone and a Pyrex window on the upper side. Methanol-saturated nitrogen was flowed through the cell for 30 min, the cell was irradiated by a 625-nm light-emitting diode beam (Luxeon LXHL-ND98, LUMILEDS, San Jose, CA, USA), modulated at 35 Hz by a function generator (DF1906, NF Corporation, Yokohama, Japan) as modulated light, and a monochromatic light beam from a Xe lamp (ASB-XE-175, Spectral Products, Putnam, CT, USA), equipped with a grating monochromator (CM110 1/8 m, Spectral Products, CT, USA) as continuous light. The continuous light was scanned from 650 to 300 nm with a 5-nm step. RDB-PA signal was detected by a digital lock-in amplifier (LI5630, NF Corporation). Obtained spectrum was differentiated from the lower-energy side and calibrated with the reported total electron-trap density in units of μmol g^{-1} measured by a photochemical method [28] to obtain an ERDT pattern. For PAS measurements, the cell window was irradiated from 650 to 300 nm by a light beam from a Xe lamp (ASB-XE-175, Spectral Products) with a grating monochromator (CM110 1/8m, Spectral Products, CT, USA) modulated at 80 Hz by a light chopper (5584A, NF Corporation, Japan) to detect the PAS signal using a digital lock-in amplifier, and then photoacoustic (PA) spectra were calibrated with a reference of a PA spectrum of graphite. The CBB, as energy from the top of the valence band (VBT), of samples was calculated from the onset wavelength corresponding to the bandgap of samples.

2.3. Activity Tests

The photocatalytic activity of samples was evaluated under UV/vis irradiation for: (1) oxidative decomposition of acetic acid (CO_2 evolution) and (2) anaerobic dehydrogenation of methanol (H_2 evolution; with/without in situ platinum deposition (2 wt% in respect to TiO_2), photodeposition details might be found here [29]), and under vis irradiation for (3) oxidation of phenol. For activity tests, (1–2) 50 mg and (3) 10 mg of photocatalyst was suspended in 5 mL of aqueous solution of (1) acetic acid (5 vol%), (2) methanol (50 vol%) and (3) 0.21 mmol/L phenol in 35-mL Pyrex test tubes. The tubes were sealed with rubber septa, the suspensions were continuously stirred in a thermostated water bath and irradiated by (1–2) Hg lamp ($\lambda > 290$ nm) and Xe lamp ($\lambda > 420$ nm; Xe lamp, water IR filter, cold mirror and cut-off filter Y45, in a reactor shown in Figure 3 of Ref. [30]). For methanol dehydrogenation, two kinds of reactions were carried out, i.e., with ("H_2 system (Pt)") and without ("H_2 system (no Pt)") platinum deposited in situ and working as co-catalyst for hydrogen formation. Suspensions containing titania and methanol and hexachloroplatinic acid (H_2PtCl_6 $6H_2O$; in the case of "H_2 system (Pt)") were pre-bubbled with Ar (100 mL/min, 15 min) to remove oxygen from the system. The amounts of (1) generated hydrogen, (2) liberated carbon dioxide and (3) phenol and benzoquinone (main degradation product) were determined by chromatography: (1–2) gas chromatography with thermal conductivity detector (GC-TCD), gas phase sampled every 20 min for (1) or 15 min for (2)) and (3) HPLC (liquid phase sampled every 30 min).

Antibacterial activity was evaluated using *Escherichia coli* K12 (ATCC29425, Manassas, VA, USA) as a model of bacteria. The procedure was described elsewhere [31]. In brief, 10 mg of sample was suspended in 7 mL of bacterial suspension (ca. 1–5 × 10^8 cells/mL) in Pyrex-glass test tube and irradiated with xenon lamp, equipped with cold mirror and cut-off filter Y45 ($\lambda > 420$ nm) or kept in the dark under continuous stirring at 25 °C. After irradiation at 0.5, 1, 2 and 3 h, portions of suspension were taken, diluted and inoculated on the plate count agar (Becton, Dickinson and Company, Franklin Lakes, NJ, USA) medium. Agar plates were incubated at 37 °C for ca. 16 h, and then formed colonies were counted.

Commercial titania P25 (Degussa/Evonik) was used as reference sample since P25 has exhibited one of the highest photocatalytic activity among various titania samples in different reaction systems (oxidation and reduction) [29,32,33], and thus is commonly used as a "standard" titania sample.

3. Results and Discussion

3.1. Characterization of Samples

SEM and TEM images of graphene, TiO_2(C) and G-TiO_2(C) samples are given in Figure 2 Original graphene sheets and graphene sheets covered with fine NPs of titania are shown in Figure 2a,b, respectively. Figure 2c,d illustrates TiO_2(C) sample, with good homogeneity of fine TiO_2 NPs of ca. 10–20 nm, corresponding well with the diameter estimated from specific surface area (SSA) measurement by BET (Brunauer, Emmet and Teller) method (ca. 25 nm).

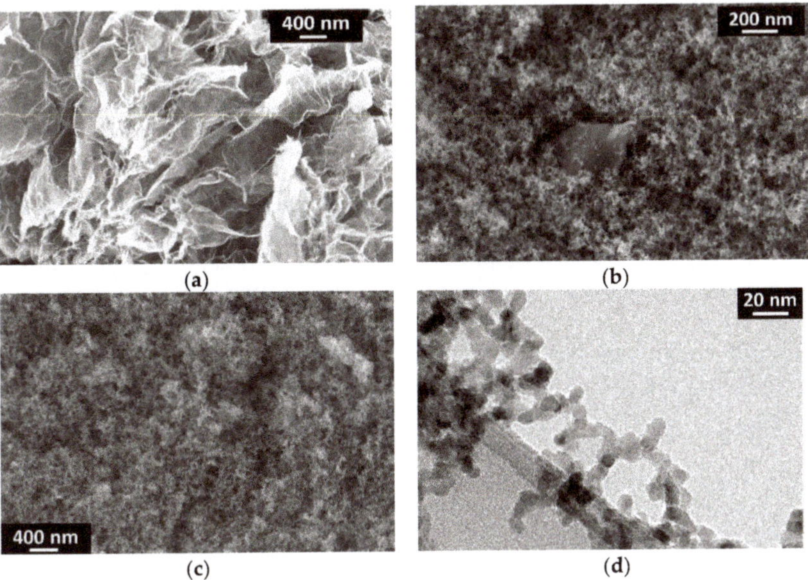

Figure 2. SEM (**a–c**) and TEM (**d**) images of: (**a**) graphene (SIMBATT); (**b**) G-TiO_2(C) and (**c**,**d**) TiO_2(C) samples.

The samples formed by the laser pyrolysis were grey, due to the presence of significant content of carbon (ca. 25% [25]) from titania precursor (TTIP) and C_2H_4 (a pyrolysis sensitizer). Although the previous study indicates the disappearance of the grey color after annealing (430 °C for 6 h; The conditions were estimated carefully to remove carbon impurities, and to avoid graphene combustion [25]), annealed samples were still greyish, probably due to different geometric conditions in a new furnace (KDF S-70; Denken-High Dental Co., Ltd.). The respective photoabsorption spectra before and after annealing are shown in Figure 3a,b.

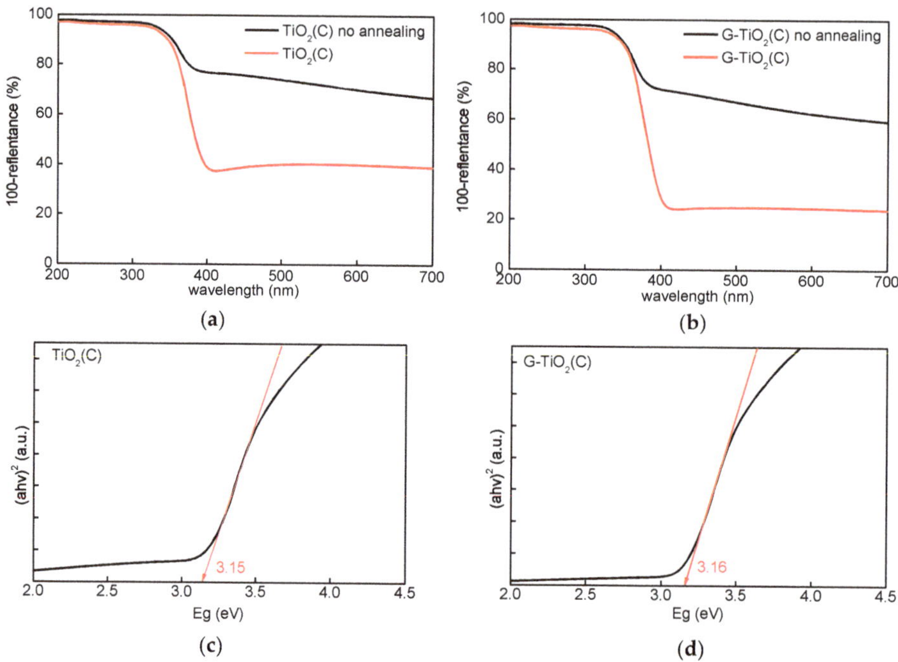

Figure 3. (**a**,**b**) Diffuse reflectance spectroscopy (DRS) spectra of samples before (black) and after (red) annealing: (**a**) TiO_2(C) and (**b**) G-TiO_2(C); (**c**,**d**) Bandgap estimation using Tauc's bandgap plots for samples after annealing: (**c**) TiO_2(C) and (**d**) G-TiO_2(C).

Similar photoabsorption spectra have already been reported for C-modified titania, prepared from TTIP by the sol–gel method and calcination at 350 °C [17]. Interestingly, it was found that photoabsorption properties at vis range were stronger for TiO_2(C) sample than for G-TiO_2(C) sample (both before and after annealing), which suggests that annealing resulted in removal of only part of the carbon used for synthesis. It is possible that adsorption of titania NPs on graphene sheets resulted in less available area of titania surface to be modified with carbon (from TTIP and C_2H_4). It should be pointed that thermal treatment (annealing) might cause stable surface modification of titania with carbon [17], whereas the pyrolysis might result in carbon doping (since carbon is introduced during titania synthesis), similar to samples prepared via vapor deposition [16]. Therefore, both kinds of modification could be expected in the case of TiO_2(C) sample, i.e., C-doping and C-modification. Whereas, in the case of G-TiO_2(C), titania was additionally modified with graphene. Although, clear bandgap narrowing was not observed, it is proposed that C-doping might be possible, considering slight shift of absorption edge from 388 nm (for pure anatase) [34] towards vis region, as calculated from bandgap energy [35] (Figure 3c,d)), i.e., ca. 394 nm (E_g = 3.15 eV) and 392 nm (E_g = 3.16 eV) for TiO_2(C) and G-TiO_2(C) samples, respectively.

To find the reason of vis absorption by TiO_2(C) sample, and to characterize both samples in detail, XRD, XPS, EDS, Raman spectroscopy and PAS/RDB-PAS analyses were performed, and obtained data are shown in Figures 4–8 XRD patterns of both samples were very similar, indicating that anatase was the main crystalline form, i.e., 78.7% in TiO_2(C) and 79.2% in G-TiO_2(C) (Figure 4 and Table 1). The minority of rutile was also detected in both samples, reaching 4.3% in TiO_2(C) and 3.6% in G-TiO_2(C), and thus the ratios of anatase to rutile were 18.5:1 and 22:1, respectively. The smaller content of rutile in G-TiO_2(C) sample suggests that co-present graphene could inhibit the anatase-to-rutile phase transition during annealing. The graphene was detected only in G-TiO_2(C) sample (ca. 3.9%). Although the signal was not intense, clear pattern could be seen after subtraction of titania peaks, as shown in Figure 4d (pink). It was found that the typical diffraction peak of graphene at $2\theta = 26.7°$ [21,36] was shifted to ca. 30° (at max. intensity) indicating possible partial hybridization with titania. Moreover, the additional, broad peaks at ca. 22° (002), 40–50° ((100), (101) and (004)) and 70–90° ((110), (112), (006) and (201)) are typical for all carbon materials, and thus might indicate the presence of both 2D nanostructures, i.e., graphene oxide (GO) and reduced graphene oxide (rGO) [23,37], and other carbon materials. Similar XRD patterns were reported for various carbon materials, e.g., carbon aerogel [38], carbon black [39], carbon nanospheres [40] and activated carbon [41]. The crystallite size changed only slightly from 9.8 to 8.99 nm for anatase and from ca. 10.5 to 10.9 nm for rutile, suggesting that graphene did not only inhibit anatase-to-rutile transition, but also influenced the formation and growing of crystals, as already reported [22,42]. Interestingly, slight shift of anatase peaks to 25.26° for TiO_2(C) and 25.27° for G-TiO_2(C) samples ((101) anatase at 25.31°) might confirm C-doping for both samples (Figure 4c; in comparison to reference titania sample: P25 for clarity).

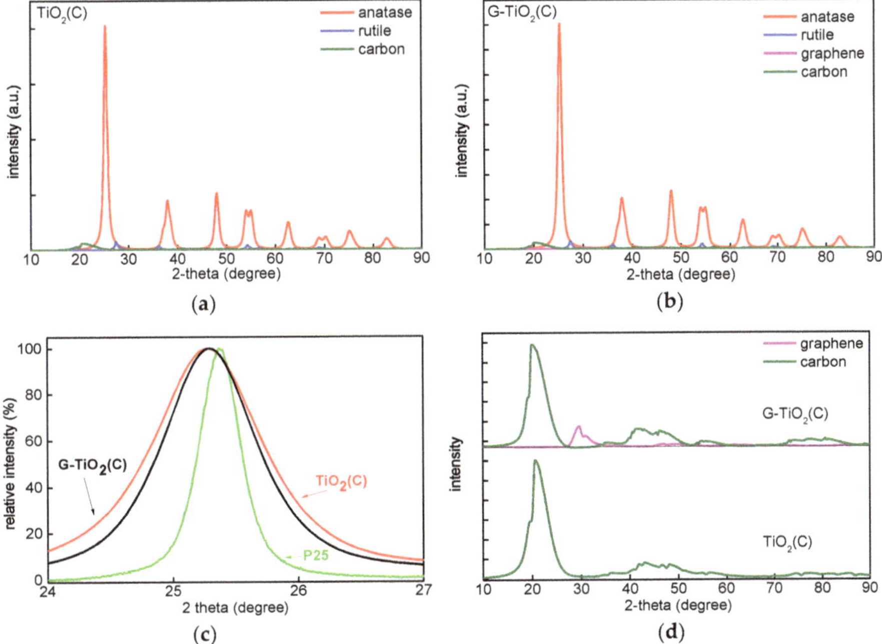

Figure 4. XRD patterns of: (**a**) TiO_2(C); (**b**) G-TiO_2(C), (**c**) TiO_2(C), G-TiO_2(C) and P25 titania at (101) anatase peak; (**d**) G-TiO_2(C) and TiO_2(C) after subtraction of titania peaks.

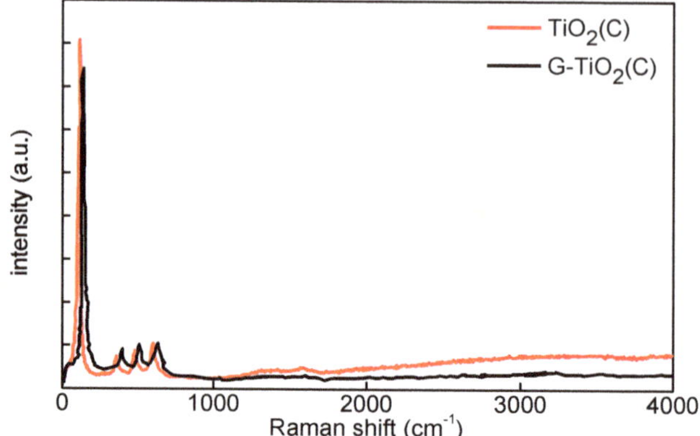

Figure 5. Raman spectra of TiO_2(C) (red) and G-TiO_2(C) (black).

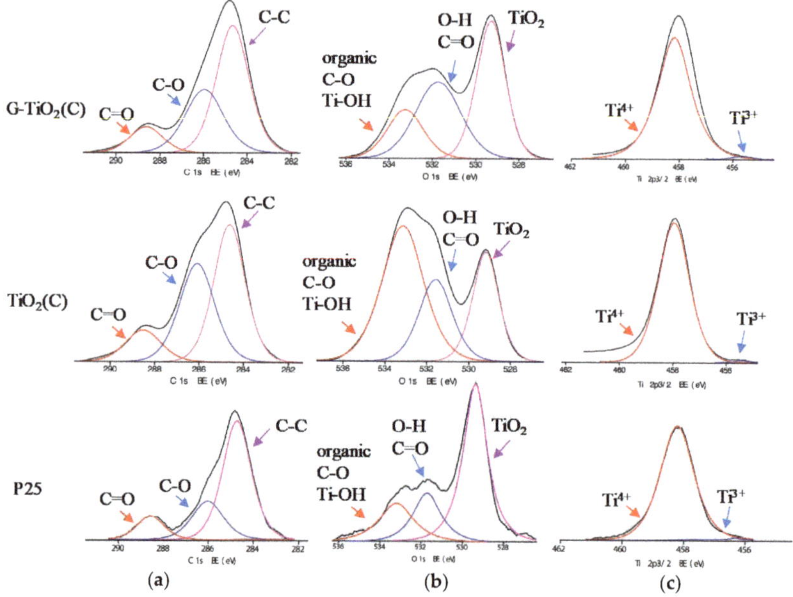

Figure 6. XPS data for G-TiO_2(C) (top) and TiO_2(C) (middle) and P25 (bottom) samples for: (**a**) C 1s, (**b**) O 1s and (**c**) Ti $2p_{3/2}$.

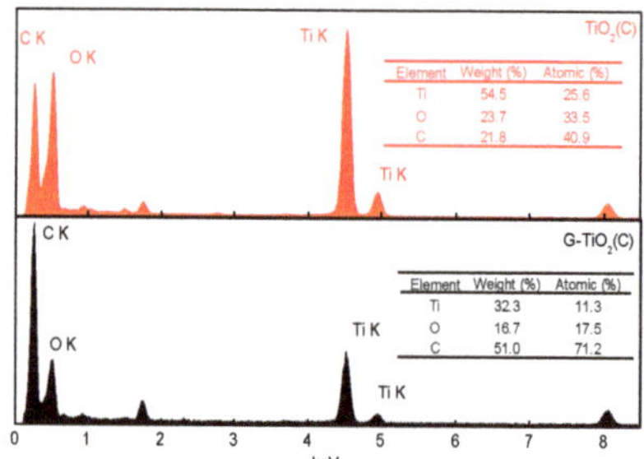

Element	Weight (%)	Atomic (%)
Ti	54.5	25.6
O	23.7	33.5
C	21.8	40.9

Element	Weight (%)	Atomic (%)
Ti	32.3	11.3
O	16.7	17.5
C	51.0	71.2

Figure 7. EDS data for TiO_2(C) (red, upper) and for G-TiO_2(C) (black, down).

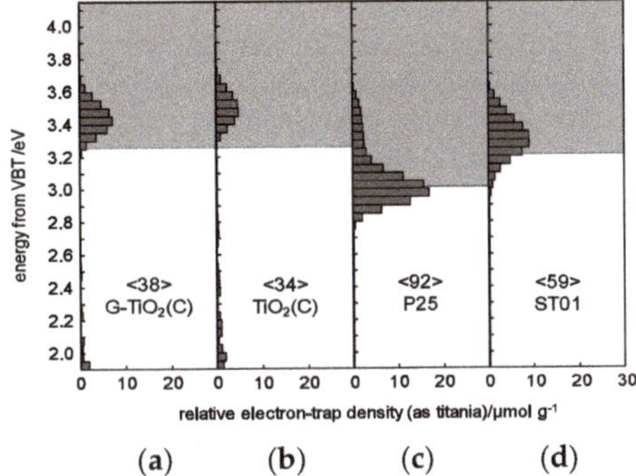

Figure 8. Energy-resolved distribution of electron traps (ERDT) pattern (bars) and conduction band bottom (CBB) position (bottom of grey box) of: (**a**) G-TiO_2(C), (**b**) TiO_2(C), (**c**) P25 and (**d**) ST01; values in < > denote the total density of electron traps (ETs) in the unit of µmol g^{-1}.

Table 1. Crystalline properties of samples.

Samples	Crystalline Content (%)				Crystallite Size (nm)			
	Anatase	Rutile	Carbon	Graphene	Anatase	Rutile	Carbon	Graphene
G-TiO_2(C)	79.2	3.6	14.5	2.7	9.0	10.9	2.2	3.9
TiO_2(C)	78.7	4.3	17.0	–	9.8	10.5	1.9	–

The Raman spectra for both samples were very similar, as shown in Figure 5, and a mapping recorded on a G-TiO_2(C) sample is provided in Figure S1. Four clear peaks appeared at ca. 136, 390, 510 and 630 cm^{-1}, corresponding well to E_g, $B_{1g(1)}$, A_{1g} + $B_{1g(2)}$ and $E_{g(2)}$ modes of anatase [19–21,43], respectively. Unfortunately, it was difficult to detect two characteristic peaks for graphene-kind structure at ca. 1304–1349 cm^{-1} and 1588–1601 cm^{-1} (as reported for graphene, GO, rGO, multiwall carbon nanotubes (MWCNT) of graphitic nature [19,20,23,44]), probably due to its low content. Interestingly, low-intensity peaks at ca. 1350 and 1560 cm^{-1} could be observed for TiO_2(C) sample (Figure 5), which might result from surface modification of titania with carbon, as suggested by DRS spectrum (Figure 3a). Similar Raman spectra were reported for anatase titania NPs uniformly distributed inside the porous carbon matrix, synthesized using furfuryl alcohol, tetrabutyltitanate, and nonionic surfactant [45]. It is known that D and G bands of carbon structures are located at ca. 1350 and 1600 cm^{-1}, and G band is common for all sp2 carbon forms giving information on the in-plane vibration of sp2 bonded carbon (tangential stretching mode of C=C bond), whereas D band is associated with structural disorder (the presence of sp3 defects) [19,44]. Therefore, it was confirmed that TiO_2(C) was surface modified with carbon species, correlating well with high absorption at vis range (Figure 3a). Moreover, slight broadening of peaks (confirming a decrease in crystallite size (XRD data; Table 1)) and frequency shift after titania modification with graphene might be attributed to photon confinement, non-stoichiometry and internal stress/surface tension effects [45], which corresponds well with higher content of Ti^{3+} and electron traps (ETs) in G-TiO_2(C) sample (as discussed further).

The surface of photocatalysts was characterized by XPS analysis, and obtained data are shown in Table 2 and Figure 6 Since both samples (TiO_2(C) and G-TiO_2(C)) were modified with carbon (during pyrolysis and annealing), pure titania sample (P25) was also analyzed for comparison and discussion. The ratio of oxygen to titanium exceeded stoichiometric value of 2.0, reaching 2.97 for P25, 5.7 for TiO_2(C) and 3.38 for G-TiO_2(C). Enrichment of titania surface with oxygen has been commonly reported as a result of adsorption of water/hydroxyl groups and carbon dioxide from air. For example, O/Ti ratios of 2.5 [46], 2.1–5.0 [47], 4.6 [48] and 7.7 [49] were reported for titania samples prepared by laser ablation, hydrothermal reaction, microemulsion and gas-phase methods, respectively. This is also confirmed by deconvolution of oxygen peak (Table 2) showing high content of hydroxyl groups on the titania surface. The binding energies of carbon, oxygen and titanium were estimated after deconvolution of C 1s, O 1s and Ti $2p_{3/2}$ peaks into three, three and two peaks, respectively, according to published reports on titania and carbon-modified titania samples [19,24,50–52]. Titanium was present mainly in Ti^{4+} form (TiO_2(C)) and only low content of reduced titanium (Ti^{3+}) was detected, i.e., 0.4%, 0.64% and 1.74% in P25, TiO_2(C) and G-TiO_2(C) samples, respectively. Almost all titania samples contain crystalline defects, observed mainly as Ti^{3+} form. It is proposed that graphene presence during titania synthesis disturbed in the formation of perfect titania crystals, and thus three times higher content of Ti^{3+} was noticed than that in TiO_2(C) sample prepared in the absence of graphene sheets. In contrast to titanium spectra, the significant differences between samples were observed for carbon and oxygen peaks. In the case of titania, oxygen peak might be usually deconvoluted into three peaks, i.e., (i) oxygen in crystal lattice of TiO_2 (at ca. 529.3 eV), (ii) C = O, Ti_2O_3 and OH groups bound to two titanium atoms (at ca. 531.7 eV) and (iii) hydroxyl groups bound to carbon or titanium (C–OH and Ti–OH; at 533.2 eV) [52]. In the case of graphene-modified samples, deconvolution of oxygen into two peaks was reported, i.e., (i) at 532.3 and 531.3 eV for lattice oxygen (Zn-O-Fe for Fe_2O_3-ZnO photocatalyst) and O–C bond in rGO, respectively [24], and (ii) at 529.7 and 531.4 eV for lattice oxygen (TiO_2) and oxygen in adsorbed hydroxyl groups, respectively [19]. It is clear that oxygen in P25 was mainly in the form of TiO_2 and water or/and carbon dioxide was also adsorbed on titania surface. However, in the case of carbon-modified samples, significant adsorption of oxygen (with different forms) on titania surfaces was noticed with different intensities, i.e., more organic carbon, C–OH and Ti–OH for TiO_2(C) sample and C=O in G-TiO_2(C) sample. Accordingly, it was confirmed that TiO_2(C) sample was surface modified with carbon (from TTIP and/or C_2H_4), whereas G-TiO_2(C) contained graphene-like forms of carbon (graphene could be partially oxidized during annealing). The presence of carbon is typical for all titania (and other oxides) samples, mainly due to carbon dioxide adsorption from surrounding air during sample preparation, which further forms bicarbonate and mono- and bidentate carbonate adsorbed on titania surface [53]. Of course, carbon from organic precursor of TiO_2 (e.g., TTIP, titanium butoxide) might be also present in the final product (as also probable in TiO_2(C) and G-TiO_2(C) samples). It was found that samples prepared by pyrolysis contained much larger content of carbon on the surface than P25, i.e., ratio of C/Ti reached 3.6, 6.39 and 15.22 for P25, G-TiO_2(C) and TiO_2(C) samples, respectively, confirming significant enrichment of titania surface with carbon, especially for TiO_2(C) sample (as suggested from DRS, XRD and Raman). In the case of titania samples, carbon might be deconvoluted also into three peaks indicating the presence of C–C, C–O and C=O at ca. 284.4, 286.1 and 288.6 eV, respectively. However, in the case of graphene-modified samples, deconvolution of carbon peak into two peaks was reported, i.e., (i) at 284.6 and 285.3 eV for graphitic carbon (C–C) and rGO (C–OH) [24], and (ii) at 284.5 and 288 eV for C–C/C–H bonds and C=O, respectively [19]. Here, carbon peak could be deconvoluted into three peaks for all samples, but the content of C–O and C=O was the highest in TiO_2(C) and G-TiO_2(C) samples, respectively. Therefore, it was concluded that adsorbed carbon (from TTIP and/or C_2H_4 or graphene) on titania surface was partly oxidized during annealing.

Table 2. Surface composition of samples determined by XPS analysis for C 1s, O 1s and Ti 2p3/2.

Samples	Content (at%)			C 1s (%)			O 1s (%)			Ti $2p_{3/2}$ (%)	
	C 1s	O 1s	Ti $2p_{3/2}$	C–C/C–H	C–O	C = O	TiO_2	=O/–OH [a]	–OH/C–O [b]	Ti^{3+}	Ti^{4+}
G-TiO_2(C)	59.30	31.42	9.28	55.85	11.35	32.80	43.70	37.11	19.20	1.74	98.26
TiO_2(C)	69.44	26.00	4.56	48.45	38.97	12.58	25.53	23.93	50.54	0.64	99.36
P25	47.08	39.73	13.19	64.87	23.27	11.86	58.37	20.94	20.69	0.40	99.60

=O/–OH [a]: Ti–(OH)–Ti, Ti_2O_3, C=O; –OH/C–O [b]: Ti–OH, C–OH, organic carbon.

Since only surface characterization and crystalline composition could be obtained from XPS and XRD analyses, respectively, EDS measurement was additionally carried out for detailed samples' characterization, and obtained data are shown in Figure 7. The large content of carbon was found in both samples, reaching 21.8 wt% and 51.0 wt% for TiO_2(C) and G-TiO_2(C) samples, respectively, confirming samples' enrichment with carbon both on the surface (XPS data; Table 2) and in the bulk (crystalline (XRD data; Table 1) and amorphous carbon). Interestingly, it was found that the molar ratios of carbon to titanium differed significantly between samples, and between EDS and XPS results for TiO_2(C) sample, i.e., C/Ti = 15.3 (XPS) and 1.6 (EDS) for TiO_2(C) and Ti/C = 6.4 (XPS) and 6.3 (EDS) for G-TiO_2(C). These differences suggest that graphene-modified sample is quite uniform in composition (bulk/surface), whereas carbon-modified sample (TiO_2(C)) is mainly surface-modified with carbon. Therefore, it has been found that graphene co-presence during titania synthesis disturbs in titania surface modification with carbon.

For detail characterization of electronic properties of samples, RDB-PAS analysis was carried out for all samples, and obtained data are shown in Figure 8. It has been clarified that ERDT/CBB patterns might be used as a fingerprint of semiconducting metal-oxides powders for their identification [27,54]. This is because the CBB position, and ERDT pattern and total density of ETs (estimated from PAS and shown inside brackets in the Figure as "<ETs>") might reflect bulk structure, surface structure and bulk/surface size, respectively. The trend of an increase in <ETs> along with an increase in specific surface area for commercial titania samples [52] suggests that measured ETs are predominantly located on the surface of particles with similar area density of ca. 1 ET per nm^2. The peak of ERDT pattern of G-TiO_2(C) was shifted to the low-energy, comparing to TiO_2(C). Moreover, it appears that larger content of deep (ca. 2–3 eV) ETs was observed in TiO_2(C) sample. On the other hand, the CBB positions were almost the same for both samples. Therefore, it is proposed that the modification with graphene could reduce the density of deep ETs (usually considered as recombination centers), but did not influence the band-gap energy. However, it must be pointed out that both samples could absorb visible light because the baseline of RDB-PA intensity in the vis range was much higher than that for commercial titania samples [27], as shown in Figure S2. Lower CBB position of P25 corresponds well with the presence of rutile of narrower bandgap than that in anatase, i.e., ca. 3.0 vs 3.2 eV, respectively. Therefore, it was concluded that P25 was not best sample for discussion of RDB-PAS data, since TiO_2(C) and G-TiO_2(C) contained mainly anatase phase. Accordingly, another titania sample (ST01; from Ishihara company) containing only anatase of fine NPs (ca. 10 nm) was analyzed. The CBB positions of G-TiO_2(C) and TiO_2(C) were nearly corresponding to that of ST01, confirming bandgap of anatase (CBB position seems to be almost constant since the bulk structure, i.e., anatase, is the same for TiO_2(C), G-TiO_2(C) and ST01). Interestingly, it was found that the total density of ETs (<ETs>)) was much lower in the samples prepared by laser synthesis than that in commercial titania samples (P25 and ST01) of similar properties (an increase in specific surface area results in an increase in the content of ETs, as discussed above [55]). Therefore, it was concluded that carbon from titania synthesis (TTIP and C_2H_4) might fill some crystalline defects, as proved by slight bandgap narrowing by DRS and XRD (Figure 3c,d and Figure 4c). Moreover, the observed higher-energy peaks at ca. 3.5 eV for TiO_2(C) and G-TiO_2(C) (inside CB), compared to those of P25 and ST01, indicate that the surfaces of TiO_2(C) and G-TiO_2(C) samples have amorphous nature, as has been observed for commercial amorphous titania and brayed samples (unpublished data), which might be induced by modification of sample with graphene.

3.2. Photocatalytic Activity

The photoactivity of samples was measured for oxidative decomposition of acetic acid under UV/vis (CO_2 system), dehydrogenation of methanol under UV/vis (H_2 system (Pt) and H_2 system (no Pt)) and phenol degradation under vis. The obtained data are shown in Figures 9 and 10. The comparison of activity under UV/vis for all systems to activity of reference sample: P25 (relative activity; activity of P25 = 100%) is shown in Figure 9a, whereas real activity data are shown in Figure 9b,c.

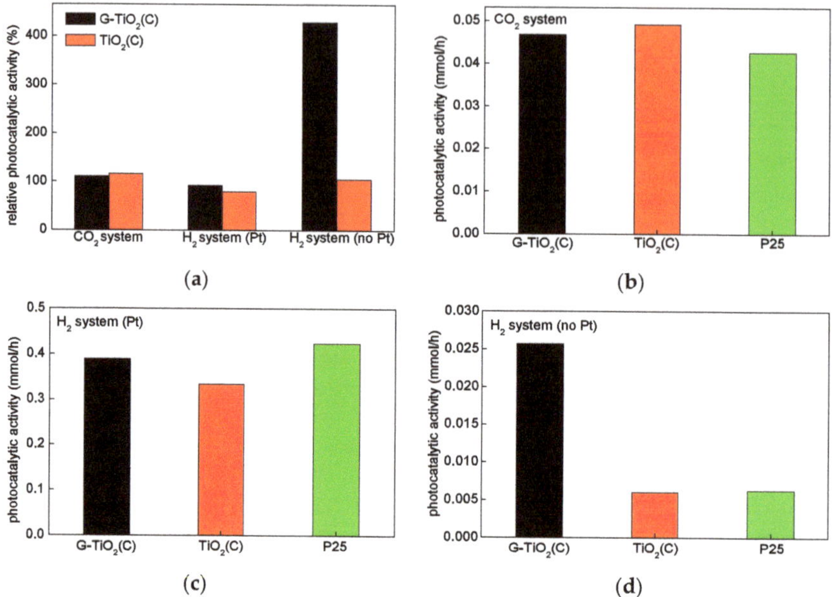

Figure 9. Photocatalytic activity data for: (**a**) relative activity in respect to that by P25 (100%) in three reaction systems; (**b**) oxidative decomposition of acetic acid (CO_2 system); (**c**) dehydrogenation of methanol with in situ platinum deposition (H_2 system (Pt)) and (**d**) dehydrogenation of methanol without in situ platinum deposition (H_2 system (no Pt)).

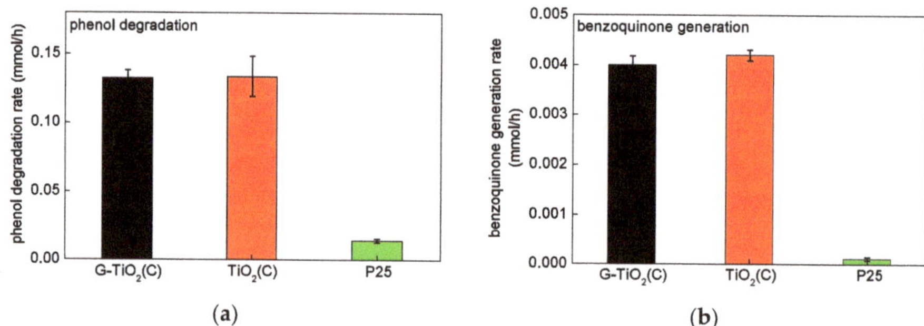

Figure 10. Photocatalytic activity under vis irradiation showing: (**a**) disappearance of phenol and (**b**) subsequent formation of benzoquinone during phenol degradation.

At first, data from Figure 9b,c were discussed, i.e., CO_2 system and H_2 system (Pt), since usually for anaerobic alcohol dehydrogenation, metallic co-catalyst (here Pt) must be used since titania is hardly active for hydrogen evolution. It was found that obtained photocatalysts exhibited high activity, similar to that by P25 in both reaction systems (oxidation and reduction), which is quite rare. Titania P25 (Degussa P25/Evonik P25/Aeroxide P25) is probably the most famous mixed-phase titania, due to extremely high photocatalytic activity in various photocatalytic reactions. For example, P25 was used for decomposition of organic and inorganic compounds present in water and wastewater [56,57], degradation of gas-phase pollutants [58], allergens' removal [59], inactivation of microorganisms [60–62], self-cleaning surfaces [63,64] and solar energy conversion [65]. P25 is a white powder with fine NPs (ca. 30 nm), the density of ca. 3.9 g cm^{-3} and specific surface area of ca. 50 m^2 g^{-1} [29,32]. The different composition of P25 might be found in the literature, i.e., 70–85% anatase and 15%–30% rutile, and the content of amorphous phase is usually not considered [66–68]. Our previous study indicated that P25 was not homogeneous samples, and the phase composition could vary between each P25 samples. Even P25 powders sampled from the same container possessed different content of anatase, rutile and amorphous phase, i.e., 73–85%, 13–17% and 0–13%, respectively [29,32], which was not surprising since the composition depended on the flame conditions (and even the location of formed particles in the flame) as P25 is produced by gas-phase flame synthesis. The comparison of P25 with 34 titania photocatalysts showed that P25 possessed one of the highest photocatalytic activity despite not the best properties (specific surface area, crystallite size, ETs content, crystallinity, etc.) [33]. It was reported that the photocatalytic efficiency did not depend only on the titania properties (specific surface area, polymorphic composition, defects' content, crystallite and particle sizes), but also on the kind of photocatalytic reaction. Accordingly, it was proposed that: (i) large particle size resulted in efficient oxygen evolution, (ii) large specific surface area (small crystallite and particle sizes) in methanol dehydrogenation (H_2 system), (iii) anatase presence in oxidative decomposition of acetic acid (CO_2 system), (iv) rutile presence and defects' content in oxidative decomposition of acetaldehyde and (v) rutile presence in organic synthesis. In our previous study, the higher activity in both reaction systems (CO_2 and H_2) than that by P25 was only obtained for faceted anatase sample of decahedral shape (decahedral anatase particles, DAPs [69]) with intrinsic properties of charge carriers' separation, i.e., the migration of electrons to {101} and holes to {001} facets [70,71].

Accordingly, similar activity of P25 to that by TiO_2(C) and G-TiO_2(C) (and even slightly lower in CO_2 system) indicates that both samples, prepared by laser synthesis, might be efficiently used for various photocatalytic reactions. Moreover, it should be pointed that even their color (greyish) does not significantly disturb in the overall activity ("inner-filter" effect). Slightly higher activity of TiO_2(C) in CO_2 system and G-TiO_2(C) in H_2 system (with Pt) suggests that the form of carbon governs photocatalytic action, and thus graphene is more recommended for hydrogen evolution. Indeed, graphene-like nanostructures have been known for high conductivity, participating in electron transfer [37], i.e., from CB of the semiconductor via G/GO/rGO to adsorbed molecules/compounds [20–23]. Interestingly, slightly higher activity of TiO_2(C) than G-TiO_2(C) and P25 for acetic acid oxidation should be discussed in consideration of photogenerated holes. Buchalska et al. reported that reactive oxygen species (ROS) generated by reaction of photogenerated holes with water or/and surface hydroxyl groups were mainly responsible for oxidation reaction (rather than ROS generated by reaction of photogenerated electrons with adsorbed oxygen) in the case of anatase samples [72]. Therefore, it is proposed that high enrichment of TiO_2(C) surface with hydroxyl groups (either C–OH or Ti–OH (Figure 6 (center)) might cause an efficient formation of ROS, and thus high activity in oxidation reactions. Interestingly, it was found that in the case of methanol dehydrogenation in the absence of platinum, G-TiO_2(C) sample was ca. five times more active than P25 and TiO_2(C) sample (Figure 9d), suggesting that graphene-like structure might work as a co-catalyst for hydrogen generation.

Since photocatalysts were colored (greyish), the photocatalytic activity under vis irradiation was also tested for oxidative decomposition of phenol as a model compound. It should be considered that a disappearance of phenol (a decrease in its concentration in water) is not always equal with its degradation (possibility of adsorption on the photocatalyst surface), and thus the determination of oxidation products is recommended. Consequently, activity data were presented for both: (i) a decrease in the phenol concentration (Figure 10a) and (ii) formation of main intermediate, i.e., benzoquinone (Figure 10b). Indeed, it was found that both samples, prepared by pyrolysis, exhibited vis activity. The most active was TiO_2(C), but an activity of G-TiO_2(C) was only slightly lower, indicating that graphene might disturb in high activity of other carbon species adsorbed on titania surface or C-doped titania. However, an activity of bare P25 was meaningless, especially considering benzoquinone formation (these results confirm that estimation of compound disappearance (here phenol) is not recommended, due to its adsorption on photocatalyst surface). In contrast to the activities under UV/vis, G-TiO_2(C) and TiO_2(C) samples possessed the superior vis-activity than P25, due to the absorption of visible light (Figure 3). However, it should be pointed that absorption of vis light by modified titania samples does not guarantee vis activity. There are many colorous titania samples, practically inactive under vis irradiation. Therefore, carbon modification during titania synthesis and post-thermal annealing results in formation of vis-active materials. For clarification of the main reason of vis activity (surface sensitization, C-doping, etc.), an action spectrum analysis should be performed (planned further study).

Since high photocatalytic activities of G-TiO_2(C) and TiO_2(C) under vis irradiation was shown, the antibacterial activity was examined under vis and in the dark condition (Figure 11). The antibacterial activities of G-TiO_2(C) and TiO_2(C) were almost same (considering experimental error) both under vis (Figure 11a) and in the dark (Figure 11b), i.e., relatively high activity under vis and low in the dark. Therefore, it is suggested that both samples showed the photocatalytic bactericidal activity, possibly due to the presence of carbon, rather than the existence of graphene. On the other hand, the activity of P25 was negligible both under vis and in the dark conditions, due to its wide band-gap and "dark" inertness. Some reports proposed that vis-responsive bactericidal activity of carbon-modified (mainly doped) titania [13,73–77] was cased by the presence of carbon and the direct interaction (e.g., redox reaction) between photocatalyst and bacteria. Interestingly, Wang et al. proposed that the electron transfer between C-doped titania nanotubes and bacteria induced the intracellular ROS formation and cell death, evaluated by charging of sample [78]. Moreover, Cruz-Ortiz et al. reported that TiO_2–rGO showed significant bactericidal activity (*E. coli*) under visible light irradiation, due to the generation of singlet oxygen [79]. Accordingly, Markowska et al. found enhanced generation of hydroxyl radicals and decrease in cell viability (via activity of antioxidant enzymes, morphology change and mineralization) on glucose-modified titania under artificial solar irradiation [15]. Therefore, our results confirm high activity of carbon-modified samples under vis irradiation for bacteria inactivation, probably via generated ROS.

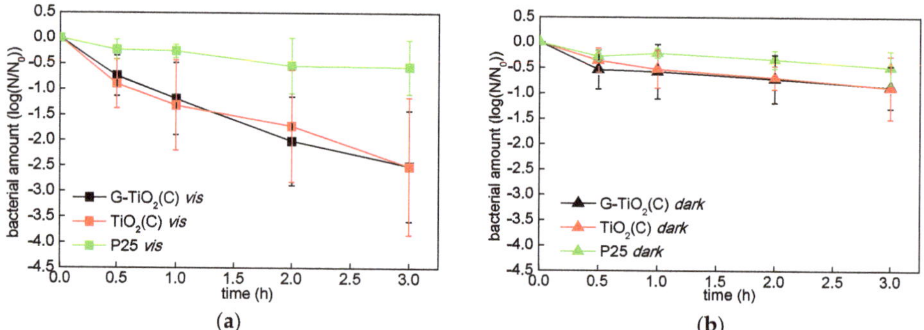

Figure 11. Bactericidal activities of G-TiO_2(C) (black), TiO_2(C) (red) and P25 (green) under: (**a**) vis irradiation and (**b**) dark conditions; data are expressed as mean ± SD (n = 3).

4. Conclusions

Laser pyrolysis with post-annealing proved to be an efficient method to obtain carbon-modified titania sample (TiO_2(C)) with high photocatalytic activity under both UV and vis irradiation. Two sources of carbon might be considered, i.e., from titanium isopropoxide (TTIP; titania precursor) and C_2H_4 (a sensitizer used for laser pyrolysis). The modification of synthesis by addition of 0.04 wt% graphene into TTIP solution resulted in preparation of graphene/carbon-modified titania (G-TiO_2(C)) with similar properties to non-modified sample (TiO_2(C)). Although, all properties and activities of these two samples were similar, G-TiO_2(C) showed ca. five times higher activity for methanol dehydrogenation without Pt co-catalyst under UV/vis irradiation than TiO_2(C) and commercial titania sample (P25), indicating that graphene might be an efficient co-catalyst for hydrogen evolution. Similar activities of both samples for other reactions, i.e., oxidative decomposition of acetic acid under UV/vis, methanol dehydrogenation under UV/vis with Pt deposition in situ, phenol oxidation under vis and *E. coli* inactivation under vis, indicate that such low content of graphene (2.7 wt%) did not have significant influence on the overall activity. In contrast, carbon-modification seemed to be highly beneficial for activities in all systems since TiO_2(C) and G-TiO_2(C) samples exhibited similar activity to one of the most active commercial titania samples (P25) under UV/vis irradiation, and few orders of magnitude higher activity than P25 under vis irradiation. Therefore, simple modification with carbon allowed preparation of highly active and cheap (carbon modified) photocatalysts, active at broad range of irradiation, and thus being attractive alternative for other titania materials for environmental application.

Supplementary Materials: The following are available online at http://www.mdpi.com/1996-1944/12/24/4158/s1, Figure S1: Raman mapping of the G-TiO_2(C) sample illustrating the presence of C species in localized places; Figure S2: RDB-PA spectra of (a) G-TiO_2(C), (b) TiO_2(C), (c) P25 (anatase/rutile), and (d) ST01 (anatase).

Author Contributions: Conceptualization, E.K. and N.H.-B., writing—original draft preparation, K.W., M.E.-K., E.K. and N.H.-B., writing-revision and editing, K.W., M.E.-K., E.K. and N.H.-B., sample preparation, K.W., A.H., R.B., sample characterization, K.W., M.E.-K., R.B., photoactivity tests, K.W., M.E.-K. and D.Z., microbiological tests, M.E.-K., supervision, writing—review and funding acquisition, B.O., J.B., E.K. and N.H.-B.

Funding: This work was financially supported by "Yugo-Sohatsu Kenkyu" for an Integrated Research Consortium on Chemical Sciences (IRCCS) from Ministry of Education, Culture, Sports, Science and Technology-Japan (MEXT). R.B. is thankful for the support of Region Ile de France of the Labex Sigma-Lim (Limoges) for PhD fellowship.

Acknowledgments: N.H.B. would like to thank Ambassade de France au Japon for the support through "decouverte Japon" program, which allowed travelling to Sapporo and establishing fruitful collaboration for this study.

Conflicts of Interest: The authors declare no conflict of interest.

References

1. Hoffmann, M.R.; Martin, S.T.; Choi, W.Y.; Bahnemann, D.W. Environmental applications of semiconductor photocatalysis. *Chem. Rev.* **1995**, *95*, 69–96. [CrossRef]
2. Fujishima, A.; Zhang, X.T.; Tryk, D.A. TiO_2 photocatalysis and related surface phenomena. *Surf. Sci. Rep.* **2008**, *63*, 515–582. [CrossRef]
3. Fujishima, A.; Rao, T.N.; Tryk, D.A. Titanium dioxide photocatalysis. *J. Photochem. Photo. C* **2000**, *1*, 1–21. [CrossRef]
4. Pichat, P. NATO ASI Series, Series C: Mathematical and Physical Sciences. *Photocatal. React.* **1985**, *146*, 425–455.
5. Minero, C.; Catozzo, F.; Pelizzetti, E. Role of adsorption in photocatalyzed reactions of organic molecules in aqueous titania suspensions. *Langmuir* **1992**, *8*, 481–486. [CrossRef]
6. Guillard, C.; Disdier, J.; Herrmann, J.M.; Lehaut, C.; Chopin, T.; Malato, S.; Blanco, J. Comparison of various titania samples of industrial origin in the solar photocatalytic detoxification of water containing 4-chlorophenol. *Catal. Today* **1999**, *54*, 217–228. [CrossRef]
7. Kraeutler, B.; Bard, A.J. Heterogeneous photocatalytic preparation of supported catalysts. Photodeposition of platinum on TiO_2 powder and other substrates. *J. Am. Chem. Soc.* **1978**, *100*, 4317–4318. [CrossRef]
8. Tian, Y.; Tatsuma, T. Mechanisms and applications of plasmon-induced charge separation at TiO_2 films loaded with gold nanoparticles. *J. Am. Chem. Soc.* **2005**, *127*, 7632–7637. [CrossRef]
9. Kowalska, E.; Abe, R.; Ohtani, B. Visible light-induced photocatalytic reaction of gold-modified titanium(IV) oxide particles: Action spectrum analysis. *Chem. Commun.* **2009**, *14*, 241–243. [CrossRef]
10. Tryba, B.; Tsumura, T.; Janus, M.; Morawski, A.W.; Inagaki, M. Carbon-coated anatase: Adsorption and decomposition of phenol in water. *Appl. Catal. B Environ.* **2004**, *50*, 177–183. [CrossRef]
11. Janus, M.; Tryba, B.; Inagaki, M.; Morawski, A.W. New preparation of a carbon-TiO_2 photocatalyst by carbonization of n-hexane deposited on TiO_2. *Appl. Catal. B Environ.* **2004**, *52*, 61–67. [CrossRef]
12. Kusiak-Nejman, E.; Janus, M.; Grzmil, B.; Morawski, A.W. Methylene blue decomposition under visible light irradiation in the presence of carbon-modified TiO_2 photocatalysts. *J. Photoch. Photobiol.* **2011**, *226*, 68–72. [CrossRef]
13. Rokicka, P.; Markowska-Szczupak, A.; Kowalczyk, L.; Kowalska, E.; Morawski, A.W. Influence of titanium dioxide modification on the antibacterial properties. *Pol. J. Chem. Technol.* **2016**, *18*, 56–64. [CrossRef]
14. Zabek, P.; Kisch, H. Polyol-derived carbon-modified titania for visible light photocatalysis. *J. Coor. Chem.* **2010**, *63*, 2715–2726. [CrossRef]

15. Markowska-Szczupak, A.; Rokicka, P.; Wang, K.L.; Endo, M.; Morawski, A.W.; Kowalska, E. Photocatalytic water disinfection under solar irradiation by D-glucose-modified titania. *Catalysts* **2018**, *8*, 316. [CrossRef]
16. Wu, G.; Nishikawa, T.; Ohtani, B.; Chen, A. Synthesis and characterization of carbon-doped TiO_2 nanostructures with enhanced visible light response. *Chem. Mater.* **2007**, *19*, 4530–4537. [CrossRef]
17. Gorska, P.; Zaleska, A.; Kowalska, E.; Klimczuk, T.; Sobczak, J.W.; Skwarek, E.; Janusz, W.; Hupka, J. TiO_2 photoactivity in vis and UV light: The influence of calcination temperature and surface properties. *Appl. Catal. B Environ.* **2008**, *84*, 440–447. [CrossRef]
18. Giovannetti, R.; Rommozzi, E.; Zannotti, M.; D'Amato, C.A. Recent advances in graphene based TiO_2 nanocomposites (gTiO_2ns) for photocatalytic degradation of synthetic dyes. *Catalysts* **2017**, *7*, 305. [CrossRef]
19. Stengl, V.; Bakardjieva, S.; Grygar, T.M.; Bludska, J.; Kormunda, M. TiO_2-graphene oxide nanocomposite as advanced photocatalytic materials. *Chem. Cent. J.* **2013**, *7*, 41. [CrossRef]
20. Nagaraju, G.; Ebeling, G.; Goncalves, R.V.; Teixeira, S.R.; Weibel, D.E.; Dupont, J. Controlled growth of TiO_2 and TiO_2-rGO composite nanoparticles in ionic liquids for enhanced photocatalytic H_2 generation. *J. Mol. Catal. A Chem.* **2013**, *378*, 213–220. [CrossRef]
21. Bai, X.; Zhang, X.; Hua, Z.; Ma, W.; Dai, Z.; Huang, X.; Gu, H. Uniformly distributed anatase TiO_2 nanoparticles on graphene: Synthesis, characterization, and photocatalytic application. *J. Alloys Comp.* **2014**, *599*, 10–18. [CrossRef]
22. Li, F.; Du, P.H.; Liu, W.; Li, X.S.; Ji, H.D.; Duan, J.; Zhao, D.Y. Hydrothermal synthesis of graphene grafted titania/titanate nanosheets for photocatalytic degradation of 4-chlorophenol: Solar-light-driven photocatalytic activity and computational chemistry analysis. *Chem. Eng. J.* **2018**, *331*, 685–694. [CrossRef]
23. Neelgund, G.M.; Oki, A. Graphene-coupled zno: A robust nir-induced catalyst for rapid photo-oxidation of cyanide. *ACS Omega* **2017**, *2*, 9095–9102. [CrossRef] [PubMed]
24. Wang, X.W.; Li, Q.C.; Zhou, C.X.; Cao, Z.Q.; Zhang, R.B. ZnO rod/reduced graphene oxide sensitized by alpha-Fe_2O_3 nanoparticles for effective visible-light photoreduction of CO_2. *J. Colloid Interf. Sci.* **2019**, *554*, 335–343. [CrossRef]
25. Belchi, R.; Habert, A.; Foy, E.; Gheno, A.; Vedraine, S.; Antony, R.; Ratier, B.; Boucle, J.; Herlin-Boime, N. One-step synthesis of TiO_2/graphene nanocomposites by laser pyrolysis with well-controlled properties and application in perovskite solar cells. *ACS Omega* **2019**, *4*, 11906–11913. [CrossRef]
26. Pignon, B.; Maskrot, H.; Ferreol, V.G.; Leconte, Y.; Coste, S.; Gervais, M.; Pouget, T.; Reynaud, C.; Tranchant, J.F.; Herlin-Boime, N. Versatility of laser pyrolysis applied to the synthesis of TiO_2 nanoparticles-application to UV attenuation. *Eur. J. Inorg. Chem.* **2008**, *2008*, 883–889. [CrossRef]
27. Nitta, A.; Takase, M.; Takashima, M.; Murakamid, N.; Ohtani, B. A fingerprint of metal-oxide powders: Energy-resolved distribution of electron traps. *Chem. Commun.* **2016**, *52*, 12096–12099. [CrossRef]
28. Ikeda, S.; Sugiyama, N.; Murakami, S.-Y.; Kominami, H.; Kera, Y.; Noguchi, H.; Uosaki, K.; Torimoto, T.; Ohtani, B. Quantitative analysis of defective sites in titanium (IV) oxide photocatalyst powders. *Phys. Chem. Chem. Phys.* **2003**, *5*, 778–783. [CrossRef]
29. Wang, K.L.; Wei, Z.S.; Ohtani, B.; Kowalska, E. Interparticle electron transfer in methanol dehydrogenation on platinum-loaded titania particles prepared from P25. *Catal. Today* **2018**, *303*, 327–333. [CrossRef]
30. Kowalska, E.; Rau, S. Photoreactors for wastewater treatment: A review. *Recent Pat. Engin.* **2010**, *4*, 242–266. [CrossRef]
31. Endo, M.; Wei, Z.S.; Wang, K.L.; Karabiyik, B.; Yoshiiri, K.; Rokicka, P.; Ohtani, B.; Markowska-Szczupak, A.; Kowalska, E. Noble metal-modified titania with visible-light activity for the decomposition of microorganisms. *Beilstein J. Nanotech.* **2018**, *9*, 829–841. [CrossRef] [PubMed]
32. Ohtani, B.; Prieto-Mahaney, O.O.; Li, D.; Abe, R. What is Degussa (Evonik) P25? Crystalline composition analysis, reconstruction from isolated pure particles and photocatalytic activity test. *J. Photoch. Photobiol. A* **2010**, *216*, 179–182. [CrossRef]
33. Prieto-Mahaney, O.O.; Murakami, N.; Abe, R.; Ohtani, B. Correlation between photocatalytic activities and structural and physical properties of titanium (IV) oxide powders. *Chem. Lett.* **2009**, *38*, 238–239. [CrossRef]
34. Rao, M.V.; Rajeshwar, K.; Verneker, V.R.P.; DuBow, J. Photosynthetic production of hydrogen and hydrogen peroxide on semiconducting oxide grains in aqueous solutions. *J. Phys. Chem.* **1980**, *84*, 1987–1991. [CrossRef]
35. Tauc, J. Optical properties and electronic structure of amorphous Ge and Si. *Mater. Res. Bull.* **1968**, *3*, 37–46. [CrossRef]

36. Zhang, Z.; He, X.R.; Zhang, J.J.; Lu, X.B.; Yang, C.H.; Liu, T.; Wang, X.; Zhang, R. Influence of graphene/ferriferrous oxide hybrid particles on the properties of nitrile rubber. *RSC Adv.* **2016**, *6*, 91798–91805. [CrossRef]
37. Gualdron-Reyes, A.F.; Melendez, A.M.; Nino-Gomez, M.E.; Rodriguez-Gonzales, V.; Carreno-Lizeano, M.I. Photoanodes modified with reduced graphene oxide to enhance photoelectrocatalytic performance of B-TiO_2 under visible light. *Rev. Acad. Colomb. Cienc. Ex. Fis. Nat.* **2015**, *39*, 77–83. [CrossRef]
38. Singh, S.; Bhatnagar, A.; Dixit, V.; Shukla, V.; Shaz, M.A.; Sinha, A.S.K.; Srivastava, O.N.; Sekkar, V. Synthesis, characterization and hydrogen storage characteristics of ambient pressure dried carbon aerogel. *Int. J. Hydrog. Energ.* **2016**, *41*, 3561–3570. [CrossRef]
39. Ungar, T.; Gubicza, J.; Ribarik, G.; Pantea, C.; Zerda, T.W. Microstructure of carbon blacks determined by X-ray diffraction profile analysis. *Carbon* **2002**, *40*, 929–937. [CrossRef]
40. Faisal, A.D.; Aljubouri, A.A. Synthesis and production of carbon nanospheres using noncatalytic CVD method. *Int. J. Adv. Mat. Res.* **2016**, *2*, 86–91.
41. Liu, X.Y.; Huang, M.A.; Ma, H.L.; Zhang, Z.Q.; Gao, J.M.; Zhu, Y.L.; Han, X.J.; Guo, X.Y. Preparation of a carbon-based solid acid catalyst by sulfonating activated carbon in a chemical reduction process. *Molecules* **2010**, *15*, 7188–7196. [CrossRef] [PubMed]
42. Akhila, A.K.; Vinitha, P.S.; Renuka, N.K. Photocatalytic activity of graphene–titania nanocomposite. *Mater. Today Proc.* **2018**, *5*, 16085–16093. [CrossRef]
43. Sekiya, T.; Ohta, S.; Kamei, S.; Hanakawa, M.; Kurita, S. Raman spectroscopy and phase transition of anatase TiO_2 under high pressure. *J. Phys. Chem. Solids* **2001**, *62*, 717–721. [CrossRef]
44. Rodriguez, L.A.A.; Pianassola, M.; Travessa, D.N. Production of tio_2 coated multiwall carbon nanotubes by the sol-gel technique. *Mater. Res.* **2017**, *20*, 96–103. [CrossRef]
45. Zakharchuk, I.; Komlev, A.A.; Soboleva, E.; Makarova, T.L.; Zherebtsov, D.A.; Galimov, D.M.; Lahderanta, E. Paramagnetic anatase titania/carbon nanocomposites. *J. Nanophotonics* **2017**, *11*, 032505. [CrossRef]
46. Siuzdak, K.; Sawczak, M.; Klein, M.; Nowaczyk, G.; Jurga, S.; Cenian, A. Preparation of platinum modified titanium dioxide nanoparticles with the use of laser ablation in water. *Phys. Chem. Chem. Phys.* **2014**, *16*, 15199–15206. [CrossRef] [PubMed]
47. Wei, Z.; Endo, M.; Wang, K.; Charbit, E.; Markowska-Szczupak, A.; Ohtani, B.; Kowalska, E. Noble metal-modified octahedral anatase titania particles with enhanced activity for decomposition of chemical and microbiological pollutants. *Chem. Eng. J.* **2017**, *318*, 121–134. [CrossRef]
48. Zielińska-Jurek, A.; Kowalska, E.; Sobczak, J.W.; Lisowski, W.; Ohtani, B.; Zaleska, A. Preparation and characterization of monometallic (Au) and bimetallic (Ag/Au) modified-titania photocatalysts activated by visible light. *Appl. Catal. B Environ.* **2011**, *101*, 504–514. [CrossRef]
49. Wei, Z.; Janczarek, M.; Endo, M.; Wang, K.L.; Balcytis, A.; Nitta, A.; Mendez-Medrano, M.G.; Colbeau-Justin, C.; Juodkazis, S.; Ohtani, B.; et al. Noble metal-modified faceted anatase titania photocatalysts: Octahedron versus decahedron. *Appl. Catal. B Environ.* **2018**, *237*, 574–587. [CrossRef]
50. Rossnagel, S.M.; Sites, J.R. X-ray photoelectron spectroscopy of ion beam sputter deposited silicon dioxide, titanium dioxide, and tantalum pentoxide. *J. Vac. Sci. Technol.* **1984**, *2*, 376–379. [CrossRef]
51. Jensen, H.; Soloviev, A.; Li, Z.S.; Sogaard, E.G. XPS and FTIR investigation of the surface properties of different prepared titania nano-powders. *Appl. Surf. Sci.* **2005**, *246*, 239–249. [CrossRef]
52. Yu, J.G.; Zhao, X.J.; Zhao, Q.N. Effect of surface structure on photocatalytic activity of TiO_2 thin films prepared by sol-gel method. *Thin Solid Films* **2000**, *379*, 7–14. [CrossRef]
53. Baltrusaitis, J.; Schuttlefield, J.; Zeitler, E.; Grassian, V.H. Carbon dioxide adsorption on oxide nanoparticle surfaces. *Chem. Eng. J.* **2011**, *170*, 471–481. [CrossRef]
54. Nitta, A.; Takashima, M.; Takase, M.; Ohtani, B. Identification and characterization of titania photocatalyst powders using their energy-resolved distribution of electron traps as a fingerprint. *Catal. Today* **2019**, *321*, 2–8. [CrossRef]
55. Murakami, N.; Mahaney, O.O.P.; Torimoto, T.; Ohtani, B. Photoacoustic spectroscopic analysis of photoinduced change in absorption of titanium(IV) oxide photocatalyst powders: A novel feasible technique for measurement of defect density. *Chem. Phys. Lett.* **2006**, *426*, 204–208. [CrossRef]
56. Ferguson, M.A.; Hering, J.G. TiO_2-photocatalyzed As (III) oxidation in a fixed-bed, flow-through reactor. *Environ. Sci. Technol.* **2006**, *40*, 4261–4267. [CrossRef]

57. Neppolian, B.; Choi, H.C.; Sakthivel, S.; Arabindoo, B.; Murugesan, V. Solar/UV-induced photocatalytic degradation of three commercial textile dyes. *J. Hazard. Mater.* **2002**, *89*, 303–317. [CrossRef]
58. Sakai, H.; Kubota, Y.; Yamaguchi, K.; Fukuoka, H.; Inumaru, K. Photocatalytic decomposition of 2-propanol and acetone in air by nanocomposites of pre-formed TiO_2 particles and mesoporous silica. *J. Porous Mat.* **2013**, *20*, 693–699. [CrossRef]
59. Nishikawa, N.; Kaneco, S.; Katsumata, H.; Suzuki, T.; Ohta, K.; Yamazaki, E.; Masuyama, K.; Hashimoto, T.; Kamiya, K. Photocatalytic degradation of allergens in water with titanium dioxide. *Fresen. Environ. Bull.* **2007**, *16*, 310–314.
60. Markowska-Szczupak, A.; Wang, K.L.; Rokicka, P.; Endo, M.; Wei, Z.S.; Ohtani, B.; Morawski, A.W.; Kowalska, E. The effect of anatase and rutile crystallites isolated from titania P25 photocatalyst on growth of selected mould fungi. *J. Photoch. Photobiol. B* **2015**, *151*, 54–62. [CrossRef]
61. Markowska-Szczupak, A.; Janda, K.; Wang, K.L.; Morawski, A.W.; Kowalska, E. Effect of water activity and titania P25 photocatalyst on inactivation of pathogenic fungi-contribution to the protection of public health. *Cent. Eur. J. Public Health* **2015**, *23*, 267–271.
62. Thabet, S.; Weiss-Gayet, M.; Dappozze, F.; Cotton, P.; Guillard, C. Photocatalysis on yeast cells: Toward targets and mechanisms. *Appl. Catal. B Environ.* **2013**, *140*, 169–178. [CrossRef]
63. Bozzi, A.; Yuranova, T.; Guasaquillo, I.; Laub, D.; Kiwi, J. Self-cleaning of modified cotton textiles by TiO_2 at low temperatures under daylight irradiation. *J. Photoch. Photobiol. A* **2005**, *174*, 156–164. [CrossRef]
64. Smits, M.; Chan, C.K.; Tytgat, T.; Craeye, B.; Costarramone, N.; Lacombe, S.; Lenaerts, S. Photocatalytic degradation of soot deposition: Self-cleaning effect on titanium dioxide coated cementitious materials. *Chem. Eng. J.* **2013**, *222*, 411–418. [CrossRef]
65. Antoniadou, M.; Lianos, P. Production of electricity by photoelectrochemical oxidation of ethanol in a photofuelcell. *Appl. Catal. B Environ.* **2010**, *99*, 307–313. [CrossRef]
66. Wang, G.H.; Xu, L.; Zhang, J.; Yin, T.T.; Han, D.Y. Enhanced photocatalytic activity of TiO_2 powders (P25) via calcination treatment. *Int. J. Photoenergy* **2012**, *9*, 265760.
67. Martin, S.T.; Herrmann, H.; Choi, W.; Hoffmann, M.R. Time-resolved microwave conductivity. Part 1. TiO_2 photoreactivity and size quantization. *J. Chem. Soc. Faraday Trans.* **1994**, *90*, 3315–3322. [CrossRef]
68. Szczepankiewicz, S.H.; Colussi, A.J.; Hoffmann, M.R. Infrared spectra of photoinduced species on hydroxylated titania surfaces. *J. Phys. Chem. B* **2000**, *104*, 9842–9850. [CrossRef]
69. Janczarek, M.; Kowalska, E.; Ohtani, B. Decahedral-shaped anatase titania photocatalyst particles: Synthesis in a newly developed coaxial-flow gas-phase reactor. *Chem. Eng. J.* **2016**, *289*, 502–512. [CrossRef]
70. Murakami, N.; Kurihara, Y.; Tsubota, T.; Ohno, T. Shape-controlled anatase titanium (IV) oxide particles prepared by hydrothermal treatment of peroxo titanic acid in the presence of polyvinyl alcohol. *J. Phys. Chem. C* **2009**, *113*, 3062–3069. [CrossRef]
71. Tachikawa, T.; Yamashita, S.; Majima, T. Evidence for crystal-face-dependent TiO_2 photocatalysis from single-molecule imaging and kinetic analysis. *J. Am. Chem. Soc.* **2011**, *133*, 7197–7204. [CrossRef]
72. Buchalska, M.; Kobielusz, M.; Matuszek, A.; Pacia, A.; Wojtyla, S.; Macyk, W. On oxygen activation at rutile- and anatase-TiO_2. *ACS Catal.* **2015**, *5*, 7424–7431. [CrossRef]
73. Koli, V.B.; Dhodamani, A.G.; Delekar, S.D.; Pawar, S.H. In situ sol-gel synthesis of anatase TiO_2-MWCNTs nanocomposites and their photocatalytic applications. *J. Photoch. Photobiol. A* **2017**, *333*, 40–48. [CrossRef]
74. Koli, V.B.; Dhodamani, A.G.; Raut, A.V.; Thorat, N.D.; Pawar, S.H.; Delekar, S.D. Visible light photo-induced antibacterial activity of TiO_2-MWCNTs nanocomposites with varying the contents of MWCNTs. *J. Photoch. Photobiol. A* **2016**, *328*, 50–58. [CrossRef]
75. Mitoraj, D.; Janczyk, A.; Strus, M.; Kisch, H.; Stochel, G.; Heczko, P.B.; Macyk, W. Visible light inactivation of bacteria and fungi by modified titanium dioxide. *Photoch. Photobiol. Sci.* **2007**, *6*, 642–648. [CrossRef]
76. Akhavan, O.; Abdolahad, M.; Abdi, Y.; Mohajerzadeh, S. Synthesis of titania/carbon nanotube heterojunction arrays for photoinactivation of *E. coli* in visible light irradiation. *Carbon* **2009**, *47*, 3280–3287. [CrossRef]
77. Cheng, C.L.; Sun, D.S.; Chu, W.C.; Tseng, Y.H.; Ho, H.C.; Wang, J.B.; Chung, P.H.; Chen, J.H.; Tsai, P.J.; Lin, N.T.; et al. The effects of the bacterial interaction with visible-light responsive titania photocatalyst on the bactericidal performance. *J. Biomed. Sci.* **2009**, *16*, 7. [CrossRef]

78. Wang, G.M.; Feng, H.Q.; Hu, L.S.; Jin, W.H.; Hao, Q.; Gao, A.; Peng, X.; Li, W.; Wong, K.Y.; Wang, H.Y.; et al. An antibacterial platform based on capacitive carbon-doped tio_2 nanotubes after direct or alternating current charging. *Nat. Commun.* **2018**, *9*, 2055. [CrossRef]
79. Cruz-Ortiz, B.R.; Hamilton, J.W.J.; Pablos, C.; Diaz-Jimenez, L.; Cortes-Hernandez, D.A.; Sharma, P.K.; Castro-Alferez, M.; Fernandez-Ibanez, P.; Dunlop, P.S.M.; Byrne, J.A. Mechanism of photocatalytic disinfection using titania-graphene composites under uv and visible irradiation. *Chem. Eng. J.* **2017**, *316*, 179–186. [CrossRef]

Article

Visible-Light Activated Titania and Its Application to Photoelectrocatalytic Hydrogen Peroxide Production

Tatiana Santos Andrade [1,2], Ioannis Papagiannis [1], Vassilios Dracopoulos [3], Márcio César Pereira [2] and Panagiotis Lianos [1,*]

1 Department of Chemical Engineering, University of Patras, 26500 Patras, Greece; tsandrade@live.com (T.S.A.); ion.papg@gmail.com (I.P.)
2 Institute of Science, Engineering, and Technology, Universidade Federal dos Vales do Jequitinhonha e Mucuri, Campus Mucuri, 39803–371 Teófilo Otoni, Minas Gerais, Brazil; mcpqui@gmail.com
3 FORTH/ICE–HT, P.O. Box 1414, 26504 Patras, Greece; indy@iceht.forth.gr
* Correspondence: lianos@upatras.gr; Tel.: +30-2610-997513

Received: 2 December 2019; Accepted: 16 December 2019; Published: 17 December 2019

Abstract: Photoelectrochemical cells have been constructed with photoanodes based on mesoporous titania deposited on transparent electrodes and sensitized in the Visible by nanoparticulate CdS or CdS combined with CdSe. The cathode electrode was an air–breathing carbon cloth carrying nanoparticulate carbon. These cells functioned in the Photo Fuel Cell mode, i.e., without bias, simply by shining light on the photoanode. The cathode functionality was governed by a two-electron oxygen reduction, which led to formation of hydrogen peroxide. Thus, these devices were employed for photoelectrocatalytic hydrogen peroxide production. Two-compartment cells have been used, carrying different electrolytes in the photoanode and cathode compartments. Hydrogen peroxide production has been monitored by using various electrolytes in the cathode compartment. In the presence of $NaHCO_3$, the Faradaic efficiency for hydrogen peroxide production exceeded 100% due to a catalytic effect induced by this electrolyte. Photocurrent has been generated by either a CdS/TiO_2 or a $CdSe/CdS/TiO_2$ combination, both functioning in the presence of sacrificial agents. Thus, in the first case ethanol was used as fuel, while in the second case a mixture of Na_2S with Na_2SO_3 has been employed.

Keywords: hydrogen peroxide; TiO_2; CdS; CdSe; photoelectrocatalysis; photocatalytic fuel cells; photo fuel cells

1. Introduction

Titanium dioxide (Titania, TiO_2) is the most popular photocatalyst, and this is justified by the fact that it is stable, it can be easily synthesized and deposited on solid substrates as mesoporous film of several types of nanostructures, it is considered non–toxic and, most of all, it has very good electronic properties [1–4]. Thus, it is characterized by relatively high charge-carrier mobility, while its valence band has a high oxidative potential capable of carrying out most oxidative reactions. Its only disadvantage is that its light absorption is limited in the UV. For this reason and in order to expand its light absorption range, titania has been combined with visible-light-absorbing sensitizers [5–17]. Dyes as well as organometal halide perovskites have been very successful as sensitizers of titania, but their functionality is limited to specific organic or solid-state environments exclusively applied to solar cells [5,6]. In aqueous environments, only inorganic sensitizers have offered acceptable performance, and among those, only a few II–VI semiconductors have led to interesting scenarios [7–17]. Indeed, sensitization of mesoporous titania by nanoparticulate CdS and CdSe has been repeatedly studied, righteously offering efficient sensitization and application in a variety of photocatalytic and photoelectrocatalytic systems [7–17]. Furthermore, formation of nanoparticulate metal sulfides within the titania mesoporous structure is very easy and necessitates only "soft" chemistry techniques. In the present work, we have followed this established route for titania sensitization in order to construct photoelectrochemical devices capable of producing a valuable solar fuel, i.e., hydrogen peroxide.

The study of hydrogen peroxide has become very popular in recent years [18]. In addition to its well-known pharmaceutical applications, hydrogen peroxide is a key substance in advanced oxidation processes for water treatment, either alone or in combination with iron ions in Fenton processes [19–21]. Hydrogen peroxide is also a means of energy storage in the form of chemical energy since it can be used as fuel in hydrogen peroxide fuel cells or as oxidant for the operation of several fuel cell types [22–26]. Hydrogen peroxide can be easily handled since it is soluble in water and its consumption simply results to oxygen or water production. It is then obvious that hydrogen peroxide is indeed a valuable fuel, and for this reason it is worth producing it in a sustainable manner [27–30]. Photoelectrocatalysis, which can directly exploit solar energy and employ biomass-derived wastes as fuel may thus offer the means for sustainable hydrogen peroxide production. This route is investigated in the present work, and in this sense it constitutes a novelty in spite of the fact that the materials used to make electrodes are well known and frequently used in the past.

The purpose then of the present work is to study the production of hydrogen peroxide by means of photoelectrocatalysis. To this goal, we have used a photoelectrochemical cell employing a photoanode carrying visible–light–sensitized nanoparticulate titania. These cells work as self-running Photo(catalytic) Fuel Cells (PFC), operating without external bias by photocatalytically oxidizing a fuel, which may as well be a waste. This offers a double environmental benefit and ensures sustainability. Hydrogen peroxide has been produced by atmospheric oxygen reduction. The photoanode produced the current while an air–breathing cathode produced hydrogen peroxide. This is schematically represented in Figure 1, while details are discussed in the following sections.

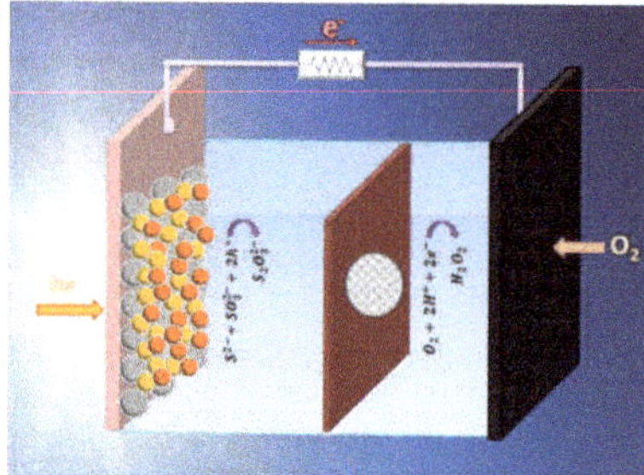

Figure 1. Schematic representation of the reactor used in the present work. The oxidation reaction in the anode compartment corresponds to the case of the S^{2-}/SO_3^{2-} electrolyte and the CdSe-enriched photoanode.

2. Materials and Methods

2.1. Materials

Unless otherwise specified, all reagents were obtained from Sigma–Aldrich and were used as received. Thus, Fluorine-doped Tin Oxide electrodes (FTO, 8 ohm/square) were purchased from Pilkington North America (Toledo, OH, USA), carbon cloth (CC) from Fuel Cell Earth (Wobum, MA, USA), carbon black (CB) from Cabot Corporation (Vulcan XC72, Billerica, MA, USA), and Nafion membrane from Ion Power, Inc (Newcastle, DE, USA).

2.2. Construction of the Photoanode

2.2.1. Deposition of the Titania Film

The photoanode electrode was constructed by the following procedure. An FTO glass was cut in the appropriate dimensions and was carefully cleaned first with soap and then by sonication in acetone, ethanol, and water. A compact titania layer was first deposited on the clean electrode by a sol gel procedure. A precursor solution was prepared by mixing 3.5 g of Triton X-100 with 19 mL of ethanol, to which 3.4 mL of glacial acetic acid and 1.8 mL of titanium isopropoxide was added under stirring. This solution was used for dipping FTO electrodes, which were patterned by covering with tapes the back side and the front side parts, which should remain clear. Then, it was calcined up to 500 °C. This was repeated once, to ensure a complete coverage of the active electrode area. Next, a mesoporous titania layer was deposited on this compact layer by doctor blading, using a paste composed of Degusa P25 nanoparticles and prepared by a standard procedure based on [31]. The mesoporous film was calcined at 550 °C. This procedure was repeated once again to ensure that a mesoporous film of around 10 μm thick was obtained. Film thickness was approximately determined by SEM. The active area of the titania film was 1 cm^2 (1 cm × 1 cm).

2.2.2. Application of CdS on the Titania Film

A fresh titania film was sensitized by CdS nanoparticles by the SILAR (Successive Ionic Layer Adsorption and Reaction) method [7,32,33] using 0.1 M cadmium nitrate as Cd^{2+} and 0.1 M sodium sulfide as S^{2-} source. Ten SILAR cycles were sufficient to cover titania with the yellow CdS layer. This method does not produce a separate CdS layer, but rather, CdS nanoparticles are formed within the titania mesoporous structure [7]. At the end, the film was first dried in a nitrogen stream and then for a few minutes in an oven at 70 °C. This electrode was either used as $CdS/TiO_2/FTO$ photoanode or as substrate for the next step of CdSe deposition.

2.2.3. Addition of the CdSe Layer and the ZnS Protective Layer

CdSe was added on the top of CdS by a chemical bath deposition (CBD), as in previous publications [8,32]. CdSe is formed in a period of about 4–5 h at low temperature (in a refrigerator). The CdS-sensitized TiO_2 film was immerged face-up in a precursor solution containing 27 mM of sodium selenosulphate (Na_2SeSO_3) as a source of Se^{2-}, and 27 mM of cadmium sulfate as a source of Cd^{2+}. The precursor solution was prepared by observing the following protocol. An aqueous solution of 80 mM Se powder was first prepared in the presence of 0.2 M Na_2SO_3 by continuous stirring and refluxing at 80 °C. The procedure lasted about 15 h and was carried out overnight. The obtained solution, denoted in the following as sol A, actually aimed at the formation of the above sodium selenosulphate (Na_2SeSO_3). Then, an aqueous solution of 0.12 M nitrilotriacetic acid trisodium salt was prepared, denoted as sol B. Finally, an aqueous 80 mM $CdSO_4{\cdot}8/3H_2O$ solution was also prepared, denoted as sol C. Sol B was mixed with an equal volume of sol C, and the obtained mixture was stirred for a few minutes. The combination of sol B with sol C leads to the formation of a complex, which is used as precursor for slow release of cadmium ions. Finally, two parts of this last mixture were mixed with one part of sol A, and the thus obtained final mixture was used for CBD. At the end of 4–5 h, the CdS/TiO_2 film was red-colored. The thus obtained film was again first dried in a N_2 stream and then in an oven at 70 °C. It is a common practice to stabilize the CdSe film by a top layer of ZnS [32], which was actually done in the present work. Thus a ZnS layer was finally deposited on the top by 2 SILAR cycles using zinc nitrate as Zn^{2+} source and N_2S as S^{2-} source, followed by drying as above. This procedure yielded $ZnS/CdSe/CdS/TiO_2/FTO$ photoanode electrodes with a broad range of light absorption.

2.3. Construction of the Counter Electrode

The cathode electrode was a carbon cloth covered on one side with two layers of carbon black (CB/CC). The deposition of the CB layer was made by using a paste made by the following recipe: 300 mg of CB was mixed with 8 mL of twice-distilled water under vigorous stirring with a mixer (more than 4000 rpm), until a viscous paste was obtained. To this mixture 0.1 mL of PTFE (60% in water) was added as a hydrophobic binder and was again vigorously stirred. This paste was applied on the carbon cloth with a spatula, dried in an oven at 80 °C, and then sintered at 340 °C. The procedure was repeated once to reach a quantity of CB equal to 1 mg per cm^2. The active area of the electrode was 1 cm^2, as in the case of the photoanode.

2.4. Description of the Reactor

The reactor was a home-made device based on Plexiglas, which was divided into two compartments by a Nafion membrane, as schematically shown in Figure 1. The capacity of each compartment was 5 mL. It had two windows which were sealed by the two electrodes (photoanode and cathode). It was filled with various electrolytes appropriate for each particular case, as described below. The active area of each window was 1 cm^2, fitting the above electrodes. Illumination of the photoanode was made with a Xe lamp, which provided approximately 100 mW cm^{-2} at the position of the catalyst. Light entered through the transparent FTO electrode.

2.5. Measurements

The quantity of produced H_2O_2 was spectroscopically determined by its concentration in the aqueous electrolyte of the cathode compartment. This was done by a standard method analytically described in the Supplementary File. Photoelectrochemical measurements were made with the support of an Autolab potentiostat PGSTAT128N (Metrohm Autolab B.V., Utrecht, The Netherlands). Reflection–Absorption spectra were recorded with a Shimadzu UV–2600 absorption spectrophotometer equipped with an integration sphere. Field-Emission Scanning Electron Microscopy images (FESEM) were obtained with a Zeiss SUPRA 35 VP device.

3. Results and Discussion

3.1. Characterization of Electrodes

As already discussed, the photoelectrochemical setup used in the present work employed three photoanode electrodes based on mesoporous TiO_2 deposited on FTO transparent electrodes: Titania alone (TiO_2/FTO), CdS-sensitized titania (CdS/TiO_2/FTO), and a film with additional CdSe sensitizer on the top (CdSe/CdS/TiO_2/FTO). The light absorption range of each film is given by the reflection–absorption spectra of Figure 2. The CdS/TiO_2/FTO film approximately absorbed photons up to about 510 nm, corresponding to a band gap of 2.43 eV. The maximum photocurrent density expected for a photoanode carrying such a film and for solar radiation equal to 1 sun (100 mW cm^{-2}) can be calculated from published charts [2,34] and is around 7 mA cm^{-2}. Correspondingly, the CdSe/CdS/TiO_2/FTO film absorbed photons up to 610 nm with a band gap of 2.03 eV and maximum current density approximately equal to 13 mA cm^{-2}. Photoanodes made with such combined semiconductors may then yield substantial current densities appropriate for practical applications.

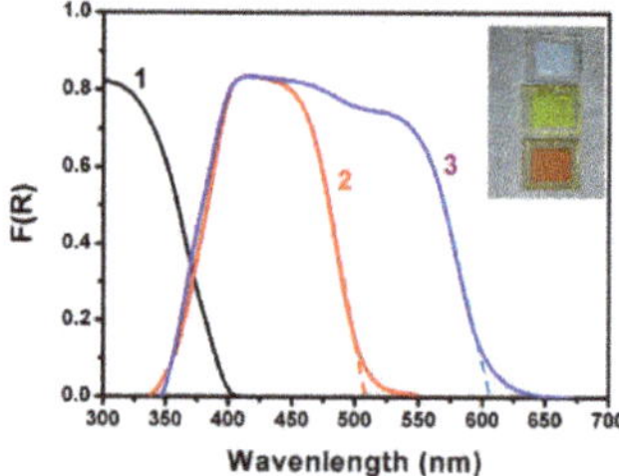

Figure 2. Reflection absorption spectra of various photoanode electrodes: (1) TiO_2/FTO; (2) CdS/TiO_2/FTO; and (3) CdSe/CdS/TiO_2/FTO. In curves (2) and (3), the background including titania absorption has been subtracted for better presentation of the spectra. The tangential dashed lines give the average value of the band gap, i.e., 510 nm (2.43 eV) for the CdS/TiO_2/FTO film and 610 nm (2.03 eV) for the CdSe/CdS/TiO_2/FTO film. Insert: Film photographs.

CdS and CdSe nanoparticles were synthesized, as explained in the Experimental section, by reaction of Cd^{2+} cations with S^{2-} anions within the pores of the mesoporous titania film. For this reason, they do not form separate layers, but they are detected inside the titania mesostructure. This is known from previous works [7,8,15,17,35,36], but can be also verified by the following FESEM images shown in Figure 3. Figure 3A shows a characteristic image of nanoparticulate Titania P25. TiO_2 nanoparticles range in sizes around 20–30 nm. Careful observation and comparison of the three images shows the formation of new species within the mesoporous titania structure in going from pure titania to CdS/TiO_2 (Figure 3B) and then to $CdSe/CdS/TiO_2$ (Figure 3C). CdS and CdSe particles grow in much smaller sizes ranging below 10 nm. This intercalated formation of the chalcogenide semiconductors ensures close proximity between nanoparticles and subsequently efficient sensitization and charge carrier mobility. Some protective role of titania on chalcogenide nanoparticles cannot be excluded either.

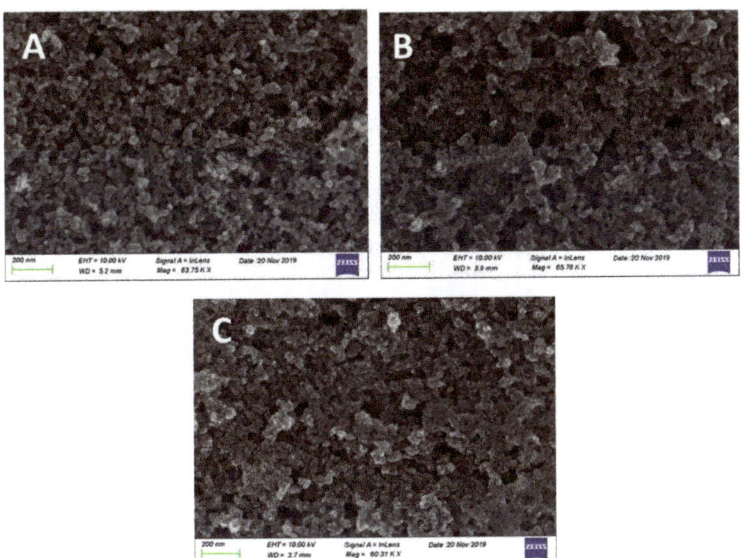

Figure 3. FESEM images of the three photoanode films: (**A**) TiO_2/FTO; (**B**) CdS/TiO_2/FTO; and (**C**) CdSe/ CdS/TiO_2/FTO. The scale bar is 200 nm in all cases.

The structure of the cathode electrodes (CB/CC) can be seen in the FESEM images of Figure 4, showing the weaving of carbon filaments making the carbon cloth and the mesostructure of the deposited carbon black. Carbon black covered all pores making an air-breathing (gas diffusion) electrode which sealed the electrolyte from any leak. This construction lasted for many hours and for many rounds of operation.

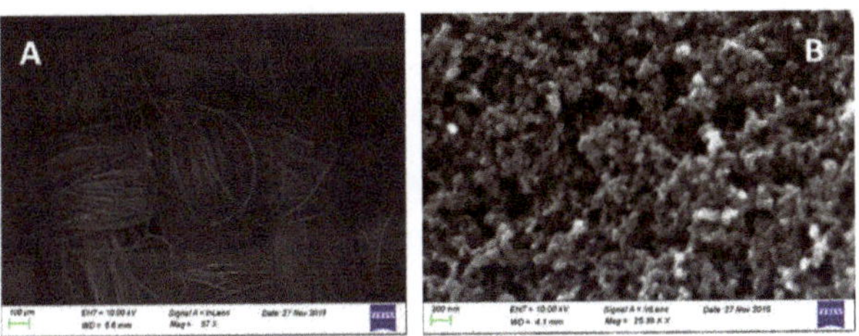

Figure 4. FESEM images of the carbon cloth (**A**) and the carbon black film (**B**). The scale bar is 100 μm and 200 nm, respectively.

The capacity of the CB/CC electrode to carry out reduction reactions may be qualitatively appreciated by the polarization curves of Figure 5. These curves were obtained in a 3-electrode configuration using CB/CC or plain CC as working, a Pt foil as counter, and Ag/AgCl as reference electrode. Curves were traced in aqueous electrolytes containing 0.5 M of either H_2SO_4, Na_2SO_4, or $NaHCO_3$. All curves have been plot vs. Reversible Hydrogen Electrode (RHE) by taking into account the pH value of each electrolyte, i.e., 1.0, 6.5, and 8.5, respectively, and by adding 0.2 V for the potential of the reference electrode. The importance of the presence of carbon black is first obvious by the fact that reduction reactions are obtained at negative potentials in its absence. In the presence of carbon black, the most favorable reduction was obtained in a carbonate electrolyte, where the reduction potential was the most positive of the three electrolytes. More positive reduction potential means higher intrinsic bias for electron flow from the photoanode to the cathode electrode, therefore, more efficient electrochemical process. Even though the data of Figure 5 are approximate and do not allow accurate quantitative results, they are sufficient to highlight $NaHCO_3$ as the most promising of the three electrolytes presently tested. As it will be seen below, hydrogen peroxide production was indeed the highest in this electrolyte.

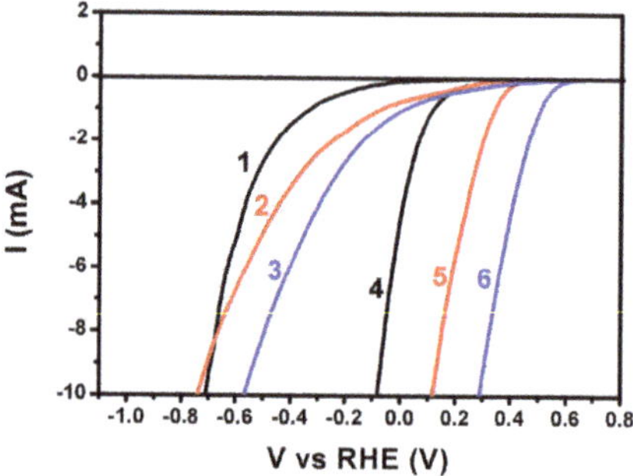

Figure 5. Polarization curves for an electrode made of carbon cloth alone (1,2,3) or carbon cloth carrying carbon black (4,5,6) in various 0.5 M aqueous electrolytes: (1,4) H_2SO_4; (2,5) Na_2SO_4; and (3,6) $NaHCO_3$. A Pt foil was used as counter and an Ag/AgCl as reference electrode.

3.2. Current–Voltage Characteristics of Various Photo Fuel Cells

The above described photoanode and cathode electrodes were used to operate various versions of a photo fuel cell. Figure 6 shows current–voltage characteristics of a two-compartment cell functioning with a CdS-sensitized photoanode ($CdS/TiO_2/FTO$) and a CB/CC cathode. A Nafion membrane separated the two compartments. The anode aqueous electrolyte was 0.5 M NaOH with added 5% w/w ethanol. The cathode compartment contained either 0.5 M aqueous H_2SO_4, Na_2SO_4, or $NaHCO_3$. Curves were plot in a two-electrode configuration in the light-chopping mode to reveal the conditions of photocurrent production. Maximum photocurrent density was obtained under forward bias and exceeded 10 mA cm^{-2} in all three cases. These values were higher than expected for the present CdS-sensitized TiO_2 photoanode (i.e., larger than 7 mA cm^{-2}, see above). This is due to the presence of ethanol and the ensuing current doubling phenomena [37]. Current doubling phenomena are always observed with photoelectrochemical cells functioning in the presence of an organic fuel, and they are reported in almost all of our related works. At zero bias, which is of interest in the present case, substantial photocurrent was produced in all three cases, therefore, production of hydrogen peroxide in a Photo Fuel Cell mode was monitored in all three cases, as will be detailed below. In the curves of Figure 6, there is interference to the photocurrent by a capacitance current due to adsorption of cations in the photoanode mesostructure [38]. This capacitance current appears only when plotting current–voltage curves and is of no importance for the rest of the measurements.

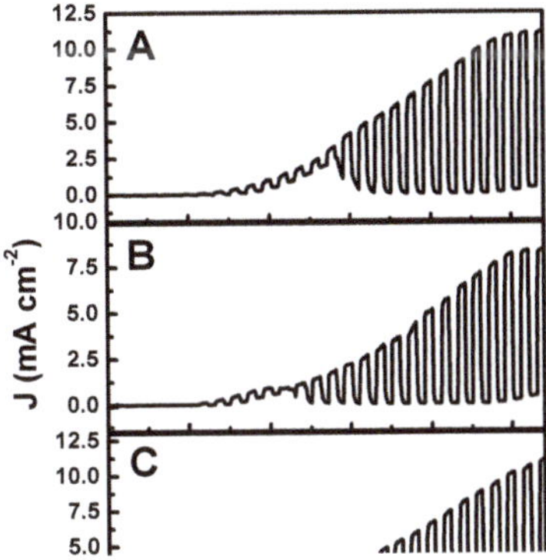

Figure 6. Photo Fuel Cell current density–voltage curves plot in the light chopping mode corresponding to the same electrolyte in the anode compartment (0.5 M aqueous NaOH + 5% w/w ethanol) and three different 0.5 M aqueous electrolytes in the cathode compartment: (**A**) H_2SO_4; (**B**) Na_2SO_4; and (**C**) $NaHCO_3$.

In another version of the Photo Fuel Cell, the photoanode carried the ternary semiconductor film, i.e., $CdSe/CdS/TiO_2/FTO$. In fact, as explained in Section 2.2.3, there was an additional protective layer of ZnS deposited on the top by 2 SILAR cycles. This layer is not sufficient to change the spectroscopic characteristics of the photoanode. ZnS absorbs in the UV, and its role is only to protect the underlying film. Its addition is a common practice for such films [32]. The ternary photoanode is not stable in the NaOH+ethanol electrolyte. For this reason, it was employed in the presence of an aqueous mixture of Na_2S with Na_2SO_3, which is frequently used with chalcogenide semiconductors and is considered a model for sulfur-containing water wastes. Figure 7 shows a current density–voltage curve obtained with a two-compartment Photo Fuel Cell comprising the above ternary semiconductor photoanode and a CB/CC cathode electrode. The two compartments were separated again by a Nafion membrane. The anode electrolyte was an aqueous 0.25 M Na_2S and 0.125 M Na_2SO_3 mixture. The cathode electrolyte was an aqueous 0.5 M $NaHCO_3$ solution. The choice of this last electrolyte was made, as it will be seen below, by the fact that it offers the highest hydrogen peroxide yield. The maximum photocurrent was within the expected range (i.e., no more than 13 mA cm^{-2}, see above). Of course, in the present case, no current doubling phenomena were observed since there was no organic additive in the anode electrolyte. It is noteworthy that the short-circuit current density in the present case was larger than 10 mA cm^{-2}, much larger than in any of the three cases of Figure 6. Here then there is an interesting case, very promising for practical applications. Both devices, i.e., the one carrying the $CdS/TiO_2/FTO$ photoanode and the one carrying the $CdSe/CdS/TiO_2/FTO$ photoanode, are useful for sustainable production of hydrogen peroxide. In the first case, a biomass-derived fuel may be used, while in the second case, a water waste containing sulfur products, for example from oil refineries, may be used.

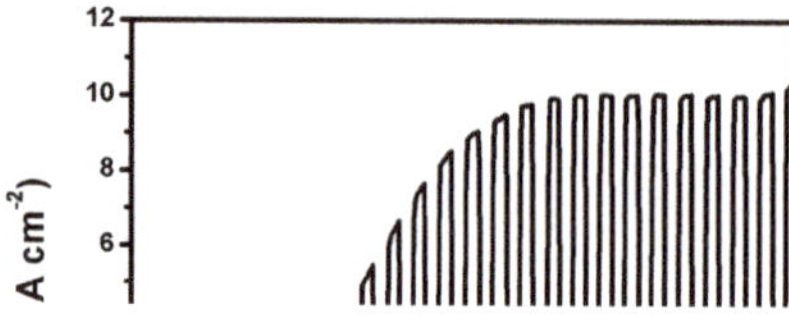

Figure 7. Photo Fuel Cell current density–voltage curve plot in the light chopping mode corresponding to 0.25 M Na_2S + 0.125 M Na_2SO_3 aqueous electrolyte in the anode compartment and 0.5 M aqueous $NaHCO_3$ electrolyte in the cathode compartment.

3.3. Photoelectrocatalytic Hydrogen Peroxide Production

Following the above characterization of electrodes and devices, hydrogen peroxide production has been monitored by PFC operation. The first system studied was the one operating with the ternary semiconductor photoanode (i.e., $CdSe/CdS/TiO_2/FTO$), which produced the highest short-circuit photocurrent. The anode compartment contained an aqueous 0.25 M Na_2S + 0.125 M Na_2SO_3 electrolyte, while the cathode electrolyte was an aqueous 0.5 M $NaHCO_3$ solution. Production of H_2O_2 was monitored under potentiostatic–amperometric conditions at V = 0.0 V. The evolution of the short-circuit photocurrent is shown in Figure 8. The current density started at 10 mA cm^{-2}. In the course of the experiment, it dropped to 7.2 mA cm^{-2}, where it was finally stabilized. Hydrogen peroxide continuously evolved in a cumulative manner within a period of 120 min, as seen in Figure 8. An analysis of the H_2O_2 as a function of time and in relation with the current flowing through the cell is presented in Table 1. The second column of this table gives the evolution of the concentration of hydrogen peroxide in the carbonate electrolyte, while the fifth column gives the average molar rate of H_2O_2 production in the period of the corresponding time (first column). To a rough approximation, the rate did not much vary in the course of the present experiment. The molar rate can be associated with the equivalent current that should flow through the cell to produce hydrogen peroxide by reduction reactions. Hydrogen peroxide production may be described by the following scheme:

$$O_2 + 2H^+ + 2e^- \rightarrow H_2O_2 \quad (1)$$

It then takes 2 electrons to form one H_2O_2 molecule. Consequently, 1 μmole min^{-1} of a substance which is formed by 2 electrons per molecule, corresponds to 10^{-6} mole × 6.023 × 10^{23} molecules $mole^{-1}$ × 2 × 1.602 × 10^{-19} C $molecule^{-1}$ × $(60\ sec)^{-1}$ = 3.21 mA. The corresponding equivalent current is then calculated by multiplying the molar rate by 3.21, and the obtained values are listed in column 6. By dividing this current by the corresponding actual average current flowing through the cell over each time period, we obtained the corresponding Faradaic efficiencies for hydrogen peroxide photoelectrocatalytic production. The Faradaic efficiency was very high and reached 100%. This means that all current generated by the present Photo Fuel Cell was consumed to produce hydrogen peroxide without losses. The carbonate environment is responsible for such a high efficiency both in view of the data of Figure 5 and the related discussion, and because $NaHCO_3$ is known to catalyze hydrogen peroxide formation [39,40]. When sulfate was used in the presence of carbonate, Faradaic efficiency was substantially lower. Indeed, Table 2 presents an equivalent analysis with that of Table 1 on data obtained by the same system as above but by substituting the $NaHCO_3$ electrolyte by aqueous 0.5 M Na_2SO_4. The current flowing through the cell was much lower, and the Faradaic efficiency for hydrogen peroxide production did not pass 65%. Apparently, $NaHCO_3$ electrolyte is much more interesting for Photo Fuel Cell operation with the ternary semiconductor photoanode in the presence of sulfide/sulfite electrolyte.

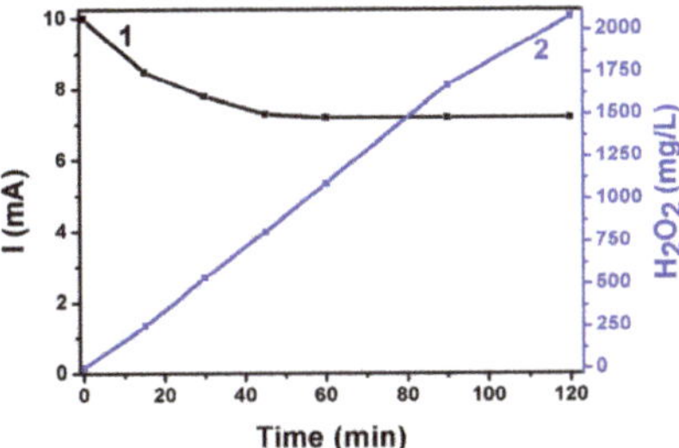

Figure 8. Evolution of the short-circuit current (1) and of the quantity of photoelectrochemically produced H_2O_2 (2) using a PFC operated with a ternary semiconductor photoanode. Anode electrolyte: Aqueous 0.25 M Na_2S + 0.125 M Na_2SO_3. Cathode electrolyte: Aqueous 0.5 M $NaHCO_3$.

Table 1. Analysis of hydrogen peroxide photoelectrocatalytic production rate in 0.5 M $NaHCO_3$. Case of the ternary semiconductor photoanode.

Time	H_2O_2 Conc.	H_2O_2 Mass	Corresp. Molarity	Molar Rate	Equivalent Current	Current at Time	Average Current	Faradaic Efficiency
(min)	(mg/L)	(mg)	(μmol)	(μmol/min)	(mA)	(mA)	(mA)	%
0	0	0	0	—	—	10	-	—
15	250	1.25	36.8	2.45	7.87	8.5	9.3	91
30	538	2.69	79.1	2.64	8.46	7.8	8.9	95
45	808	4.04	119	2.64	8.48	7.3	8.7	98
60	1092	5.46	161	2.68	8.59	7.2	8.6	100
90	1672	8.36	246	2.73	8.77	7.2	8.6	102
120	2077	10.4	305	2.54	8.17	7.2	8.6	95

Table 2. Analysis of hydrogen peroxide photoelectrocatalytic production rate in 0.5 M Na_2SO_4. Case of the ternary semiconductor photoanode.

Time	H_2O_2 Conc.	H_2O_2 Mass	Corresp. Molarity	Molar Rate	Equivalent Current	Current at time	Average Current	Faradaic Efficiency
(min)	(mg/L)	(mg)	(μmol)	(μmol/min)	(mA)	(mA)	(mA)	%
0	0	0	0	-	-	6.5	-	-
30	201	1.01	29.6	0.98	3.16	6.1	6.3	50
47	329	1.65	48.4	1.03	3.30	5.1	5.8	57
77	594	2.97	87.4	1.13	3.64	4.7	5.6	65

Hydrogen peroxide was also produced by the PFC, which used the binary semiconductor photoanode, i.e., CdS/TiO_2/FTO, and ethanol as a fuel. In that case, data have been obtained for three different electrolytes in the cathode compartment and the results are presented in Tables 3–5. The first case involved the carbonate electrolyte, which led to the highest H_2O_2 production. The results are shown in Table 3. The current flowing through the cell was now lower, as expected in accordance to the data of Figure 3. As seen in the 7th column, the original current was 3.8 mA cm^{-2} and settled at 3.2 mA cm^{-2}. The H_2O_2 concentration continuously increased and so did the molar rate. It is interesting that the Faradaic efficiency was very high and grew beyond 100%, verifying the catalytic effect that the carbonate electrolyte has on the electrocatalytic production of hydrogen peroxide [39,40]. In these works, this effect was studied mainly for cases of hydrogen peroxide production by water oxidation and is related with the intermediate formation of HCO_3^- and its catalytic effect on water oxidation. Apparently, equivalent processes may take place during oxygen and water reduction.

This phenomenon necessitates further study. It must be noted at this point that no hydrogen peroxide was detected in the anode compartment, suggesting that for the present cells hydrogen peroxide is an exclusive product of reduction processes. Likewise, it is also a product of photoelectrochemical processes since no hydrogen peroxide was detected in the dark. In the presence of Na_2SO_4 instead of $NaHCO_3$, according to the data of Table 4, the catalytic effect was not present anymore and the Faradaic efficiency was substantially lower. Finally, when sulfuric acid was used as cathode electrolyte, the system showed a very poor behavior, as seen in Table 5, so this electrolyte was excluded from any further measurements. The higher short-circuit current recorded in the case of H_2SO_4 and Na_2SO_4 (7th column of Tables 4 and 5) is in accordance with the data of Figure 6. The PFC in that case had an alkaline electrolyte (i.e., NaOH) in the anode compartment with a pH value around 13, while the pH in the cathode compartment was 1, 6.5, and 8.5 in the case of H_2SO_4, Na_2SO_4, and $NaHCO_3$, respectively. **In the case then of the first two electrolytes, a strong forward bias of chemical nature developed between the two electrodes, which justifies the higher currents at V = 0.**

Table 3. Analysis of hydrogen peroxide photoelectrocatalytic production rate in 0.5 M $NaHCO_3$. Case of the binary semiconductor photoanode.

Time	H_2O_2 Conc.	H_2O_2 Mass	Corresp. Molarity	Molar Rate	Equivalent Current	Current at Time	Average Current	Faradaic Efficiency
(min)	(mg/L)	(mg)	(μmol)	(μmol/min)	(mA)	(mA)	(mA)	%
0	0	0	0	–	–	3.8	–	–
15	113	0.57	16.7	1.11	3.56	3.4	3.6	99
30	247	1.23	36.3	1.21	3.89	3.2	3.5	111
47	415	2.07	61.0	1.30	4.17	3.2	3.5	119
60	556	2.78	81.8	1.36	4.38	3.2	3.5	125
75	728	3.64	107	1.43	4.59	3.2	3.5	131
95	944	4.72	139	1.46	4.69	3.2	3.5	134

Table 4. Analysis of hydrogen peroxide photoelectrocatalytic production rate in 0.5 M Na_2SO_4. Case of the binary semiconductor photoanode.

Time	H_2O_2 Conc.	H_2O_2 Mass	Corresp. Molarity	Molar Rate	Equivalent Current	Current at Time	Average Current	Faradaic Efficiency
(min)	(mg/L)	(mg)	(μmol)	(μmol/min)	(mA)	(mA)	(mA)	%
0	0	0	0	–	–	5.1	–	–
18	115	0.58	16.9	0.94	3.00	5.1	5.1	60
40	279	1.40	41.0	1.03	3.28	5.2	5.1	64
47	352	1.76	51.8	1.10	3.52	5.2	5.2	68
85	705	3.53	104	1.22	3.90	4.9	5.0	78

Table 5. Analysis of hydrogen peroxide photoelectrocatalytic production rate in 0.5 M H_2SO_4. Case of the binary semiconductor photoanode.

Time	H_2O_2 Conc.	H_2O_2 Mass	Corresp. Molarity	Molar Rate	Equivalent Current	Current at Time	Average Current	Faradaic Efficiency
(min)	(mg/L)	(mg)	(μmol)	(μmol/min)	(mA)	(mA)	(mA)	%
0	0	0	0	–	–	5.1	–	–
30	20	0.10	2.9	0.10	0.31	6.2	6.3	5
83	70	0.35	10.3	0.12	0.39	5.6	5.9	7
125	125	0.63	18.4	0.15	0.47	4.9	5.3	9
142	152	0.76	22.3	0.16	0.50	4.6	4.7	11

In conclusion, the above data show that a photoelectrochemical cell operating as a Photo Fuel Cell, i.e., without any external bias, can produce substantial quantities of hydrogen peroxide with high Faradaic efficiency reaching values higher than 100% in the presence of a carbonate electrolyte. Hydrogen peroxide was produced at the cathode electrode, which was a Pt-free inexpensive combination of a carbon cloth with carbon black. Both anode and cathode electrodes were characterized after use. Some of the deposited material leached off the electrodes, but their nanostructure remained the same as imaged in Figures 3 and 4. This materials leach is mainly responsible for the drop of current during cell operation.

4. Conclusions

This work has shown that a Photo Fuel Cell can be constructed with a visible light responsive photoanode based on chalcogenide-semiconductors-sensitized mesoporous titania and a simple Pt-free cathode made of carbon cloth with deposited nanoparticulate carbon (carbon black). This cell functioned without any bias producing substantial current. The cathode functionality is based on atmospheric oxygen reduction, which leads to hydrogen peroxide production. This functionality was exploited in order to photoelectrochemically produce hydrogen peroxide. High Faradaic efficiencies have been reached for the production of hydrogen peroxide, which in the presence of $NaHCO_3$ rose beyond 100% due to the catalytic effect of the latter.

Supplementary Materials: The following are available online at http://www.mdpi.com/1996-1944/12/24/4238/s1, Detailed description of the determination of hydrogen peroxide in solution.

Author Contributions: T.S.A. investigation; I.P. investigation; V.D. investigation; M.C.P. supervision; P.L. conceptualization.

Funding: This study was partially supported by the Coordenação de Aperfeiçoamento de Pessoal de Nível Superior—Brasil (Capes)—Finance Code 001.

Acknowledgments: T.S.A. acknowledges support through a scholarship provided by the Coordenação de Aperfeiçoamento de Pessoal de Nível Superior—Brasil (Capes)—Finance Code 001 that allowed her stay in the University of Patras.

Conflicts of Interest: The authors declare no conflict of interest.

References

1. Singh, R.; Dutta, S. A review on H_2 production through photocatalytic reactions using TiO_2/TiO_2–assisted catalysts. *Fuel* **2018**, *220*, 607–620. [CrossRef]
2. Lianos, P. Review of recent trends in photoelectrocatalytic conversion of solar energy to electricity and hydrogen. *Appl. Catal. B Environ.* **2017**, *210*, 235–254. [CrossRef]
3. Daghrir, R.; Drogui, P.; Robert, D. Modified TiO_2 for environmental photocatalytic applications: A review. *Ind. Eng. Chem. Res.* **2013**, *52*, 3581–3599. [CrossRef]
4. Humayun, M.; Raziq, F.; Khan, A.; Luo, W. Modification strategies of TiO_2 for potential applications in photocatalysis: A critical review. *Green Chem. Lett. Rev.* **2018**, *11*, 86–102. [CrossRef]
5. Yeoh, M.-E.; Chan, K.-Y. Recent advances in photo-anode for dye-sensitized solar cells: A review. *Int. J. Energy Res.* **2017**, *41*, 2446–2467. [CrossRef]
6. Wang, Z.; Lang, X. Visible light photocatalysis of dye-sensitized TiO_2: The selective aerobic oxidation of amines to imines. *Appl. Catal. B Environ.* **2018**, *224*, 404–409. [CrossRef]
7. Sfaelou, S.; Sygellou, L.; Dracopoulos, V.; Travlos, A.; Lianos, P. Effect of the nature of cadmium salts on the effectiveness of CdS SILAR deposition and its consequences on the performance of sensitized solar cells. *J. Phys. Chem. C* **2014**, *118*, 22873–22880. [CrossRef]
8. Antoniadou, M.; Sfaelou, S.; Dracopoulos, V.; Lianos, P. Platinum-free photoelectrochemical water splitting. *Catal. Commun.* **2014**, *43*, 72–74. [CrossRef]
9. Mezzetti, A.; Balandeh, M.; Luo, J.; Bellani, S.; Tacca, A.; Divitini, G.; Cheng, C.; Ducati, C.; Meda, L.; Hongjin Fan, H.; et al. Hyperbranched TiO_2-CdS nanoheterostructures for highly efficient photoelectrochemical photoanodes. *Nanotechnology* **2018**, *29*, 335404–335415. [CrossRef]
10. Zhao, D.; Yang, C.-F. Recent advances in the TiO_2/CdS nanocomposite used for photocatalytic hydrogen production and quantum-dot-sensitized solar cells. *Renew. Sust. Energ. Rev.* **2016**, *54*, 1048–1059. [CrossRef]

11. Chen, Y.; Chuang, C.-H.; Qin, Z.; Shen, S.; Doane, T.; Burda, C. Electron-transfer dependent photocatalytic hydrogen generation over cross-linked CdSe/TiO_2 type-II heterostructure. *Nanotechnology* **2017**, *28*, 084002–084012. [CrossRef] [PubMed]
12. Pawar, S.A.; Patilb, D.S.; Junga, H.R.; Parka, J.Y.; Malic, S.S.; Hongc, C.K.; Shinc, J.-C.; Patild, P.S.; Kim, J.-H. Quantum dot sensitized solar cell based on TiO_2/CdS/CdSe/ZnS heterostructure. *Electrochim. Acta.* **2016**, *203*, 74–83. [CrossRef]
13. Ai, G.; Li, H.; Liu, S.; Mo, R.; Zhong, J. Solar water splitting by TiO_2/CdS/Co-Pi nanowire array photoanode enhanced with Co–Pi as hole transfer relay and CdS as light absorber. *Adv. Funct. Mater.* **2015**, *25*, 5706–5713. [CrossRef]
14. Lin, K.-H.; Chuang, C.-Y.; Lee, Y.-Y.; Li, F.-C.; Chang, Y.-M. Charge transfer in the heterointerfaces of CdS/CdSe cosensitized TiO_2 photoelectrode. *J. Phys. Chem. C* **2012**, *116*, 1550–1555. [CrossRef]
15. Liu, J.; Xia, M.; Chen, R.; Zhu, X.; Liao, Q.; Ye, D.; Zhang, B.; Zhang, W.; Yu, Y. A membrane-less visible–light responsive micro photocatalytic fuel cell with the laterally-arranged CdS/ZnS-TiO_2 photoanode and air-breathing CuO photocathode for simultaneous wastewater treatment and electricity generation. *Sep. Purif. Technol.* **2019**, *229*, 115821. [CrossRef]
16. Chen, Y.-L.; Chen, Y.-H.; Chen, J.-W.; Cao, F.; Li, L.; Luo, Z.-M.; Leu, I.-C.; Pu, Y.-C. New insights into the electron collection efficiency improvement of CdS–sensitized TiO_2 nanorod photoelectrodes by interfacial seed-layer mediation. *ACS Appl. Mater. Interfaces* **2019**, *11*, 8126–8137. [CrossRef]
17. Li, L.; Chen, R.; Zhu, X.; Liao, Q.; Ye, D.; Zhang, B.; He, X.; Jiao, L.; Feng, H.; Zhang, W. A ternary hybrid CdS/SiO_2/TiO_2 photoanode with enhanced photoelectrochemical activity. *Renew. Energy* **2018**, *127*, 524–530. [CrossRef]
18. Ciriminna, R.; Albanese, L.; Meneguzzo, F.; Pagliaro, M. Hydrogen peroxide: A key chemical for today's sustainable development. *ChemSusChem* **2016**, *9*, 3374–3381. [CrossRef]
19. Asghar, A.; Raman, A.A.A.; Daud, W.M.A.W. Advanced oxidation processes for in-situ production of hydrogen peroxide/hydroxyl radical for textile wastewater treatment: A review. *J. Clean. Prod.* **2015**, *87*, 826–838. [CrossRef]
20. Patton, S.; Romano, M.; Naddeo, V.; Ishida, K.P.; Liu, H. Photolysis of mono- and dichloramines in UV/hydrogen peroxide: Effects on 1,4–dioxane removal and relevance in water reuse. *Environ. Sci. Technol.* **2018**, *52*, 11720–11727. [CrossRef]
21. Zhao, H.; Chen, Y.; Peng, Q.; Wang, Q.; Zhao, G. Catalytic activity of MOF(2Fe/Co)/carbon aerogel for improving H_2O_2 and OH generation in solar photo-electro-Fenton process. *Appl. Catal. B Environ.* **2017**, *203*, 127–137. [CrossRef]
22. Fujiwara, K.; Akita, A.; Kawano, S.; Fujishima, M.; Tada, H. Hydrogen peroxide-photofuel cell using TiO_2 photoanode. *Electrochem. Commun.* **2017**, *84*, 71–74. [CrossRef]
23. Onishi, T.; Fujishima, M.; Tada, H. Solar-driven one-compartment hydrogen peroxide-photofuel cell using bismuth vanadate photoanode. *ACS Omega* **2018**, *3*, 12099–12105. [CrossRef] [PubMed]
24. Yamazaki, S.; Siroma, Z.; Senoh, H.; Ioroi, T.; Fujiwara, N.; Yasuda, K. A fuel cell with selective electrocatalysts using hydrogen peroxide as both an electron acceptor and A fuel. *J. Power Sources* **2008**, *178*, 20–25. [CrossRef]
25. Mcdonnell–Worth, C.J.; Macfarlane, D.R. Progress towards direct hydrogen peroxide fuel cells (DHPFCS) as an energy storage concept. *Aust. J. Chem.* **2018**, *71*, 781–788. [CrossRef]
26. Han, L.; Guo, S.; Wang, P.; Dong, S. Light-driven, membraneless, hydrogen peroxide based fuel cells. *Adv. Energy Mater.* **2015**, *5*, 1400424. [CrossRef]
27. Papagiannis, I.; Doukasm, E.; Kalarakis, A.; Avgouropoulos, G.; Lianos, P. Photoelectrocatalytic H_2 and H_2O_2 production using visible-light-absorbing photoanodes. *Catalysts* **2019**, *9*, 243. [CrossRef]
28. Shiraishi, Y.; Kanazawa, S.; Kofuji, Y.; Sakamoto, H.; Ichikawa, S.; Tanaka, S.; Hirai, T. Sunlight-driven hydrogen peroxide production from water and molecular oxygen by metal-free photocatalysts. *Angew. Chem. Int. Ed. Engl.* **2014**, *53*, 13454–13459. [CrossRef]
29. Shi, X.; Zhang, Y.; Siahrostami, S.; Zheng, X. Light-driven $BiVO_4$-C fuel cell with simultaneous production of H_2O_2. *Adv. Energy Mater.* **2018**, *8*, 1801158. [CrossRef]
30. Xiao, K.; Liang, H.; Chen, S.; Yang, B.; Zhang, J.; Li, J. Enhanced photoelectrocatalytic degradation of bisphenol A and simultaneous production of hydrogen peroxide in saline wastewater treatment. *Chemosphere* **2019**, *222*, 141–148. [CrossRef]
31. Ito, S.; Chen, P.; Comte, P.; Nazeeruddin, M.K.; Liska, P.; Pechy, P.; Gratzel, M. Fabrication of screen–printing pastes from TiO_2 powders for dye-sensitised solar cells. *Prog. Photovolt. Res. Appl.* **2007**, *15*, 603–612. [CrossRef]

32. Sfaelou, S.; Kontos, A.G.; Falaras, P.; Lianos, P. Micro-Raman, Photoluminescence and Photocurrent studies on the photostability of quantum dot sensitized photoanodes. *J. Photochem. Photobiol. A* **2014**, *275*, 127–133. [CrossRef]
33. Nicolau, Y.F. Solution deposition of thin solid compound films by A successive ionic–layer adsorption and reaction process. *Appl. Surf. Sci.* **1985**, *22/23*, 1061–1074. [CrossRef]
34. Li, Z.; Luo, W.; Zhang, M.; Feng, J.; Zou, Z. Photoelectrochemical cells for solar hydrogen production: Current state of promising photoelectrodes, methods to improve their properties, and outlook. *Energy Environ. Sci.* **2013**, *6*, 347. [CrossRef]
35. Sfaelou, S.; Antoniadou, M.; Dracopoulos, V.; Bourikas, K.; Kondarides, D.I.; Lianos, P. Quantum Dot Sensitized Titania as Visible-light Photocatalyst for Solar Operation of Photofuel Cells. *J. Adv. Oxid. Technol.* **2014**, *17*, 59–65. [CrossRef]
36. Pop, C.L.; Sygellou, L.; Dracopoulos, V.; Andrikopoulos, K.S.; Sfaelou, S.; Lianos, P. One-step electrodeposition of CdSe on nanoparticulate titania filmsand their use as sensitized photoanodes for photoelectrochemicalhydrogen production. *Catal. Today* **2015**, *252*, 157–161. [CrossRef]
37. Kalamaras, E.; Lianos, P. Current Doubling effect revisited: Current multiplication in A Photo Fuel Cell. *J. Electroanal. Chem.* **2015**, *751*, 37–42. [CrossRef]
38. Pop, L.C.; Sfaelou, S.; Lianos, P. Cation adsorption by mesoporous titania photoanodes and its effect on the current-voltage characteristics of photoelectrochemical cells. *Electrochim. Acta* **2015**, *156*, 223–227. [CrossRef]
39. Sayama, K. Production of High-Value-Added Chemicals on Oxide Semiconductor Photoanodes under Visible Light for Solar Chemical-Conversion Processes. *ACS Energy Lett.* **2018**, *3*, 1093–1101. [CrossRef]
40. Shi, X.; Siahrostami, S.; Li, G.L.; Zhang, Y.; Chakthranont, P.; Studt, F.; Jaramillo, T.F.; Zheng, X.; Nørskov, J.K. Understanding activity trends in electrochemical water oxidation to form hydrogen peroxide. *Nat. Commun.* **2017**, *8*, 701. [CrossRef]

MDPI
St. Alban-Anlage 66
4052 Basel
Switzerland
Tel. +41 61 683 77 34
Fax +41 61 302 89 18
www.mdpi.com

Materials Editorial Office
E-mail: materials@mdpi.com
www.mdpi.com/journal/materials

www.ingramcontent.com/pod-product-compliance
Lightning Source LLC
LaVergne TN
LVHW070904170726
843515LV00005B/698

* 9 7 8 3 0 3 6 5 1 0 3 2 3 *